SCIENCE RULES

Science Rules

A Historical Introduction to Scientific Methods

Edited by

Peter Achinstein

The Johns Hopkins University Press

Baltimore & London

Printed in the United States of America on acid-free paper
2 4 6 8 9 7 5 3 1

The Johns Hopkins University Press
2715 North Charles Street
Baltimore, Maryland 21218-4363
www.press.jhu.edu

Library of Congress Cataloging-in-Publication Data

Achinstein, Peter.
Science rules : a historical introduction to scientific methods /
Peter Achinstein.
p. cm.
Includes bibliographical references and index.
ISBN 0-8018-7943-4 (hardcover : alk. paper)
ISBN 0-8018-7944-2 (pbk. : alk. paper)
1. Science—Methodology—History. I. Title.
Q175.A2685 2004
501—dc22
2003024784

A catalog record for this book is available from the British Library.

For my Students,

Past and Future

Contents

General Introduction 1

PART I

Descartes' Rationalism & Laws of Motion

Introduction 9

1 / Descartes' Methodological Rules 17

From Rules for the Direction of the Mind / René Descartes

2 / Descartes' Ontological Proof of God 35

From Meditations on First Philosophy / René Descartes

3 / Descartes' Laws of Motion 40

From Principles of Philosophy / René Descartes

4 / A Discussion of Descartes' Methodology 48

From Descartes' Metaphysical Physics / Daniel Garber

PART II

Newton's Inductivism & the Law of Gravity

Introduction 69

5 / Newton's Methodological Rules 78

From The Principia, Book 3 / Isaac Newton

6 / Newton's "Phenomena" and Derivation of the Law of Gravity 81

From The Principia, Book 3 / Isaac Newton

7 / Newton on "Hypotheses," God, and Gravity 95

From The Principia, General Scholium / Isaac Newton

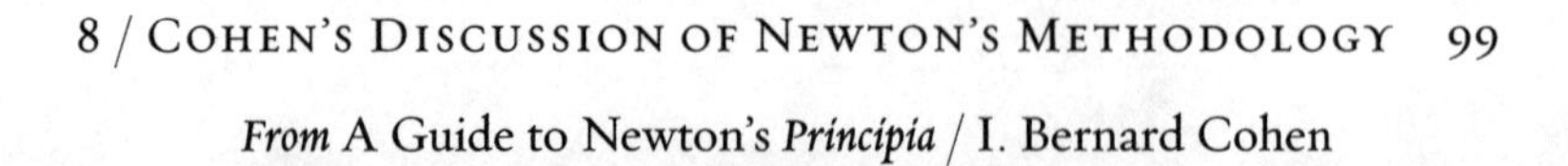

8 / Cohen's Discussion of Newton's Methodology 99

From A Guide to Newton's *Principia* / I. Bernard Cohen

9 / Whewell's Critique of Newton's Methodology 112

From The Philosophy of the Inductive Sciences / William Whewell

PART III

Hypothetico-Deductivism, the Mill–Whewell Debate, & the Wave Theory of Light

Introduction 127

10 / Young's Wave Theory of Light 137

From A Course of Lectures on Natural Philosophy / Thomas Young

11 / Whewell's Hypothetico-Deductivism 150

From The Philosophy of the Inductive Sciences / William Whewell

12 / Popper's Falsificationism 168

From The Logic of Scientific Discovery / Karl R. Popper

13 / Mill's Inductivism and Debate with Whewell 173

From A System of Logic, Book 3 / John Stuart Mill

14 / The Wave Theory of Light and the Mill–Whewell Debate 234

Waves and Scientific Method / Peter Achinstein

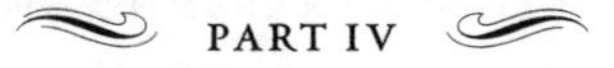 PART IV

Realism vs. Antirealism & Molecular Reality

Introduction 251

15 / Duhem's Antirealism 258

From The Aim and Structure of Physical Theory / Pierre Duhem

16 / Van Fraassen's Antirealism 281

From The Scientific Image / Bas C. van Fraassen

17 / Perrin's Realism and Argument for Molecules 298

From Atoms / Jean Perrin

18 / Salmon's Empirical Defense of Realism 312

From Scientific Explanation and the Causal Structure of the World / Wesley C. Salmon

19 / Realism and Perrin's Argument for Molecules 327

Is There a Valid Experimental Argument for Scientific Realism? / Peter Achinstein

 PART V

Galileo's Tower Argument & Rejections of Universal Rules of Method

Introduction 355

20 / Galileo's Refutation of the Tower Argument 361

From Dialogue Concerning the Two Chief World Systems / Galileo Galilei

21 / Feyerabend's Rejection of Universal Rules 372

From Against Method: Outline of an Anarchistic Theory of Knowledge / Paul K. Feyerabend

22 / A Critique of Feyerabend's Anarchism 389

Proliferation: Is It a Good Thing? / Peter Achinstein

23 / Kuhn's Rejection of Universal Rules 402

From The Structure of Scientific Revolutions / Thomas S. Kuhn

24 / A Discussion of Kuhn's "Values" 412

Subjective Views of Kuhn / Peter Achinstein

Suggested Further Reading 423

Index 425

Science Rules

General Introduction

Scientific methods are rules for constructing and testing scientific theories. This volume introduces the reader to scientific methods that have played a prominent role in the practice of science and have generated lively debates about scientific theorizing. The methods selected are ones advocated by some scientists, methodologists, and philosophers of science and rejected by others. The presentation of each method will include (1) a formulation of the method by the scientist, methodologist, or philosopher advocating it; (2) a scientific selection intended by the proponent of the method to show how the method is applicable to a particular scientific example; and (3) a critical discussion of the method and its application. The volume includes historical and contemporary discussions. Each topic is preceded by an introduction by the editor. Because the book contains scientific as well as philosophical material, it is intended to serve as a new type of introduction to issues about scientific method.

Five topics are included. The first is the method of rationalism illustrated by the seventeenth-century philosopher-scientist-mathematician René Descartes. This method, which is based on the idea that scientific truths must be established with certainty by pure thought rather than empirical observation, is expressed in his *Rules for the Direction of the Mind* and illustrated in his derivation of his three laws of motion in the *Principles of Philosophy*. In proving what he takes to be the laws of motion, Descartes invokes a principle of conservation of motion, which, he asserts, follows from God's existence and the claim that God created matter and motion in the universe. One of Descartes' proofs for the existence of God from his *Meditations* is also included. The main question for the reader will be to understand Descartes' rules of reasoning, which appeal to ideas he develops about "intuitions," "deductions," and "simplification," and to see to what extent these rules are followed in the fundamental part of his physics.

The second topic is the inductive empiricism of Isaac Newton, as expressed in his "Rules for the Study of Natural Philosophy" at the beginning of book 3 of the *Principia* and in his remarks about "hypotheses" at the end of that work. Newton formulates four rules of reasoning to be used in science and employs them in the derivation of his universal law of gravity. The rules and the derivation are included in the volume. Although Newton does not mention Descartes explicitly, it is clear that in formulating these rules he is rejecting Descartes' method of rationalism and defending the idea that scientific truths are to be established by appeal to observation and experiment. Included as well are a historical introduction to Newton's rules by the contemporary Newton scholar I. Bernard Cohen and a critical discussion of these rules by the nineteenth-century scientist, philosopher, and historian of science William Whewell.

The third topic is the method of hypothesis, or hypothetico-deductivism, which is an empirical method that rejects both Descartes' rationalism and Newton's inductivism. Two versions of the view will be considered. The first is developed by the aforementioned William Whewell, the second, very differently, by the twentieth-century philosopher of science Karl Popper. Whewell argues that his version is well illustrated by the development of the wave theory of light during the nineteenth century. That theory was defended at the beginning of the nineteenth century by the British physician and scientist Thomas Young, a selection from whose argument for the wave theory is included. Whewell considers the wave theory established on the grounds that it satisfies his three rules for testing a scientific theory, which appeal to the idea that the theory should correctly predict new phenomena, that it should explain and predict phenomena of a type different from those that prompted the theory in the first place, and that as the theory develops over time, it should be "simple" and "coherent." The philosopher John Stuart Mill, an inductivist contemporary of Whewell, engaged in an important debate with Whewell over the nature of scientific reasoning. Mill rejected Whewell's rules of testing. He also rejected the wave theory of light, particularly the luminiferous (light-bearing) ether that it presupposes, on the grounds that the theory is not supported by an inductive inference, which is the first step required in what Mill calls the "deductive method." The present volume contains selections from Whewell's *Philosophy of the Inductive Sciences* and from Mill's *A System of Logic,* in which the debate between the two is joined, Mill's inductivism and Whewell's hypothetico-deductivism are formulated, and the wave theory of light is discussed. Also included is the editor's critical analysis of this debate and its application to the wave theory. In addition, there is a selection from Popper's influential work *The Logic of Scientific Discovery,* in which he argues for a different version of hypothetico-deductivism. He defends the idea that the only inferences allowable in science are deductive ones of the sort found in mathematical proofs and because of this scientific hypotheses can never be proved but only disproved.

Although each of the scientific methods noted above is different, their defenders agree that the rules are universal, in two important senses. First, whatever the rules are, they apply to theories about the unobservable world as well as to theories about the observable world. If one can use Descartes' rationalism, or Newton's inductivism, or Whewell's hypothetico-deductivism, to infer the truth of claims about the planets and the sun, which are observable, then one can use them to infer the truth of claims about molecules, atoms, and electrons, which are not. Defenders of these methodologies claim that their rules apply to the entire universe of objects. A second point of agreement is that the rules of method hold for all scientific reasoning; they do not vary from one science, or scientist, or scientific period, to another. These assumptions about the universality of scientific rules are the subject of the final two parts of the present volume.

The first assumption is the basis for a crucial debate among scientists and philosophers of science between so-called realists and antirealists. Realists, including Descartes, Newton, and Whewell, believe that, using methodological rules, scientists can learn the truth about both the unobservable and the observable world. Antirealists, including various positivistic and instrumentalist scientists and philosophers in the nineteenth and twentieth centuries, deny this and maintain that scientific knowledge is possible only with respect to the observable world. Part IV of the present volume contains selections from two important antirealists, Pierre Duhem and Bas van Fraassen. Duhem, a physicist, defended antirealism in his 1906 work *The Aim and Structure of Physical Theory.* Van Fraassen, a philosopher of science, revived the debate in 1980 with his book *The Scientific Image.* This part of the volume also contains a selection from the book *Atoms* by the French physical chemist Jean Perrin, whose experiments on Brownian motion earned him a Nobel prize. From these experiments Perrin argued for the existence of unobservable molecules which cause the observable Brownian motion. Included in this selection by Perrin is his philosophical discussion of a method he calls the "intuitive method," a realist position that sanctions inferences from observables to the truth of claims about unobservables, and material from his argument from Brownian motion to the existence of molecules. Whether Perrin's argument is sufficient to establish scientific realism empirically is a question discussed by the philosopher of science Wesley Salmon in a selection from his 1984 book *Scientific Explanation and the Causal Structure of the World,* and by the editor in a selection from an article on scientific realism published in 2002.

The final topic is the question of whether there are rules of scientific method that are and are supposed to be used by all scientists at all times. Two very influential twentieth-century philosophers of science deny this. The more extreme of the two is Paul Feyerabend, who defends a view he calls "anarchism." His idea is that there are no universal rules that scientists should follow, except a rule he calls "proliferation." The latter directs scientists to introduce hypotheses that are incompatible with accepted scientific theories even when those theories are highly confirmed by experiments and observations. To illustrate his view, Feyerabend discusses Galileo's response to the "tower argument"—an argument intended by Galileo's opponents to demonstrate that the earth is stationary and does not rotate on its axis.

Feyerabend claims that Galileo was practicing proliferation, since he was introducing hypotheses that are refuted by observation. Part V of the present volume includes selections from Galileo's *Dialogues Concerning the Two Chief World Systems,* in which the tower argument is presented and examined. It also includes Feyerabend's discussion of this argument and his philosophy of anarchism, as well as the editor's critical examination of Feyerabend's doctrine of proliferation.

Thomas Kuhn also rejects the idea of universal rules, but in a somewhat less radical way than Feyerabend. For such rules Kuhn substitutes what he calls "values" (such as accuracy, consistency, and simplicity), which he thinks all scientists accept but apply in quite different ways. A selection from Kuhn's Postscript to his *Structure of Scientific Revolutions* is included, as well as a critical essay by the editor.

Questions about scientific method of the sort discussed in this volume fall most appropriately under the general rubric of philosophy of science. Because both philosophical and scientific material are included, as well as historical and contemporary works, this volume cannot cover all topics within the philosophy of science. The selections here include central issues that would be studied in most philosophy of science courses. Another restriction should be noted. The scientific examples pertain exclusively to physics. There are three reasons for this: historically much philosophy of science has been derived from concerns about physics; most of the scientists who have advocated scientific methods or rules that are supposed to be applicable to all the sciences have been physicists; and this focus reflects the interests of the editor.

This book could be used in several ways: (1) as a text in an undergraduate or graduate introduction to the philosophy of science (perhaps in addition to an anthology of contemporary readings); (2) as a text in an undergraduate "introduction to scientific thinking" core-course offered by a philosopher, scientist, historian of science, or some combination of the three; (3) as a supplementary text in a history of science course that includes historical questions about scientific method; (4) as a volume of interest to scientists with philosophical or historical concerns about science. With respect to (1)–(3) the text might help to shape or reshape the way such courses are taught.

I am very much indebted to the following people: Trevor Lipscombe, editor in chief of the Johns Hopkins University Press, for his interest in philosophy of science and support of this project; Sean Greenberg, John Norton, Richard Richards, Sherri Roush, Kent Staley, and Maura Tumulty for very helpful comments and corrections pertaining to my section introductions; Christopher Altermatt, Matthew Holtzman, Crystal L'Hote, and Cherie McGill for reading proofs; and L. Suzanne Brown for refelction and all the other good things of life.

PART I

Descartes' Rationalism & Laws of Motion

René Descartes—mathematician, scientist, and philosopher—was born in 1596 in La Haye, France. He was educated by Jesuits at the Collège Royal de la Flèche, receiving training in mathematics, natural science, philosophy, and the classics. Descartes is the father of what today is called analytical (or "Cartesian") geometry, which was based on mathematical investigations that he began in 1618 and that culminated in 1637 in his *Geometry.* In physics he formulated early versions of the law of conservation of momentum and the law of inertia (*Principles of Philosophy,* 1644), as well as derivations of the laws of reflection and refraction of light (*Dioptric,* 1637). The most important and influential of his philosophical works—in metaphysics, epistemology, and scientific method—that were published during his lifetime were *Discourse on Method* (1637) and *Meditations on First Philosophy* (1641). Descartes died in 1650 in Stockholm, where he was serving at the court of Queen Christina.

Descartes' Methodological Rules

The first of the three selections from Descartes included in this volume is taken from his *Rules for the Direction of the Mind,* which was written between 1619 and 1628 but first published posthumously in Latin in 1701. In this work Descartes formulates a set of twenty-one rules (of which the first thirteen are reproduced here) that are intended to constitute a method for obtaining knowledge about the world. These rules are supposed to apply to all the sciences, which, he writes, "are nothing other than human wisdom" (rule 1).[1] Today we might call these rules an expression of Descartes' scientific method: his ideas of how to proceed to gain scientific knowledge. The general view he expresses in these rules is a form of what is called rationalism. On his version of this doctrine, although experience may suggest certain ideas to the scientist, whether these ideas are true can be determined only by pure thought of a sort that is characteristic of mathematics.

Rule 1 asserts that the aim of scientific studies is to direct the mind to form judgments that are both true and sound—in other words, that are justified. For Descartes the scientist should not accept theories or hypotheses that are false, however useful they may be for providing suggestive pictures, or for making predictions, or for generating practical applications. Nor does forming a judgment that is true suffice, unless that truth can be appropriately established. Rule 2 expresses Descartes' belief that science should aim at certainty, and not mere probability. Without certainty there is frequent disagreement, so scientific progress is hindered. Descartes claims that in his time only arithmetic and geometry have the desired certainty. These sciences are the exemplars that other sciences should emulate.

Rule 3, perhaps the most famous, indicates that the only way to satisfy the first two rules, and thus obtain truth that is certain, is to employ "intuition" and "deduction," terms Descartes employs in special ways. By "intuition" Descartes is not referring to what people often mean by that word today: a basic, subjective feeling that something is the case that is not self-evident and that one cannot really prove or justify. Rather, he means "the indubitable conception of a clear and attentive mind which proceeds solely from the light of reason." His examples include one's thoughts that one exists, that one is thinking, that a triangle has three sides, that a sphere has a single surface, and that 2 + 2 = 4. These are thoughts whose truth is immediately evident to us just by our thinking them; they cannot be doubted. By "deduction," Descartes means an inference, or train of reasoning, to some proposition that follows necessarily from other propositions known with certainty, provided that the inference involves a continuous, uninterrupted movement of thought. (For Descartes, proofs must be short enough to be taken in at one sitting.) His example is the inference from 2 + 2 = 4 and 3 + 1 = 4 to 2 + 2 = 3 + 1. Again, his models are arithmetic and geometry, whose certainty other sciences should try to achieve.

Rule 4 declares that to investigate and obtain truth, we need a method. Rules 5–8 discuss features of this method. The method is to consist in reducing "complicated and obscure propositions step by step to simpler ones, and then, starting with the intuition of the simplest ones of all, . . . ascend through the same steps to a knowledge of all the rest" (rule 5). Descartes' discussion here is somewhat vague, but, following the lead of Daniel Garber, a leading contemporary Descartes scholar, it may be useful to think of a scientific investigation as beginning with a problem or question. One of Descartes' examples (given in rule 8) is a problem in optics of determining the anaclastic line: a surface at which parallel rays of light are refracted so that they meet at a point. The "reduction," Garber suggests, consists in raising a series of more and more basic questions. For example, what is the relationship between the angles of incidence and refraction? How does this relationship depend on differences in the media in which the refraction occurs? What is the nature of light and its activity, and, more generally, what is a natural power (of which light is a particular example)? The idea is to raise increasingly fundamental questions in such a way that one can answer the most fundamental questions of all by means of one or more Cartesian intuitions. The investigator should then try to answer the other questions by constructing "deductions" from the basic intuitions.

In rule 12 Descartes declares that although "only the intellect . . . is capable of perceiving the truth" (by intuition and deduction), the intellect is aided by other faculties of the mind, including sense perception, memory, and imagination. Sense perception and memory may suggest the idea that between two points one and only one straight line can be drawn. But only an intuition of this fact—only a clear and distinct conception proceeding from the light of reason—can guarantee the truth of this geometrical proposition. In this rule Descartes also suggests that all knowledge is based on intuitions about what he calls "simple natures." These are few in number and in the case of the material world include extension, shape, and motion. As will be seen, the laws of motion he develops in his physics pertain to bodies that have these simple natures.

Ontological Proof of God

The second selection is Descartes' *Fifth Meditation,* in which he offers a proof that God exists. This is included in the present volume because in developing his laws of motion Descartes assumes that the general cause of all motion is God, who created all matter and motion and who preserves the total quantity of motion in the universe.

Descartes' proof of God's existence in the *Fifth Meditation* is a version of what is called the "ontological argument": the idea that God's existence follows from his essence. The argument is very simple. It is part of God's essence to be a supremely perfect being, that is, one with all perfections. But a God that does not exist lacks a perfection, namely, existence. (A God that exists is more perfect than one that does not.) Existence, then, is part of the essence of God, just as having angles equal to two right angles is part of the essence of a triangle. However, there is an important difference between the two cases. From the fact that it is part of the essence of a triangle to have angles equal to two right angles it does not necessarily follow that any triangle in fact exists. (Triangles do not have all perfections; *if* one exists, then it has angles equal to two right angles.) But from the fact that it is part of the essence of God to have all perfections, it does necessarily follow that God exists, since existence is a perfection.

Although the proof, as Descartes presents it, might seem to be a "deduction" from intuitions about perfection and God's essence, Descartes does claim that it is self-evident that God exists, and that one can see immediately that God's existence is part of his essence. If so, then although the claim that God exists is presented as the conclusion of an argument, Descartes is thinking of the claim itself as an "intuition."

Laws of Motion

The third selection from Descartes is taken from his *Principles of Philosophy* and deals with laws of motion in the physical world. In part I, section 28, of the

Principles, Descartes writes, "We should . . . consider him [God] as the efficient cause of all things; and starting from the divine attributes which by God's will we have some knowledge of, we shall see, with the aid of our God-given natural light, what conclusions should be drawn concerning those effects which are apparent to our senses." One of these effects apparent to our senses is motion. Since God is the efficient cause of all things, and since motion exists, God must be the cause of motion, a claim Descartes makes in part 2, section 36.

Now, says Descartes, perfection, which is part of God's essence, involves immutability. (If something changes, then either it was not perfect before, or it is not perfect afterward.) God's perfection, and hence immutability, must hold not only for God himself but also for his operations. Accordingly, Descartes concludes, God not only causes the motion in the universe but also sustains the total amount of motion in the universe, even if the motions of individual bodies can vary. This conclusion, which Descartes seems to be treating as a "deduction" from "intuitions" concerning God and motion, is an ancestor of what is now called the principle of conservation of momentum. However, Descartes' "momentum," or what he calls "quantity of motion," is not the later Newtonian idea of mass times velocity but rather volume (which is purely spatial) times speed (which has magnitude but, unlike Newton's "velocity," no direction).

Next Descartes formulates three "laws of nature," which, he says, follow (by "deduction") from God's immutability. God is what he calls the "primary" cause of motion; these laws are the "secondary, particular causes of the various motions we see in different bodies" (pt. 2, sec. 37). While the previously cited law of conservation of momentum applies to the totality of motion in the universe (it must be constant), the three laws Descartes now introduces determine the motions of individual bodies.

The first of these laws is extremely general and is applicable not just to motion: each thing, insofar as it is simple and undivided, always remains in the same state, as far as it can, and never changes except as a result of external causes. Descartes gives three examples: if a piece of matter is square, it will remain square, unless some external cause changes its shape; if it is at rest, it will remain so, unless it is pushed; if it is moving, it will continue to move, unless something intervenes. He concludes with this formulation of the first law for the case of motion: "A moving body, so far as it can, goes on moving" (pt. 2, sec. 37). With this simple statement Descartes provides a revolutionary answer to the age-old question of why projectiles such as stones continue to move after they leave the hand that throws them, when no known force is acting on them. The answer is not that there is some (hitherto unknown) external (or internal) force causing them to continue in motion. The answer, provided by the first law, is that they continue to move because they are already in motion, and there is no force causing them to change that motion.

Descartes' second "law of nature" is that any piece of matter that is moving tends to continue to move in a straight line. This law goes beyond the first law in two important ways. Unlike the first law, it gives a natural direction to motion: rectilinear. Moreover, it introduces the idea of a *tendency* to move in a straight line, which is what a body would do if it did not interact with other bodies. Descartes cites the example of a stone moving in a circle in a sling held by a hand. At each moment when it is still in the sling the stone has a *tendency* to move in a straight line, even though it is actually moving in a circle. Observation confirms this, says Descartes, because when the stone is released from the sling it does move in a straight line.

Despite this experiential confirmation, Descartes regards the second law as derivable by deduction from the immutability of God and (he adds) the simplicity of the operation by which God preserves motion. A body may in fact change its motion in many different ways, depending on encounters with other bodies. But God causes motion, and because of his immutability and the immutability of his effects, he must cause a state of motion to continue unchanged. This state is the body's tendency to move. For Descartes, God not only creates the world but sustains it by recreating it at each moment. Therefore, Descartes concludes, God preserves the motion of a body at each moment, and does so without regard to what that motion was previously. Descartes asserts that the only direction possible for a tendency to move that does not depend on previous motion is rectilinear: at each moment a body tends to move in a straight line tangent to the point on the actual curve which that body occupies at that moment.

The first two "laws of nature" are concerned with what happens if bodies are left to themselves and do not interact. The third law introduces impact. Suppose that a moving body A collides with another body B. Descartes envisages two possibilities. One is that A's power or force for continuing in a straight line is less than the force of resistance of B. In this case A is deflected so that its quantity of motion (size times speed) is conserved even though its direction of motion changes. The second possibility is that A's power for continuing in a straight line is greater than that of B. Then, says Descartes, A carries B along with it, and A loses a quantity of motion equal to that which it gives to B. (At this point Descartes does not consider the possibility that the two opposing forces are equal, although he does so later when he states seven more specific rules of impact.) Impact, according to Descartes, is a "winner-take-all" phenomenon. If B's power to resist is greater than A's power to continue, then B "wins" and A is deflected. If A's power to continue is greater than B's to resist, then A "wins" and carries B with it. In both cases the quantity of motion is conserved. In the first case, A's quantity of motion remains the same even though its direction changes; B's quantity of motion is unaltered, so that the total remains the same. In the second case, A loses a quantity of motion equal to that which B gains, so again the total is conserved.

Descartes offers a separate proof for each part of this law. The basic idea behind the first proof derives from his concept of quantity of motion, which, being size times speed, is distinct from direction. So the direction of motion can change even if the quantity of motion does not. And, according to Descartes' first law, motion tends to persist. (Also, according to his law of conservation of motion, the quantity of motion in the universe, and presumably in any isolated two-body system in the universe, remains constant.) And, in the first case, when A is deflected and B is unaltered, the total quantity of motion remains constant, since the speed of each body remains what it was before collision, even though A's direction changes. The reason A's direction changes is that B's force on A is overpowering.

Descartes' proof of the second part of the law reverts once more to God's immutability and the fact that God created and continually preserves the world. While the quantity of motion in individual bodies can change as a result of impacts, the total quantity of motion that God created in the universe must be constant because of God's immutability. For this to occur, motion must be transferred during impact. This is just what happens in the second case, when body A loses a quantity of motion equal to that which it imparts to B.

Descartes scholars raise many interpretative and critical questions about these laws and their "proofs." A discussion of some of the issues appears in the selection reprinted here from Daniel Garber's 1992 book, *Descartes' Metaphysical Physics*. As noted earlier, it should also be remembered that although Descartes did formulate a universal law of conservation of motion, his concept of motion (or momentum), unlike one employed by physicists since Newton, is nondirectional. Newton would agree that in a collision between two isolated bodies the total quantity of motion is conserved, even though he rejected Descartes' third law (regarding impact). For Newton and later physicists, what is conserved is the quantity mass times velocity, not size times speed. The laws of motion that Newton first published in 1687 can be shown to require rejecting Descartes' mistaken "winner-take-all" idea.

It seems pretty clear that Descartes' "proofs" of his laws of motion provide examples of what, in his *Rules for the Direction of the Mind,* he calls "deductions" founded on "intuitions," rather than examples of empirical generalizations based entirely on observations and experiments. Indeed, having described these laws in part 2 of the *Principles,* he begins part 3 by saying, "We have thus discovered certain principles as regards material objects, derived not from the prejudices of our senses but from the light of reason, so that their truth is indubitable" (sec. 1).

Nevertheless, in actual practice Descartes does not always proceed by using this method, even if he regards it as the surest way to truth. For example, in a brief selection that follows from part 3 of the *Principles,* Descartes indicates that, having derived the laws of motion, his task is now to discern the general structure of the visible world. He indicates that here he cannot definitively claim to have discovered the truth, so what he will introduce will be in the form of a hypothesis: "Even if this be thought to be false, I shall think my achievement is sufficiently worth while if all inferences from it agree with observation . . . ; for in that case we shall get as much practical benefit from it as we should from the knowledge of the actual truth" (sec. 44).

For example, he indicates (in sec. 46) that although we can establish by reason (intuition and deduction) alone that matter is in fact divided into a large number of parts that move in different directions, we cannot determine by reason alone how big these parts are or how fast they move. We know by reason alone that God created matter and motion, but there are many different configurations of the parts God might have selected. We are free to propose a hypothesis here, says Descartes, subject to the condition that the consequences of the hypothesis agree with experience. For this purpose Descartes assumes that the visible world was originally divided by God into particles of approximately equal size, which had two motions: a circular motion about their centers, and a collective motion in groups around other points, so as to comprise a set of vortices. Descartes asserts that from these assumptions, together with his laws of motion, he can derive all the observed effects in the universe. However, even if this were doable, Descartes would not conclude that the hypothesis is true. He recognizes that it is also possible to derive observed effects from false hypotheses.

Later in the volume we shall consider *hypothetico-deductivism,* the view that from the fact that one can derive true observational conclusions from one's hypothesis one may infer the truth of the hypothesis (depending on how many, varied, and novel these conclusions are). Descartes rejects this view. Hypotheses can be useful for various purposes, including making predictions about what can be observed, but from their successful use one must not infer that they are true.

Finally, given Descartes' rationalism, one might ask what role experience plays in his methodology. One has just been noted, namely, the practical benefit he sees in determining whether inferences from hypotheses conform to observation when those hypotheses cannot be deduced with certainty. Indeed, even when he regards a general proposition as having been deduced, experience can be useful in offering an empirical confirmation of that proposition (as when he notes that observation of the motion of a stone released from a sling confirms his second law of motion). This notion of empirical "confirmation," however, is not to be understood as establishing the truth of a proposition, at least not with the kind of certainty Descartes demands.

Daniel Garber argues that an appeal to experience is also important for Descartes in posing questions leading to a reduction of complex statements to simple ones and finally to an intuition.[2] Thus, in determining the shape of the anaclastic line in optics, experience informs us that light is refracted when going from one medium to another, and that this can happen in such a way that parallel rays are refracted to the same point. Again, however, this appeal to experience does not establish with certainty the fundamental scientific propositions being sought—

those that permit the deduction of others that can be used in answering the questions posed.

NOTES

1. Unless otherwise indicated, all quotations are from the editions from which the following selections are drawn.

2. Daniel Garber, *Descartes Embodied* (Cambridge: Cambridge University Press, 2001), 85–110.

1

Descartes' Methodological Rules

From

Rules for the Direction of the Mind / René Descartes

Rule 1

The aim of our studies should be to direct the mind with a view to forming true and sound judgements about whatever comes before it.

The sciences as a whole are nothing other than human wisdom, which always remains one and the same, however different the subjects to which it is applied, it being no more altered by them than sunlight is by the variety of the things it shines on. Hence there is no need to impose any restrictions on our mental powers; for the knowledge of one truth does not, like skill in one art, hinder us from discovering another; on the contrary it helps us. . . . It must be acknowledged that all the sciences are so closely interconnected that it is much easier to learn them all together than to separate one from the other. If, therefore, someone seriously wishes to investigate the truth of things, he ought not to select one science in particular, for they are all interconnected and interdependent. . . .

Reprinted from *Descartes: Selected Philosophical Writings,* trans. John Cottingham, Robert Stoothoff, and Dugald Murdoch (Cambridge: Cambridge University Press, 1988), 1–19.

Although the title of this work is usually translated as given here, the translators of this edition render it as *Rules for the Direction of Our Native Intelligence.* Translators' notes have been omitted. Elisions are theirs.

Rule 2

We should attend only to those objects of which our minds seem capable of having certain and indubitable cognition.

All knowledge is certain and evident cognition. Someone who has doubts about many things is no wiser than one who has never given them a thought; indeed, he appears less wise if he has formed a false opinion about any of them. Hence it is better never to study at all than to occupy ourselves with objects which are so difficult that we are unable to distinguish what is

18
true from what is false, and are forced to take the doubtful as certain; for in such matters the risk of diminishing our knowledge is greater than our hope of increasing it. So, in accordance with this Rule, we reject all such merely probable cognition and resolve to believe only what is perfectly known and incapable of being doubted. . . .

Nevertheless, if we adhere strictly to this Rule, there will be very few things which we can get down to studying. For there is hardly any question in the sciences about which clever men have not frequently disagreed. . . . Accordingly, if my reckoning is correct, out of all the sciences so far devised, we are restricted to just arithmetic and geometry if we stick to this Rule. . . . So if we seriously wish to propose rules for ourselves which will help us scale the heights of human knowledge, we must include, as one of our primary rules, that we should take care not to waste our time by neglecting easy tasks and occupying ourselves only with difficult matters. . . .

Of all the sciences so far discovered, arithmetic and geometry alone are, as we said above, free from any taint of falsity or uncertainty. If we are to give a careful estimate of the reason why this should be so, we should bear in mind that there are two ways of arriving at a knowledge of things—through experience and through deduction. Moreover, we must note that while our experiences of things are often deceptive, the deduction or pure inference of one thing from another can never be performed wrongly by an intellect which is in the least degree rational, though we may fail to make the inference if we do not see it. Furthermore, those chains with which dialecticians suppose they regulate human reason seem to me to be of little use here, though I do not deny that they are very useful for other purposes. In fact none of the errors to which men—men, I say, not the brutes—are liable is ever due to faulty inference; they are due only to the fact that men take for granted certain poorly understood observations, or lay down rash and groundless judgements.

These considerations make it obvious why arithmetic and geometry prove to be much more certain than other disciplines: they alone are concerned with an object so pure and simple that they make no assumptions that experience might render uncertain; they consist entirely in deducing conclusions by means of rational arguments. They are therefore the easiest and clearest of all the sciences and have just the sort of object we are looking for. . . .

RULE 3

Concerning objects proposed for study, we ought to investigate what we can clearly and evidently intuit or deduce with certainty, and not what other people have thought or what we ourselves conjecture. For knowledge can be attained in no other way.

19

We would be well-advised not to mix any conjectures into the judgements we make about the truth of things. It is most important to bear this point in mind. The main reason why we can find nothing in ordinary philosophy which is so evident and certain as to be beyond dispute is that students of the subject first of all are not content to acknowledge what is clear and certain, but on the basis of merely probable conjectures venture also to make assertions on obscure matters about which nothing is known; they then gradually come to have complete faith in these assertions, indiscriminately mixing them up with others that are true and evident. The result is that the only conclusions they can draw are ones which apparently rest on some such obscure proposition, and which are accordingly uncertain.

But in case we in turn should slip into the same error, let us now review all the actions of the intellect by means of which we are able to arrive at a knowledge of things with no fear of being mistaken. We recognize only two: intuition and deduction.

By 'intuition' I do not mean the fluctuating testimony of the senses or the deceptive judgement of the imagination as it botches things together, but the conception of a clear and attentive mind, which is so easy and distinct that there can be no room for doubt about what we are understanding. Alternatively, and this comes to the same thing, intuition is the indubitable conception of a clear and attentive mind which proceeds solely from the light of reason. Because it is simpler, it is more certain than deduction, though deduction, as we noted above, is not something a man can perform wrongly. Thus everyone can mentally intuit that he exists, that he is thinking, that a triangle is bounded by just three lines, and a sphere by a single surface, and the like. Perceptions such as these are more numerous than most people realize, disdaining as they do to turn their minds to such simple matters. . . .

The self-evidence and certainty of intuition is required not only for apprehending single propositions, but also for any train of reasoning whatever. Take for example, the inference that 2 plus 2 equals 3 plus 1: not only must we intuitively perceive that 2 plus 2 makes 4, and that 3 plus 1 makes 4, but also that the original proposition follows necessarily from the other two.

There may be some doubt here about our reason for suggesting another mode of knowing in addition to intuition, *viz.* deduction, by which we mean the inference of something as following necessarily from some other propositions which are known with certainty. But this distinction had to be made, since very many facts which are not self-evident are known with certainty, provided they are inferred from true and known principles through a continuous and uninterrupted movement of thought in which each individual proposition is clearly intuited. This is similar to the way in which we

20

know that the last link in a long chain is connected to the first: even if we cannot take in at one glance all the intermediate links on which the connection depends, we can have knowledge of the connection provided we survey the links one after the other, and keep in mind that each link from first to last is attached to its neighbour. Hence we are distinguishing mental intuition from certain deduction on the grounds that we are aware of a movement or a sort of sequence in the latter but not in the former, and also because immediate self-evidence is not required for deduction, as it is for intuition; deduction in a sense gets its certainty from memory. It follows that those propositions which are immediately inferred from first principles can be said to be known in one respect through intuition, and in another respect through deduction. But the first principles themselves are known only through intuition, and the remote conclusions only through deduction. . . .

Rule 4

We need a method if we are to investigate the truth of things.

By 'a method' I mean reliable rules which are easy to apply, and such that if one follows them exactly, one will never take what is false to be true or fruitlessly expend one's mental efforts, but will gradually and constantly increase one's knowledge till one arrives at a true understanding of everything within one's capacity. . . . The method cannot go so far as to teach us how to perform the actual operations of intuition and deduction, since these are the simplest of all and quite basic. If our intellect were not already able to perform them, it would not comprehend any of the rules of the method, however easy they might be. . . .

So useful is this method that without it the pursuit of learning would, I think, be more harmful than profitable. Hence I can readily believe that the great minds of the past were to some extent aware of it, guided to it even by nature alone. For the human mind has within it a sort of spark of the divine, in which the first seeds of useful ways of thinking are sown, seeds which, however neglected and stifled by studies which impede them, often bear fruit of their own accord. This is our experience in the simplest of sciences, arithmetic and geometry: we are well aware that the geometers of antiquity employed a sort of analysis which they went on to apply to the solution of every problem, though they begrudged revealing it to posterity. . . .

I shall have much to say below about figures and numbers, for no other disciplines can yield illustrations as evident and certain as these. But if one attends closely to my meaning, one will readily see that ordinary mathematics is far from my mind here, that it is quite another discipline I am expounding, and that these illustrations are more its outer garments than inner parts. This discipline should contain the primary rudiments of human reason and extend to the discovery of truths in any field whatever. . . .

In the present age some very gifted men have tried to revive this method, for the method seems to me to be none other than the art which goes by the outlandish name of 'algebra'—or at least it would be if algebra were divested of the multiplicity of numbers and incomprehensible figures which overwhelm it and instead possessed that abundance of clarity and simplicity which I believe the true mathematics ought to have. It was these thoughts which made me turn from the particular studies of arithmetic and geometry to a general investigation of mathematics. I began my investigation by inquiring what exactly is generally meant by the term 'mathematics' and why it is that, in addition to arithmetic and geometry, sciences such as astronomy, music, optics, mechanics, among others, are called branches of mathematics. To answer this it is not enough just to look at the etymology of the word, for, since the word 'mathematics' has the same meaning as 'discipline', these subjects have as much right to be called 'mathematics' as geometry has. Yet it is evident that almost anyone with the slightest education can easily tell the difference in any context between what relates to mathematics and what to the other disciplines. When I considered the matter more closely, I came to see that the exclusive concern of mathematics is with questions of order or measure and that it is irrelevant whether the measure in question involves numbers, shapes, stars, sounds, or any other object whatever. This made me realize that there must be a general science which explains all the points that can be raised concerning order and measure irrespective of the subject-matter, and that this science should be termed *mathesis universalis* [universal mathematics]—a venerable term with a well-established meaning—for it covers everything that entitles these other sciences to be called branches of mathematics. . . .

Aware how slender my powers are, I have resolved in my search for knowledge of things to adhere unswervingly to a definite order, always starting with the simplest and easiest things and never going beyond them till there seems to be nothing further which is worth achieving where they are concerned. Up to now, therefore, I have devoted all my energies to this universal mathematics, so that I think I shall be able in due course to tackle the somewhat more advanced sciences, without my efforts being premature. . . .

Rule 5

The whole method consists entirely in the ordering and arranging of the objects on which we must concentrate our mind's eye if we are to discover some truth. We shall be following this method exactly if we first reduce complicated and obscure propositions step by step to simpler

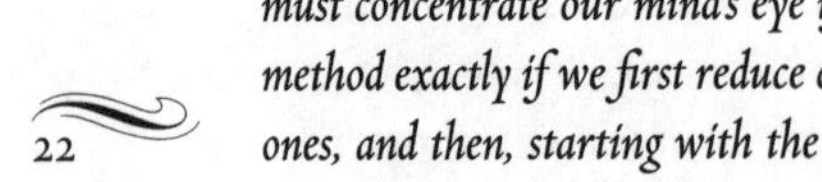

22

ones, and then, starting with the intuition of the simplest ones of all, try to ascend through the same steps to a knowledge of all the rest.

This one Rule covers the most essential points in the whole of human endeavour. Anyone who sets out in quest of knowledge of things must follow this Rule as closely as he would the thread of Theseus if he were to enter the Labyrinth. . . . But the order that is required here is often so obscure and complicated that not everyone can make out what it is; hence it is virtually impossible to guard against going astray unless one carefully observes the message of the following Rule.

Rule 6

In order to distinguish the simplest things from those that are complicated and to set them out in an orderly manner, we should attend to what is most simple in each series of things in which we have directly deduced some truths from others, and should observe how all the rest are more, or less, or equally removed from the simplest.

Although the message of this Rule may not seem very novel, it contains nevertheless the main secret of my method; and there is no more useful Rule in this whole treatise. For it instructs us that all things can be arranged serially in various groups, not in so far as they can be referred to some ontological genus (such as the categories into which philosophers divide things), but in so far as some things can be known on the basis of others. Thus when a difficulty arises, we can see at once whether it will be worth looking at any others first, and if so which ones and in what order.

In order to be able to do this correctly, we should note first that everything, with regard to its possible usefulness to our project, may be termed either 'absolute' or 'relative'—our project being, not to inspect the isolated natures of things, but to compare them with each other so that some may be known on the basis of others.

I call 'absolute' whatever has within it the pure and simple nature in question; that is, whatever is viewed as being independent, a cause, simple, universal, single, equal, similar, straight, and other qualities of that sort. I call this the simplest and the easiest thing when we can make use of it in solving problems.

The 'relative', on the other hand, is what shares the same nature, or at least something of the same nature, in virtue of which we can relate it to the absolute and deduce it from the absolute in a definite series of steps. The concept of the 'relative' involves other terms besides, which I call 'relations': these include whatever is said to be dependent, an effect, composite, particular, many, unequal, dissimilar, oblique, etc. The further removed from the absolute such relative attributes are, the more mutually dependent relations of this sort they contain. This Rule points out that all these relations should be distinguished, and the interconnections between them, and their natural order, should be noted, so that given the last term we should be able to reach the one that is absolute in the highest degree, by passing through all the intermediate ones. . . .

We should note, secondly, that there are very few pure and simple natures which we can intuit straight off and *per se* (independently of any others) either in our sensory experience or by means of a light innate within us. We should, as I said, attend carefully to the simple natures which can be intuited in this way, for these are the ones which in each series we term simple in the highest degree. As for all the other natures, we can apprehend them only by deducing them from those which are simple in the highest degree, either immediately and directly, or by means of two or three or more separate inferences. . . .

The third and last point to note is that we should not begin our studies by investigating difficult matters. Before tackling any specific problems we ought first to make a random selection of truths which happen to be at hand, and ought then to see whether we can deduce some other truths from them step by step, and from these still others, and so on in logical sequence. This done, we should reflect attentively on the truths we have discovered and carefully consider why it was we were able to discover some of these truths sooner and more easily than others, and what these truths are. . . .

Rule 7

In order to make our knowledge complete, every single thing relating to our undertaking must be surveyed in a continuous and wholly uninterrupted sweep of thought, and be included in a sufficient and well-ordered enumeration.

It is necessary to observe the points proposed in this Rule if we are to admit as certain those truths which, we said above, are not deduced immediately from first and self-evident principles. For this deduction sometimes requires such a long chain of inferences that when we arrive at such a truth it is not easy to recall the entire route which led us to it. That is why we say that a continuous movement of thought is needed to make good any weakness of memory. . . .

In addition, this movement must nowhere be interrupted. Frequently those who attempt to deduce something too swiftly and from remote initial premisses do not go over the entire chain of intermediate conclusions very

carefully, but pass over many of the steps without due consideration. But, whenever even the smallest link is overlooked, the chain is immediately broken, and the certainty of the conclusion entirely collapses.

We maintain furthermore that enumeration is required for the completion of our knowledge. The other Rules do indeed help us resolve most questions, but it is only with the aid of enumeration that we are able to make a true and certain judgement about whatever we apply our minds to. By means of enumeration nothing will wholly escape us and we shall be seen to have some knowledge on every question.

In this context enumeration, or induction, consists in a thorough investigation of all the points relating to the problem at hand, an investigation which is so careful and accurate that we may conclude with manifest certainty that we have not inadvertently overlooked anything. So even though the object of our inquiry eludes us, provided we have made an enumeration, we shall be wiser at least to the extent that we shall perceive with certainty that it could not possibly be discovered by any method known to us. . . .

We should note, moreover, that by 'sufficient enumeration' or 'induction' we must mean the kind of enumeration which renders the truth of our conclusions more certain than any other kind of proof (simple intuition excepted) allows. . . .

The enumeration should sometimes be complete, and sometimes distinct, though there are times when it need be neither. That is why I said only that the enumeration must be sufficient. For if I wish to determine by enumeration how many kinds of corporeal entity there are or how many are in some way perceivable by the senses, I shall not assert that there are just so many and no more, unless I have previously made sure I have included them all in my enumeration and have distinguished one from another. But if I wish to show in the same way that the rational soul is not corporeal, there is no need for the enumeration to be complete; it will be sufficient if I group all bodies together into several classes so as to demonstrate that the rational soul cannot be assigned to any of these. . . .

Rule 8

If in the series of things to be examined we come across something which our intellect is unable to intuit sufficiently well, we must stop at that point, and refrain from the superfluous task of examining the remaining items.

The most useful inquiry we can make at this stage is to ask: What is human knowledge and what is its scope? We are at present treating this as one single question, which in our view is the first question of all that should be examined by means of the Rules described above. . . . In order to see how the above points apply to the problem before us, we shall first divide into two parts whatever is relevant to the question; for the question ought to relate either to us, who have the capacity for knowledge, or to the actual things it is possible to know. We shall discuss these two parts separately.

25

Within ourselves we are aware that, while it is the intellect alone that is capable of knowledge, it can be helped or hindered by three other faculties, *viz.* imagination, sense-perception, and memory. We must therefore look at these faculties in turn, to see in what respect each of them could be a hindrance, so that we may be on our guard, and in what respect an asset, so that we may make full use of their resources. We shall discuss this part of the question by way of a sufficient enumeration, as the following Rule will make clear.

We should then turn to the things themselves; and we should deal with these only in so far as they are within the reach of the intellect. In that respect we divide them into absolutely simple natures and complex or composite natures. Simple natures must all be either spiritual or corporeal, or belong to each of these categories. As for composite natures, there are some which the intellect experiences as composite before it decides to determine anything about them: but there are others which are put together by the intellect itself. All these points will be explained at greater length in Rule Twelve, where it will be demonstrated that there can be no falsity save in composite natures which are put together by the intellect. In view of this, we divide natures of the latter sort into two further classes, *viz.* those that are deduced from natures which are the most simple and self-evident (which we shall deal with throughout the next book), and those that presuppose others which experience shows us to be composite in reality. . . .

Rule 9

We must concentrate our mind's eye totally upon the most insignificant and easiest of matters, and dwell on them long enough to acquire the habit of intuiting the truth distinctly and clearly.

We have given an account of the two operations of our intellect, intuition and deduction, on which we must, as we said, exclusively rely in our acquisition of knowledge. In this and the following Rule we shall proceed to explain how we can make our employment of intuition and deduction more skilful and at the same time how to cultivate two special mental faculties, *viz.* perspicacity in the distinct intuition of particular things and discernment in the methodical deduction of one thing from another.

We can best learn how mental intuition is to be employed by comparing it with ordinary vision. If one tries to look at many objects at one glance, one sees none of them distinctly. Likewise, if one is inclined to attend to many

26

things at the same time in a single act of thought, one does so with a confused mind. Yet craftsmen who engage in delicate operations, and are used to fixing their eyes on a single point, acquire through practice the ability to make perfect distinctions between things, however minute and delicate. The same is true of those who never let their thinking be distracted by many different objects at the same time, but always devote their whole attention to the simplest and easiest of matters: they become perspicacious. . . .

Everyone ought therefore to acquire the habit of encompassing in his thought at one time facts which are very simple and very few in number—so much so that he never thinks he knows something unless he intuits it just as distinctly as any of the things he knows most distinctly of all. . . . There is, I think, one point above all others which I must stress here, which is that everyone should be firmly convinced that the sciences, however abstruse, are to be deduced only from matters which are easy and highly accessible, and not from those which are grand and obscure. . . .

Rule 10

In order to acquire discernment we should exercise our native intelligence by investigating what others have already discovered, and methodically survey even the most insignificant products of human skill, especially those which display or presuppose order.

Since not all minds have such a natural disposition to puzzle things out by their own exertions, the message of this Rule is that we must not take up the more difficult and arduous issues immediately, but must first tackle the simplest and least exalted arts, and especially those in which order prevails—such as weaving and carpet-making, or the more feminine arts of embroidery, in which threads are interwoven in an infinitely varied pattern. Number-games and any games involving arithmetic, and the like, belong here. It is surprising how much all these activities exercise our native intelligence, provided of course we discover them for ourselves and not from others. . . .

Rule 11

If, after intuiting a number of simple propositions, we deduce something else from them, it is useful to run through them in a continuous and completely uninterrupted train of thought, to reflect on their relations to one another, and to form a distinct and, as far as possible, simultaneous conception of several of them. For in this way our knowledge becomes much more certain, and our intellectual capacity is enormously increased.

Two things are required for mental intuition: first, the proposition intuited must be clear and distinct; second, the whole proposition must be understood all at once, and not bit by bit. But when we think of the process of deduction as we did in Rule Three, it does not seem to take place all at once: inferring one thing from another involves a kind of movement of our mind. In that passage, then, we were justified in distinguishing intuition from deduction. But if we look on deduction as a completed process, as we did in Rule Seven, then it no longer signifies a movement but rather the completion of a movement. That is why we are supposing that the deduction is made through intuition when it is simple and transparent, but not when it is complex and involved. When the latter is the case, we call it 'enumeration' or 'induction', since the intellect cannot simultaneously grasp it as a whole, and its certainty in a sense depends on memory, which must retain the judgements we have made on the individual parts of the enumeration if we are to derive a single conclusion from them taken as a whole. . . .

As we have said, conclusions which embrace more than we can grasp in a single intuition depend for their certainty on memory, and since memory is weak and unstable, it must be refreshed and strengthened through this continuous and repeated movement of thought. Say, for instance, in virtue of several operations, I have discovered the relation between the first and the second magnitude of a series, then the relation between the second and the third and the third and fourth, and lastly the fourth and fifth: that does not necessarily enable me to see what the relation is between the first and the fifth, and I cannot deduce it from the relations I already know unless I remember all of them. That is why it is necessary that I run over them again and again in my mind until I can pass from the first to the last so quickly that memory is left with practically no role to play, and I seem to be intuiting the whole thing at once. . . .

Rule 12

Finally we must make use of all the aids which intellect, imagination, sense-perception, and memory afford in order, firstly, to intuit simple propositions distinctly; secondly, to combine correctly the matters under investigation with what we already know, so that they too may be known; and thirdly, to find out what things should be compared with each other so that we make the most thorough use of all our human powers.

This Rule sums up everything that has been said above, and sets out a general lesson the details of which remain to be explained as follows.

Where knowledge of things is concerned, only two factors need to be considered: ourselves, the knowing subjects, and the things which are the objects of knowledge. As for ourselves, there are only four faculties which we can use for this purpose, *viz.* intellect, imagination, sense-perception and memory. It is of course only the intellect that is capable of perceiving the

truth, but it has to be assisted by imagination, sense-perception and memory if we are not to omit anything which lies within our power. As for the objects of knowledge, it is enough if we examine the following three questions: What presents itself to us spontaneously? How can one thing be known on the basis of something else? What conclusions can be drawn from each of these? This seems to me to be a complete enumeration and to omit nothing which is within the range of human endeavour. . . .

With respect to the second factor, our aim is to distinguish carefully the notions of simple things from those which are composed of them, and in both cases to try to see where falsity can come in, so that we may guard against it, and to see what can be known with certainty, so that we may concern ourselves exclusively with that. To this end, as before, certain assumptions must be made in this context which perhaps not everyone will accept. But even if they are thought to be no more real than the imaginary circles which astronomers use to describe the phenomena they study, this matters little, provided they help us to pick out the kind of apprehension of any given thing that may be true and to distinguish it from the kind that may be false.

We state our view, then, in the following way. First, when we consider things in the order that corresponds to our knowledge of them, our view of them must be different from what it would be if we were speaking of them in accordance with how they exist in reality. If, for example, we consider some body which has extension and shape, we shall indeed admit that, with respect to the thing itself, it is one single and simple entity. For, viewed in that way, it cannot be said to be a composite made up of corporeal nature, extension and shape, since these constituents have never existed in isolation from each other. Yet with respect to our intellect we call it a composite made up of these three natures, because we understood each of them separately before we were in a position to judge that the three of them are encountered at the same time in one and the same subject. That is why, since we are concerned here with things only in so far as they are perceived by the intellect, we term 'simple' only those things which we know so clearly and distinctly that they cannot be divided by the mind into others which are more distinctly known. Shape, extension and motion, etc. are of this sort; all the rest we conceive to be in a sense composed out of these. This point is to be taken in a very general sense, so that not even the things that we occasionally abstract from these simples are exceptions to it. We are abstracting, for example, when we say that shape is the limit of an extended thing, conceiving by the term 'limit' something more general than shape, since we can talk of the limit of a duration, the limit of a motion, etc. But, even if the sense of the term 'limit' is derived by abstraction from the notion of shape, that is no reason to regard it as simpler than shape. On the contrary, since the term 'limit' is also applied to other things—such as the limit of a duration or a motion, etc., things totally different in kind from shape—it must have been abstracted from these as well. Hence, it is something compounded out of many quite different natures, and the term 'limit' does not have a univocal application in all these cases.

Secondly, those things which are said to be simple with respect to our intellect are, on our view, either purely intellectual or purely material, or common to both. Those simple natures which the intellect recognizes by means of a sort of innate light, without the aid of any corporeal image, are purely intellectual. That there is a number of such things is certain: it is impossible to form any corporeal idea which represents for us what knowledge or doubt or ignorance is, or the action of the will, which may be called 'volition', and the like; and yet we have real knowledge of all of these, knowledge so easy that in order to possess it all we need is some degree of rationality. Those simple natures, on the other hand, which are recognized to be present only in bodies—such as shape, extension and motion, etc.—are purely material. Lastly, those simples are to be termed 'common' which are ascribed indifferently, now to corporeal things, now to spirits—for instance, existence, unity, duration and the like. To this class we must also refer those common notions which are, as it were, links that connect other simple natures together, and whose self-evidence is the basis for all the rational inferences we make. Examples of these are: 'Things that are the same as a third thing are the same as each other'; 'Things that cannot be related in the same way to a third thing are different in some respect.' These common notions can be known either by the pure intellect or by the intellect as it intuits the images of material things.

Moreover, it is as well to count among the simple natures the corresponding privations and negations, in so far as we understand these. For when I intuit what nothing is, or an instant, or rest, my apprehension is as much genuine knowledge as my understanding what existence is, or duration, or motion. This way of conceiving of things will be helpful later on in enabling us to say that all the rest of what we know is put together out of these simple natures. Thus, if I judge that a certain shape is not moving, I shall say that my thought is in some way composed of shape and rest; and similarly in other cases.

Thirdly, these simple natures are all self-evident and never contain any falsity. This can easily be shown if we distinguish between the faculty by which our intellect intuits and knows things and the faculty by which it makes affirmative or negative judgements. For it can happen that we think we are ignorant of things we really know, as for example when we suspect that they contain something else which eludes us, something beyond what we intuit or reach in our thinking, even though we are mistaken in thinking

30

this. For this reason, it is evident that we are mistaken if we ever judge that we lack complete knowledge of any one of these simple natures. For if we have even the slightest grasp of it in our mind—which we surely must have, on the assumption that we are making a judgement about it—it must follow that we have complete knowledge of it. Otherwise it could not be said to be simple, but a composite made up of that which we perceive in it and that of which we judge we are ignorant.

Fourthly, the conjunction between these simple things is either necessary or contingent. The conjunction is necessary when one of them is somehow implied (albeit confusedly) in the concept of the other so that we cannot conceive of either of them distinctly if we judge them to be separate from each other. It is in this way that shape is conjoined with extension, motion with duration or time, etc., because we cannot conceive of a shape which is completely lacking in extension, or a motion wholly lacking in duration. Similarly, if I say that 4 and 3 make 7, the composition is a necessary one, for we do not have a distinct conception of the number 7 unless in a confused sort of way we include 3 and 4 in it. In the same way, whatever we demonstrate concerning figures or numbers necessarily links up with that of which it is affirmed. This necessity applies not just to things which are perceivable by the senses but to others as well. If, for example, Socrates says that he doubts everything, it necessarily follows that he understands at least that he is doubting, and hence that he knows that something can be true or false, etc.; for there is a necessary connection between these facts and the nature of doubt. The union between such things, however, is contingent when the relation conjoining them is not an inseparable one. This is the case when we say that a body is animate, that a man is dressed, etc. Again, there are many instances of things which are necessarily conjoined, even though most people count them as contingent, failing to notice the relation between them: for example the proposition, 'I am, therefore God exists', or 'I understand, therefore I have a mind distinct from my body.' Finally, we must note that very many necessary propositions, when converted, are contingent. Thus from the fact that I exist I may conclude with certainty that God exists, but from the fact that God exists I cannot legitimately assert that I too exist.

Fifthly, it is not possible for us ever to understand anything beyond those simple natures and a certain mixture or compounding of one with another. Indeed, it is often easier to attend at once to several mutually conjoined natures than to separate one of them from the others. For example, I can have knowledge of a triangle, even though it has never occurred to me that this knowledge involves knowledge also of the angle, the line, the number three, shape, extension, etc. But that does not preclude our saying that the nature of a triangle is composed of these other natures and that they are better known than the triangle, for it is just these natures that we understand to be present in it. Perhaps there are many additional natures implicitly contained in the triangle which escape our notice, such as the size of the angles being equal to two right angles, the innumerable relations between the sides and the angles, the size of its surface area, etc.

Sixthly, those natures which we call 'composite' are known by us either because we learn from experience what sort they are, or because we ourselves put them together. Our experience consists of whatever we perceive by means of the senses, whatever we learn from others, and in general whatever reaches our intellect either from external sources or from its own reflexive self-contemplation. We should note here that the intellect can never be deceived by any experience, provided that when the object is presented to it, it intuits it in a fashion exactly corresponding to the way in which it possesses the object, either within itself or in the imagination. Furthermore, it must not judge that the imagination faithfully represents the objects of the senses, or that the senses take on the true shapes of things, or in short that external things always are just as they appear to be. In all such cases we are liable to go wrong, as we do for example when we take as gospel truth a story which someone has told us; or as someone who has jaundice does when, owing to the yellow tinge of his eyes, he thinks everything is coloured yellow; or again, as we do when our imagination is impaired (as it is in depression) and we think that its disordered images represent real things. But the understanding of the wise man will not be deceived in such cases: while he will judge that whatever comes to him from his imagination really is depicted in it, he will never assert that it passes, complete and unaltered, from the external world to his senses, and from his senses to the corporeal imagination, unless he already has some other grounds for claiming to know this. But whenever we believe that an object of our understanding contains something of which the mind has no immediate perceptual experience, then it is we ourselves who are responsible for its composition. In the same way, when someone who has jaundice is convinced that the things he sees are yellow, this thought of his will be composite, consisting partly of what his corporeal imagination represents to him and partly of the assumption he is making on his own account, *viz.* that the colour looks yellow not owing to any defect of vision but because the things he sees really are yellow. It follows from this that we can go wrong only when we ourselves compose in some way the objects of our belief.

Seventhly, this composition can come about in three ways: through impulse, through conjecture or through deduction. It is a case of composition through impulse when, in forming judgements about things, our native intelligence leads us to believe something, not because good reasons con-

32

vince us of it, but simply because we are caused to believe it, either by some superior power, or by our free will, or by a disposition of the corporeal imagination. The first cause is never a source of error, the second rarely, the third almost always; but the first of these is irrelevant in this context, since it does not come within the scope of method. An example of composition by way of conjecture would be our surmising that above the air there is nothing but a very pure ether, much thinner than air, on the grounds that water, being further from the centre of the globe than earth, is a thinner substance than earth, and air, which rises to greater heights than water, is thinner still. Nothing that we put together in this way really deceives us, so long as we judge it to be merely probable, and never assert it to be true; nor for that matter does it make us any the wiser.

Deduction, therefore, remains as our sole means of compounding things in a way that enables us to be certain of their truth. Yet even with deduction there can be many drawbacks. If, say, we conclude that a given space full of air is empty, on the grounds that we do not perceive anything in it, either by sight, touch, or any other sense, then we are incorrectly conjoining the nature of a vacuum with the nature of this space. This is just what happens when we judge that we can deduce something general and necessary from something particular and contingent. But it is within our power to avoid this error, *viz.* by never conjoining things unless we intuit that the conjunction of one with the other is wholly necessary, as we do for example when we deduce that nothing which lacks extension can have a shape, on the grounds that there is a necessary connection between shape and extension, and so on.

From all these considerations we may draw several conclusions. First, we have explained distinctly and, I think, by an adequate enumeration, what at the outset we were able to present only in a confused and rough-and-ready way, *viz.* that there are no paths to certain knowledge of the truth accessible to men save manifest intuition and necessary deduction. . . .

Second, we need take no great pains to discover these simple natures, because they are self-evident enough. What requires effort is distinguishing one from another, and intuiting each one separately with steadfast mental gaze. . . .

Third, the whole of human knowledge consists uniquely in our achieving a distinct perception of how all these simple natures contribute to the composition of other things. This is a very useful point to note, since whenever some difficulty is proposed for investigation, almost everyone gets stuck right at the outset, uncertain as to which thoughts he ought to concentrate his mind on, yet quite convinced that he ought to seek some new kind of entity previously unknown to him. . . .

Lastly, from what has been said it follows that we should not regard some branches of our knowledge of things as more obscure than others, since they are all of the same nature and consist simply in the putting together of self-evident facts. . . .

For the rest, in case anyone should fail to see the interconnection between our Rules, we divide everything that can be known into simple propositions and problems. As for simple propositions, the only rules we provide are those which prepare our cognitive powers for a more distinct intuition of any given object and for a more discerning examination of it. For these simple propositions must occur to us spontaneously; they cannot be sought out. We have covered simple propositions in the preceding twelve Rules, and everything that might in any way facilitate the exercise of reason has, we think, been presented in them. As for problems, however, some can be understood perfectly, even though we do not know the solutions to them, while others are not perfectly understood. . . . We must note that a problem is to be counted as perfectly understood only if we have a distinct perception of the following three points: first, what the criteria are which enable us to recognize what we are looking for when we come upon it; second, what exactly is the basis from which we ought to deduce it; third, how it is to be proved that the two are so mutually dependent that the one cannot alter in any respect without there being a corresponding alteration in the other. So now that we possess all the premisses, the only thing that remains to be shown is how the conclusion is to be found. This is not a matter of drawing a single deduction from a single, simple fact, for, as we have already pointed out, that can be done without the aid of rules; it is, rather, a matter of deriving a single fact which depends on many interconnected facts, and of doing this in such a methodical way that no greater intellectual capacity is required than is needed for the simplest inference. Problems of this sort are for the most part abstract, and arise almost exclusively in arithmetic and geometry, which is why they will seem to ignorant people to be of little use. . . .

Rule 13

If we perfectly understand a problem we must abstract it from every superfluous conception, reduce it to its simplest terms and, by means of an enumeration, divide it up into the smallest possible parts.

We view the whole matter in the following way. First, in every problem there must be something unknown; otherwise there would be no point in posing the problem. Secondly, this unknown something must be delineated in some way, otherwise there would be nothing to point us to one line of investigation as opposed to any other. Thirdly, the unknown something can be delineated only by way of something else which is already known. These conditions hold also for imperfect problems. If, for example, the problem concerns the nature of the magnet, we already understand what is meant by the words 'magnet' and 'nature', and it is this knowledge which determines us to adopt one line of inquiry rather than another, etc. But if the problem is to be perfect, we want it to be determinate in every respect, so that we are not looking for anything beyond what can be deduced from the data. . . .

34

Furthermore, the problem should be reduced to the simplest terms according to Rules Five and Six, and it should be divided up according to Rule Seven. Thus if I carry out many observations in my research on the magnet, I shall run through them separately one after another. Again, if the subject of my research is sound, as in the case above, I shall make separate comparisons between strings A and B, then between A and C, etc., with a view to including all of them together in a sufficient enumeration. With respect to the terms of a given problem, these three points are the only ones which the pure intellect has to observe before embarking on the final solution of the problem, for which the following eleven Rules may be required. . . .

In every problem, of course, there has to be something unknown—otherwise the inquiry would be pointless. Nevertheless this unknown something must be delineated by definite conditions, which point us decidedly in one direction of inquiry rather than another. These conditions should, in our view, be gone into right from the very outset. We shall do this if we concentrate our mind's eye on intuiting each individual condition distinctly, looking carefully to see to what extent each condition delimits the unknown object of our inquiry. For in this context the human mind is liable to go wrong in one or other of two ways: it may assume something beyond the data required to define the problem, or on the other hand it may leave something out.

. . . .

2

Descartes' Ontological Proof of God

From Meditations on First Philosophy / René Descartes

Fifth Meditation. The Essence of Material Things, and the Existence of God Considered a Second Time

. . . .

I find within me countless ideas of things which even though they may not exist anywhere outside me still cannot be called nothing; for although in a sense they can be thought of at will, they are not my invention but have their own true and immutable natures. When, for example, I imagine a triangle, even if perhaps no such figure exists, or has ever existed, anywhere outside my thought, there is still a determinate nature, or essence, or form of the triangle which is immutable and eternal, and not invented by me or dependent on my mind. This is clear from the fact that various properties can be demonstrated of the triangle, for example that its three angles equal two right angles, that its greatest side subtends its greatest angle, and the like; and since these properties are ones which I now clearly recognize whether I want to or not, even if I never thought of them at all when I previously imagined the triangle, it follows that they cannot have been invented by me.

Translators' notes have been omitted.

It would be beside the point for me to say that since I have from time to time seen bodies of triangular shape, the idea of the triangle may have come to me from external things by means of the sense organs. For I can think up countless other shapes which there can be no suspicion of my ever having encountered through the senses, and yet I can demonstrate various properties of these shapes, just as I can with the triangle. All these properties are certainly true, since I am clearly aware of them, and therefore they are something, and not merely nothing; for it is obvious that whatever is true is something; and I have already amply demonstrated that everything of which I am clearly aware is true. And even if I had not demonstrated this,

36

the nature of my mind is such that I cannot but assent to these things, at least so long as I clearly perceive them. I also remember that even before, when I was completely preoccupied with the objects of the senses, I always held that the most certain truths of all were the kind which I recognized clearly in connection with shapes, or numbers or other items relating to arithmetic or geometry, or in general to pure and abstract mathematics.

But if the mere fact that I can produce from my thought the idea of something entails that everything which I clearly and distinctly perceive to belong to that thing really does belong to it, is not this a possible basis for another argument to prove the existence of God? Certainly, the idea of God, or a supremely perfect being, is one which I find within me just as surely as the idea of any shape or number. And my understanding that it belongs to his nature that he always exists is no less clear and distinct than is the case when I prove of any shape or number that some property belongs to its nature. Hence, even if it turned out that not everything on which I have meditated in these past days is true, I ought still to regard the existence of God as having at least the same level of certainty as I have hitherto attributed to the truths of mathematics.

At first sight, however, this is not transparently clear, but has some appearance of being a sophism. Since I have been accustomed to distinguish between existence and essence in everything else, I find it easy to persuade myself that existence can also be separated from the essence of God, and hence that God can be thought of as not existing. But when I concentrate more carefully, it is quite evident that existence can no more be separated from the essence of God than the fact that its three angles equal two right angles can be separated from the essence of a triangle, or than the idea of a mountain can be separated from the idea of a valley. Hence it is just as much of a contradiction to think of God (that is, a supremely perfect being) lacking existence (that is, lacking a perfection), as it is to think of a mountain without a valley.

However, even granted that I cannot think of God except as existing, just as I cannot think of a mountain without a valley, it certainly does not follow from the fact that I think of a mountain with a valley that there is any mountain in the world; and similarly, it does not seem to follow from the fact that I think of God as existing that he does exist. For my thought does not impose any necessity on things; and just as I may imagine a winged horse even though no horse has wings, so I may be able to attach existence to God even though no God exists.

But there is a sophism concealed here. From the fact that I cannot think of a mountain without a valley, it does not follow that a mountain and valley exist anywhere, but simply that a mountain and a valley, whether they exist or not, are mutually inseparable. But from the fact that I cannot think of God except as existing, it follows that existence is inseparable from God, and hence that he really exists. It is not that my thought makes it so, or imposes any necessity on any thing; on the contrary, it is the necessity of the thing itself, namely the existence of God, which determines my thinking in this respect. For I am not free to think of God without existence (that is, a supremely perfect being without a supreme perfection) as I am free to imagine a horse with or without wings.

And it must not be objected at this point that while it is indeed necessary for me to suppose God exists, once I have made the supposition that he has all perfections (since existence is one of the perfections), nevertheless the original supposition was not necessary. Similarly, the objection would run, it is not necessary for me to think that all quadrilaterals can be inscribed in a circle; but given this supposition, it will be necessary for me to admit that a rhombus can be inscribed in a circle—which is patently false. Now admittedly, it is not necessary that I ever light upon any thought of God; but whenever I do choose to think of the first and supreme being, and bring forth the idea of God from the treasure house of my mind as it were, it is necessary that I attribute all perfections to him, even if I do not at that time enumerate them or attend to them individually. And this necessity plainly guarantees that, when I later realize that existence is a perfection, I am correct in inferring that the first and supreme being exists. In the same way, it is not necessary for me ever to imagine a triangle; but whenever I do wish to consider a rectilinear figure having just three angles, it is necessary that I attribute to it the properties which license the inference that its three angles equal no more than two right angles, even if I do not notice this at the time. By contrast, when I examine what figures can be inscribed in a circle, it is in no way necessary for me to think that this class includes all quadrilaterals. Indeed, I cannot even imagine this, so long as I am willing to admit only what I clearly and distinctly understand. So there is a great difference between this kind of false supposition and the true ideas which are innate in me, of which the first and most important is the idea of God. There are many ways in which I understand that this idea is not something fictitious which is dependent on my thought, but is an image of a true and immutable nature. First of all, there is the fact that, apart from God, there is nothing else of which I am capable of thinking such that existence belongs to its essence. Second, I cannot understand how there could be two or more Gods of this kind; and after supposing that one God exists, I plainly see that it is necessary that he has existed from eternity and will abide for eternity. And finally, I perceive many other attributes of God, none of which I can remove or alter.

38

But whatever method of proof I use, I am always brought back to the fact that it is only what I clearly and distinctly perceive that completely convinces me. Some of the things I clearly and distinctly perceive are obvious to everyone, while others are discovered only by those who look more closely and investigate more carefully; but once they have been discovered, the latter are judged to be just as certain as the former. In the case of a right-angled triangle, for example, the fact that the square on the hypotenuse is equal to the square on the other two sides is not so readily apparent as the fact that the hypotenuse subtends the largest angle; but once one has seen it, one believes it just as strongly. But as regards God, if I were not overwhelmed by preconceived opinions, and if the images of things perceived by the senses did not besiege my thought on every side, I would certainly acknowledge him sooner and more easily than anything else. For what is more self-evident than the fact that the supreme being exists, or that God, to whose essence alone existence belongs, exists?

Although it needed close attention for me to perceive this, I am now just as certain of it as I am of everything else which appears most certain. And what is more, I see that the certainty of all other things depends on this, so that without it nothing can ever be perfectly known.

Admittedly my nature is such that so long as I perceive something very clearly and distinctly I cannot but believe it to be true. But my nature is also such that I cannot fix my mental vision continually on the same thing, so as to keep perceiving it clearly; and often the memory of a previously made judgement may come back, when I am no longer attending to the arguments which led me to make it. And so other arguments can now occur to me which might easily undermine my opinion, if I were unaware of God; and I should thus never have true and certain knowledge about anything, but only shifting and changeable opinions. For example, when I consider the nature of a triangle, it appears most evident to me, steeped as I am in the principles of geometry, that its three angles are equal to two right angles; and so long as I attend to the proof, I cannot but believe this to be true. But as soon as I turn my mind's eye away from the proof, then in spite of still remembering that I perceived it very clearly, I can easily fall into doubt about its truth, if I am unaware of God. For I can convince myself that I have a natural disposition to go wrong from time to time in matters which I think I perceive as evidently as can be. This will seem even more likely when I remember that there have been frequent cases where I have regarded things as true and certain, but have later been led by other arguments to judge them to be false.

Now, however, I have perceived that God exists, and at the same time I have understood that everything else depends on him, and that he is no deceiver; and I have drawn the conclusion that everything which I clearly and distinctly perceive is of necessity true. Accordingly, even if I am no longer attending to the arguments which led me to judge that this is true, as long as I remember that I clearly and distinctly perceived it, there are no counter-arguments which can be adduced to make me doubt it, but on the contrary I have true and certain knowledge of it. And I have knowledge not just of this matter, but of all matters which I remember ever having demonstrated, in geometry and so on. For what objections can now be raised? That the way I am made makes me prone to frequent error? But I now know that I am incapable of error in those cases where my understanding is transparently clear. Or can it be objected that I have in the past regarded as true and certain many things which I afterwards recognized to be false? But none of these were things which I clearly and distinctly perceived: I was ignorant of this rule for establishing the truth, and believed these things for other reasons which I later discovered to be less reliable. So what is left to say? Can one raise the objection I put to myself a while ago, that I may be dreaming, or that everything which I am now thinking has as little truth as what comes to the mind of one who is asleep? Yet even this does not change anything. For even though I might be dreaming, if there is anything which is evident to my intellect, then it is wholly true.

Thus I see plainly that the certainty and truth of all knowledge depends uniquely on my awareness of the true God, to such an extent that I was incapable of perfect knowledge about anything else until I became aware of him. And now it is possible for me to achieve full and certain knowledge of countless matters, both concerning God himself and other things whose nature is intellectual, and also concerning the whole of that corporeal nature which is the subject-matter of pure mathematics.

3

Descartes' Laws of Motion

From

Principles of Philosophy / René Descartes

Part 2. Principles of Material Things

. . . .

XXXVI. After considering the nature of motion, we must treat of its cause; in fact, of two sorts of cause. First, the universal and primary cause—the general cause of all the motions in the universe; secondly the particular cause that makes any given piece of matter assume a motion that it had not before.

As regards the general cause, it seems clear to me that it can be none other than God himself. He created matter along with motion and rest in the beginning; and now, merely by his ordinary co-operation, he preserves just the quantity of motion and rest in the material world that he put there in the beginning. Motion, indeed, is only a state of the moving body; but it has a certain definite quantity, and it is readily conceived that this quantity may be constant in the universe as a whole, while varying in any given part. (We must reckon the quantity of motion in two pieces of matter as equal if one moves twice as fast as the other, and this in turn is twice as big as the first; again, if the motion of one piece of matter is retarded, we must assume

Reprinted from *Descartes: Philosophical Writings,* trans. and ed. Elizabeth Anscombe and Peter Thomas Geach (Indianapolis: Bobbs-Merrill, 1971), 215–26.

Translators' notes have been omitted. Elisions and bracketed interpolations are theirs.

an equal acceleration of some other body of the same size.) Further, we conceive it as belonging to God's perfection, not only that he should in himself be unchangeable, but also that his operation should occur in a supremely constant and unchangeable manner. Therefore, apart from the changes of which we are assured by manifest experience or by divine revelation, and about which we can see, or believe [by faith], that they take place without any change in the Creator, we must not assume any others in the works of God, lest they should afford an argument for his being inconstant. Consequently it is most reasonable to hold that, from the mere fact that God gave pieces of matter various movements at their first creation, and that he now preserves all this matter in being in the same way as he first created it, he must likewise always preserve in it the same quantity of motion.

XXXVII. From God's immutability we can also know certain rules or natural laws which are the secondary, particular causes of the various motions we see in different bodies. The *first* law is: *Every reality, in so far as it is simple and undivided, always remains in the same condition so far as it can, and never changes except through external causes.* Thus if a piece of matter is square, one readily convinces oneself that it will remain square for ever, unless something comes along from elsewhere to change its shape. If it is at rest, one thinks it will never begin to move, unless impelled by some cause. Now there is equally no reason to believe that if a body is moving its motion will ever stop, spontaneously that is, and apart from any obstacle. So our conclusion must be: *A moving body, so far as it can, goes on moving.*

We, however, live on the Earth, and the constitution of the Earth is such that all motions in her neighbourhood are soon arrested—often by insensible causes. Thus from our earliest years we have held the view that these motions (which in fact are brought to rest by causes unknown to us) come to an end spontaneously. And we tend to hold in all cases what we think we have observed in many cases—that motion ceases, or tends towards rest, by its very nature. Now this is in fact flatly opposed to the laws of nature; for rest is the opposite of motion, and nothing can by its own nature tend towards its opposite, towards its own destruction.

XXXVIII. Our everyday observation of projectiles completely confirms this rule. The reason why projectiles persist in motion for some time after leaving the hand that throws them is simply that when they once move they go on moving, until their motion is retarded by bodies that get in the way. Obviously the air, or other fluid in which they are moving, gradually retards their motion, so that it cannot last long. The resistance of air to the movement of other bodies may be verified by the sense of touch if we beat it with a fan; and the flight of birds confirms this. And the resistance of any fluid other than air to the motion of projectiles is even more obvious.

XXXIX. The *second* natural law is: *Any given piece of matter considered by itself tends to go on moving, not in any oblique path, but only in straight lines.* (Of course many pieces of matter are constantly being compelled to swerve by meeting with others; and, as I said, any motion involves a kind of circulation of matter all moving simultaneously.) The reason for this rule, like that for the last one, is the immutability and simplicity of the operation by which God preserves motion in matter. For he preserves the motion in the precise form in which it occurs at the moment when he preserves it, without regard to what it was a little while before. *In* the instant, of course, no motion can take place; but obviously the motion of any moving body is determined *at* any assigned instant of its duration as capable of being continued in a given direction; continued, that is, in a straight line, not some sort of curve. For example, a stone A is moving in a sling EA in a circle ABF. At the moment when it is at the point A, it has motion in a definite direction, viz. in a straight line towards C, where the straight line AC is a tangent to the circle. It cannot be imagined that the stone has any definite curvilinear motion; it is true that it arrived at A from L along a curved path, but none of this curvature can be conceived as inherent in its motion when it is at the point A. Observation confirms this; for if the stone leaves the sling just then, it goes on towards C, not towards B. . . .

42

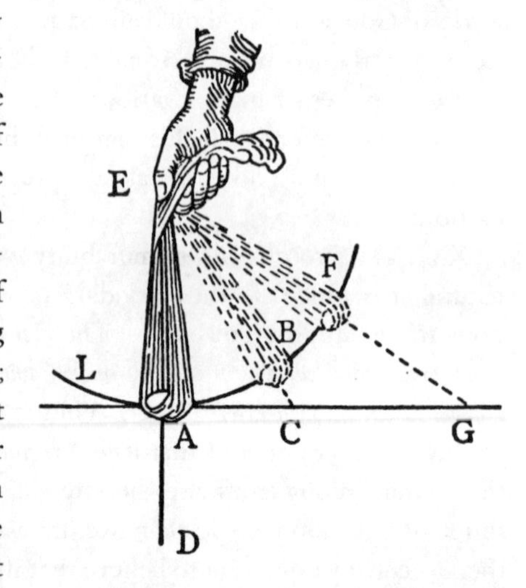

XL. The *third* natural law is this. *When a moving body collides with another, then if its own power of going on in a straight line is less than the resistance of the other body, it is reflected in another direction and retains the same amount of motion, with only a change in its direction; but if its power of going on is greater than the resistance, it carries the other body along with it, and loses a quantity of motion equal to what it imparts to the other body.* Thus we observe that hard projectiles, when they strike some other hard body, do not stop moving but are reflected in the opposite direction; on the other hand, when they collide with a soft body, they readily transfer all their motion to it, and are thus at once stopped. This third law covers all the particular causes of corporeal change—so far as they are themselves corporeal; I am not now considering whether, or how, human or angelic minds have the power to move bodies. . . .

XLI. To prove the first part of this law: there is a difference between a motion as such and its determinate direction; it is thus possible for the direction to change while the motion remains unaltered. Now, as I said, any given reality which, like motion, is not complex but simple, persists in being so long as it is not destroyed by any external cause. In a collision with a hard body, there is an obvious reason why the motion of the other body that collides with it should not continue in the same direction; but there is no obvious reason why this motion should be stopped or lessened, for one motion is not the opposite of another motion; so the motion ought not to be diminished.

XLII. The second part is proved from the immutability of the divine operation; God preserving the world by the same activity by which he once created it. For all places are filled with body, and at the same time the motion of every body is rectilinear in tendency; so clearly, when God first created the world, he must not only have assigned various motions to its various parts, but also have caused their mutual impulses and the transference of motion from one to another; and since he now preserves motion by the same activity and according to the same laws, as when he created it, he does not preserve it as a constant inherent property of given pieces of matter, but as something passing from one piece to another as they collide. Thus the very fact that creatures are thus continually changing argues the immutability of God.

XLIII. It must be carefully observed what it is that constitutes the power of a body to act on another body or resist its action; it is simply the tendency of everything to persist in its present state so far as it can (according to the first law). Thus what is joined to another thing has some power of resisting separation from it; what is separate has some power of remaining separate; what is at rest has some power of remaining at rest, and consequently resisting everything that might change its state of rest; what is moving has some power of persisting in its motion—in a motion constant as regards velocity and direction. This last power must be estimated according to the size of the body, and of its surface, which separates it from others, and the velocity of the motion, and the kind and degrees of opposition of state involved in the collision of bodies.

XLIV. Here we must observe that one motion is in no way opposite to another of equal velocity. Properly speaking, there are two sorts of opposition. First, motion is opposite to rest, and likewise a swift motion to a slow one, since slowness has something of the nature of rest. Secondly, the determinate direction of a motion is opposed to the body's meeting with another that lies in that direction and is at rest or is moving differently. The degree of this opposition depends on the direction in which a body is moving when it collides with another.

XLV. To determine from this how collision increases or diminishes the amounts of motion in bodies, or how it alters the direction of their motion, we need only calculate the power of each body to move or to resist motion, and use the principle that the greater power always produces its effect. The calculation would be easy if there were just two bodies colliding, and these were perfectly solid and entirely separated from all others, so that no surrounding bodies impeded or assisted their motion. . . .

LIII. In fact, no bodies in the universe can be thus separated from all others; and in our environment we do not ordinarily get perfectly solid bodies; so it is much harder to calculate how the motions of bodies are changed by collision with others. For we have to take into account all the

44

bodies that touch a body on every side; and the effect of these is very different according as they are solid or fluid.

We must now consider what constitutes this distinction.

LIV. According to the evidence of the senses, the only distinction we can discern is that the parts of fluids readily leave their place, and so offer no resistance when our hands move towards them; whereas the parts of solids cohere together so that they can be separated only by a force sufficient to overcome their cohesion. If we inquire further why some bodies readily abandon their place for others, while other bodies do not, we can easily see that what is already in motion does not hinder the occupation by another body of the place that it is in any case leaving, whereas what is at rest cannot be driven from its place except by some force. Hence we may infer that bodies divided into many small particles that are agitated by a variety of motions are *fluids;* while those whose particles are at rest relatively to their neighbours are *solids.*

LVI. As regards fluids, we cannot observe any sensible motion of their particles, because they are too small; but such motion is readily inferred from its effects, especially in the case of air and water. For these fluids corrupt many other bodies; and no corporeal activity such as corruption is possible apart from local motion. . . .

LXIV. I will not here say anything further about [geometrical] figures, or as to how there follow from their infinite variety countless varieties of kinds of motion; these points will be sufficiently clear in themselves when we have to treat of them. I presuppose in my readers either a familiarity with elementary geometry, or at least a mental aptitude for following mathematical proofs. I must here make it clear that I recognise no kind of 'matter' in corporeal objects except that 'matter' susceptible of every sort of division, shape, and motion, which geometers call quantity, and which they presuppose as the subject-matter of their proofs. Further, the only properties I am considering in it are these divisions, shapes and motions; and about them I assume only what can be derived in a self-evident way from indubitably true axioms so that it can be counted as a mathematical proof. All natural phenomena, as I shall show, can be explained in this way; I therefore do not think any other principles need be admitted in physics or are to be desired.

Part 3. The Visible World

I. We have thus discovered certain principles as regards material objects, derived not from the prejudices of our senses but from the light of reason, so that their truth is indubitable; we must now consider whether they suffice to explain all natural phenomena. We must begin with the most general facts on which the rest depend—the construction of the visible universe as a whole. For correct theorising about this, two cautions are needed. First, we must consider the infinite power and goodness of God, and not be afraid that we are imagining his works to be too vast, too beautiful, too perfect; what we must beware of is, on the contrary, the supposition of any bounds to God's works that we do not certainly know, lest we may seem not to have a sufficiently grand conception of the power of the Creator.

II. Secondly, we must beware of thinking too proudly of ourselves. We should be doing this, not merely if we imagined any limits to the universe, when none are known to us either by reason or by divine revelation (as if our powers of thought could extend beyond what God has actually made); but also, and that in a special degree, if we imagined everything had been created by God for our sake; or even if we thought our minds had the power to comprehend the ends God set before himself in creating the world.

III. In ethics indeed it is an act of piety to say that God made everything for our sake, that we may be the more impelled to thank him, and the more on fire with love of him; and in a sense this is true; for we can make *some* use of all things—at least we can employ our mind in contemplating them, and in admiring God for his wonderful works. But it is by no means probable that all things were made for our sake in the sense that they have no other use. In physical theory this supposition would be wholly ridiculous and absurd; for undoubtedly many things exist (or did exist formerly and now do so no longer) that have never been seen or thought of by any man, and have never been any use to anybody.

IV. The principles we have discovered so far are so vast and so fertile, that their consequences are far more numerous than the observable contents of the visible universe; far too numerous, indeed, to be ever exhaustively considered. For an investigation of causes, I here present a brief account of the principal phenomena of nature. Not that we should use these as grounds for proving anything; for our aim is to deduce an account of the effects from the causes, not to deduce an account of the causes from the effects. It is just a matter of turning our mind to consider some effects rather than others out of an innumerable multitude; all producible, on our view, by a single set of causes.

XLII. . . . To discern the real nature of this visible universe, it is not enough to find causes in terms of which we may explain what we see far away in the heavens; we must also deduce from the same causes everything that we see close at hand on earth. We need not indeed consider all of these phenomena in order to determine the causes of more general effects; but *ex postfacto* we shall know that we have determined these causes correctly only when we see that we can explain in terms of them, not merely the effects we had originally in mind, but also all other phenomena of which we did not previously think.

XLIII. But assuredly, if the only principles we use are such as we see to be self-evident; if we infer nothing from them except by mathematical deduction; and if these inferences agree accurately with all natural phenomena: then we should, I think, be wronging God if we were to suspect this discovery of the causes of things to be delusive. God would, so to say, have made us so imperfectly that by using reason rightly we nevertheless went wrong.

XLIV. However, to avoid the apparent arrogance of asserting that the actual truth has been discovered in such an important subject of speculation, I prefer to waive this point; I will put forward everything that I am going to write just as a hypothesis. Even if this be thought to be false, I shall think my achievement is sufficiently worth while if all inferences from it agree with observation; for in that case we shall get as much practical benefit from it as we should from the knowledge of the actual truth.

XLV. Moreover, in order to explain natural objects the better, I shall pursue my inquiry into their causes further back than I believe the causes ever in fact existed. There is no doubt that the world was first created in its full perfection; there were in it a Sun, an Earth, a Moon, and the stars; and on the Earth there were not only the seeds of plants, but also the plants themselves; and Adam and Eve were not born as babies, but made as full-grown human beings. This is the teaching of the Christian faith; and natural reason convinces us that it was so; for considering the infinite power of God, we cannot think he ever made anything that was not peerless. Nevertheless, in order to understand the stature of plants or man, it is far better to consider how they may now gradually develop from seed, rather than the way they were created by God at the beginning of the world; and in just the same way we may conceive certain elements, very simple and very easily understood, and from these seeds, so to say, we may prove that there could have arisen stars, and an Earth, and in fact everything we observe in this visible universe; and although we know perfectly well they never did arise in this way, yet by this method we shall give a far better account of their nature than if we merely describe what they now are. . . .

XLVI. From what has already been said it is established that all bodies in the universe consist of one and the same matter; that this is divisible arbitrarily into parts, and is actually divided into many pieces with various motions; that their motion is in a way circular, and that the same quantity of motion is constantly preserved in the universe. We cannot determine by reason how big these pieces of matter are, how quickly they move, or what circles they describe. God might have arranged these things in countless different ways; which way he in fact chose rather than the rest is a thing we must learn from observation. Therefore, we are free to make any assumption we like about them, so long as all the consequences agree with experience. So, by your leave, I shall suppose that all the matter constituting the visible world was originally divided by God into unsurpassably equal particles of medium size—that is, of the average size of those that now form the heavens and the stars; that they had collectively just the quantity of motion now found in the world; that . . . each turned round its own centre, so that they formed a fluid body, such as we take the heavens to be; and that many revolved together around various other points . . . and thus constituted as many different vortices as there now are stars in the world.

XLVII. These few assumptions are, I think, enough to supply causes from which all effects observed in our universe would arise by the laws of nature previously stated; and I think one cannot imagine any first principles that are more simple, or easier to understand, or indeed more likely. The actual arrangement of things might perhaps be inferable from an original Chaos, according to the laws of nature; and I once undertook to give such an explanation. But confusion seems less in accord with the supreme perfection of God the Creator of all things than proportion or order, and we can form a less distinct notion of it. . . . In any case, it matters very little what supposition we make; for change must subsequently take place according to the laws of nature; and it is hardly possible to make a supposition that does not allow of our inferring the same effects (perhaps with more labour) according to the same laws of nature. For according to these, matter must successively assume all the forms of which it admits; and if we consider these forms in order, we can at last come to that which is found in this universe. So no error is to be apprehended from a false supposition at this point.

LV. It is a law of nature that *all bodies moving in a circle move away from the centre of their motion so far as they can.* I shall at this point explain as accurately as I can this force by which [bodies] try to move away from these centres. . . .

LVI. When I say [they] 'try' to move away from the centres around which they revolve, I must not, therefore, be thought to be fancying that they have some consciousness from which this 'effort' proceeds; I just mean that their positions, and the forces that impel them to motion, are such that they would in fact go in that direction, if no other cause hindered them.

4

A Discussion of Descartes' Methodology

From

Descartes' Metaphysical Physics / Daniel Garber

Chapter 2. Descartes' Project: Method and Order in the Rules

. . . .

The order of knowledge is a basic theme in the *Rules.* But to understand the conception of the order of knowledge that underlies the *Rules,* it will be helpful to begin with an account of the method. And to understand method in the *Rules,* we must understand something of what Descartes takes the goal of inquiry to be.

From the earliest portions of the *Rules,* portions thought to date from as early as mid-November 1619, shortly after the dreams of November 10, Descartes is clear that the goal of method is certainty. Thus he wrote in the very first rule: "The goal [*finis*] of studies ought to be the direction of one's mind [*ingenium*] toward making solid and true judgments about everything

---

Reprinted from *Descartes' Metaphysical Physics* (Chicago: University of Chicago Press, 1992), 31–37, 211–23.

Garber's endnotes have been omitted. The abbreviation AT refers to the original-language text of Descartes' writings in Charles Adam and Paul Tannery, eds., *Oeuvres de Descartes* (Paris: J. Vrin, 1964–74).

which comes before it" (AT X 359). And in the second, probably from the same period, he wrote: "We should concern ourselves only with those objects for which our minds seem capable of certain and indubitable cognition" (AT X 362).

Descartes' conception of certainty in the *Rules* has two principal elements, the mental processes that result in certainty, intuition, and deduction, and the primary objects of certainty, what Descartes calls simple natures. In Rule 3 he announces that it is only intuition and deduction that give us the sort of knowledge which he seeks: "Concerning things proposed, one ought to seek not what others have thought, nor what we conjecture, but what we can clearly and evidently intuit or deduce with certainty; for in no other way is knowledge [*scientia*] acquired" (AT X 366). Similarly, he says that intuition and deduction are the only acts of intellect that lead to a knowledge of things "without any fear of being deceived" (AT X 368). He defines intuition as follows:

> By *intuition* [*intuitus*] I understand, not the fluctuating faith in the senses, nor the deceitful judgment of a poorly composed imagination; but a conception of a pure and attentive mind [*mens*], so easy and distinct that concerning that which we understand [*intelligere*] no further doubt remains; or, what is the same, the undoubted conception of a pure and attentive mind, which arises from the light of reason alone. (AT X 368)

Deduction is characterized as a chain of successive intuitions. Descartes writes:

> Many things are known with certainty, even though they are not themselves evident, only because they are deduced from true and known principles through a continuous and uninterrupted movement of thought, perspicuously intuiting each individual thing. . . . We can therefore distinguish an intuition of the mind from a certain deduction, by the fact that in the one, we conceive a certain motion or succession, but not in the other. (AT X 369–70)

Or, as he puts it a bit more clearly in Rule 11: "Deduction . . . involves a certain movement of our mind, inferring one thing from another" (AT X 407).

Certainty as characterized in terms of intuition and deduction has a number of salient properties. Certainty, first of all, seems to entail indubitability in at least some weak sense; intuition, the ground of certainty, is, after all, the "undoubted conception of a pure and attentive mind." Furthermore, certainty involves a particular faculty of the mind; intuition,

Descartes tells us, is not derived from the senses or the imagination, but "arises from the light of reason alone" (AT X 368). And finally, he claims, for the mind to have an intuition, and thus, for it to perform a deduction, it must be in a special state, it must be both "pure" and "attentive."

Descartes also thinks certain knowledge has its own special set of objects, a particular collection of notions or concepts; certain knowledge, strictly speaking, is knowledge of these simple natures and their combinations. Simple natures are first introduced in Rule 6, which most likely dates from mid-November 1619, the same time he drafted the passages on intuition and deduction that we have been discussing. When introduced in Rule 6, they are connected not so much with the characterization of certainty as with the method as sketched in Rule 5, a question to which we shall shortly turn. In Rule 6 he explains how "things" (the Latin *res,* noncommittal in the extreme) can be arranged in a series, "not, indeed, insofar as they are referred to some genus of being, as the philosophers divide them up into their categories, but insofar as some can be known from others" (AT X 381). At the beginning of the series are those things he calls complete (*absolutus*), things that "contain in themselves a pure and simple nature" (AT X 381). The "pure and simple natures" these complete things contain are, according to Rule 6, few in number and can be grasped (*intueri,* i.e., intuited) "first and through themselves, not dependently on the grasping of others, but through experience itself or through a certain light inherent in us" (AT X 383). Later in the series come relative things, things understood through the simple natures or absolute things (AT X 382–83). Descartes' examples here suggest that simple natures include cause, the one, and the equal, for example, and it is through these notions that we comprehend the effect, the many, and the unequal (AT X 381–83).

Simple natures reappear in Rule 12, in a passage written some years later, probably between 1626 and 1628. There the list has changed; instead of the grab bag of diverse and supposedly basic notions or concepts found in Rule 6, Descartes recognizes three kinds: those purely mental [*intellectualis*] (thought, doubt, will, etc.), those purely material (shape, extension, motion, etc.), and those that pertain to both (existence, unity, duration, etc.; see AT X 419). According to Rule 12, these simple natures are the primary objects of intuition and, indeed, the primary objects of all knowledge: "We can never understand anything but these simple natures and a certain mixture or composition of them"; "all human knowledge [*scientia*] consists in this one thing, that we distinctly see how these simple natures come together at the same time in the composition of other things" (AT X 422, 427).

This conception of certainty constitutes a kind of epistemology that underlies the methodological project of the *Rules.* Though unargued and unjustified until the very latest strata of the *Rules,* as I shall later discuss, it tells us what knowledge is, where it derives from, and what it comprehends. And insofar as it defines knowledge, it sets a goal for method: the construction of a body of knowledge grounded in intuition and deduction, derived from the intellect's acquaintance with simple natures, derived from the light of reason.

Descartes declares in Rule 4 that "method is necessary for seeking after the truth of things" (AT X 371). He continues:

> By method, moreover, I understand certain and easy rules which are such that whoever follows them exactly will never take that which is false to be true, and without consuming any mental effort uselessly, but always step by step increasing knowledge [*scientia*], will arrive at the true knowledge [*vera cognitio*] of everything of which he is capable. (AT X 371–72)

It is important to understand here that Descartes' idea is not that method will somehow help us in actually performing intuitions or deductions. Rather, method is supposed to be a tool by virtue of which the mind is led toward the *discovery* of genuine knowledge, knowledge based in intuitions and deductions, knowledge that the mind is in principle, though not perhaps in practice, capable of having independently of the method itself (AT X 372).

But what are the "certain and easy rules" that Descartes promises? Descartes summarizes them as follows in Rule 5:

> The whole of method consists in the order and disposition of those things toward which the mental insight [*mentis acies*] is to be directed so that we discover some truth. And this [rule] is observed exactly if we reduce involved and obscure propositions step by step to simpler ones, and thus from an intuition of the simplest we try to ascend by those same steps to a knowledge of all the rest. (AT X 379)

Descartes' rule of method has two steps: a *reductive* step, in which "involved and obscure propositions" are reduced to simpler ones, and a *constructive* step, in which we proceed from the simpler propositions back to the more complex. But the rule makes little sense, nor does it connect very clearly with the account of knowledge and certainty in terms of intuition and deduction unless we know what he means here by the reduction to simples, and the construction, or *re*construction of the complex from the simple.

The account of the workings of the method given in the earliest strata of the *Rules* is obscure and very difficult to translate into concrete terms. It is quite possible that Descartes' vision in those texts is still somewhat cloaked in poetic enthusiasm and that he himself may not have had a clear and distinct idea of precisely how the method was to work in actual practice. But matters are clarified considerably in an example of methodological investigation he gave late in the composition of the *Rules.* The example I have in mind is that of the anaclastic line, which Descartes gives in the commentary to Rule 8. The example he chooses is closely connected with the optical investigations that he undertook between 1626 and 1628, and it is highly probable that the example was added to the *Rules* sometime in that period. . . . As such it may well represent an understanding of the method that he simply did not have when he gave his first exposition of the method in 1619. Also, it is difficult to connect the account of method in Rule 8 with the account of method he gives in some other passages of the *Rules,* particularly in Rule 4, where the method seems closely connected with the *mathesis universalis,* or in Rule 14, where it seems more obviously mathematical. But this example is still useful for understanding one central theme in Descartes' thought on method, as it is developed in the Paris years of the late 1620s, when he was doing some of his most important scientific work, and just before he began to compose the first version of his system.

The problem that Descartes poses for himself is that of finding the anaclastic line, that is, the shape of a surface "in which parallel rays are refracted in such a way that they all intersect in a single point after refraction." Now, he notices—and this seems to be the first step in the reduction—that "the determination of this [anaclastic] line depends on the relation between the angle of incidence and the angle of refraction." But, he notes, this question is still "composite and relative," that is, not sufficiently simple, and we must proceed further in the reduction. He suggests that we must next ask how the relation between angles of incidence and refraction is caused by the difference between the two media (e.g., air and glass) which in turn raises the question as to "how the ray penetrates the whole transparent thing, and the knowledge of this penetration presupposes that the nature of illumination is also known." But, he claims, in order to understand what illumination is, we must know what a natural power (*potentia naturalis*) is. This is where the reductive step ends. At this point, Descartes seems to think that we can "clearly see through an intuition of the mind" what a natural power is. While Rule 8 is not entirely clear on this, it is plausible to connect this intuition to the conception of simple natures he had come to hold in the late 1620s. As I noted earlier, in Rule 12, written shortly after the completion of Rule 8, Descartes lists the simple natures pertaining to body as "shape, extension, motion, etc." (AT X 419). Rule 9, again dated from the same stage of composition, tells us, in turn, that in order to understand the notion of a natural power, "I will reflect on the local motions of bodies" (AT X 402). This suggests that the understanding of illumination is, somehow, an intuitive judgment about the simple nature, motion, though it is not clear how exactly he thought this would work. Once we have an intuition, we can begin the constructive step, and follow, in order, through the questions raised until we have answered the original question. That would involve understanding the nature of illumination from the nature of a natural power, understanding the way rays penetrate transparent bodies from the nature of illumination, and the relation between angle of incidence and angle of refraction from all that precedes. And finally, once we know how the angle of incidence and angle of refraction are related, we can solve the problem of the anaclastic line.

This example develops the programmatic statement of the method as given in Rule 5 in a fairly concrete way. If we take the anaclastic line example as our guide, then methodical investigation begins with a question, a question which, in turn, is reduced to questions whose answers are presupposed for the resolution of the original question posed (i.e., q1 is reduced to q2 if we must answer q2 before we can answer q1). The reductive step of the method thus involves, as Descartes suggested earlier in Rule 6, ordering things "insofar as some can be known from others, so that whenever some difficulty arises, we will immediately be able to perceive whether it will be helpful to examine some other [questions] first, and what, and in what order" (AT X 381). And so, in a sense, the reduction leads us to more basic and fundamental questions, from the anaclastic line, to the law of refraction, and back eventually to the nature of a natural power and to the motion of bodies. Ultimately, he thinks, when we follow out this series of questions, from the one that first interests us, to the "simpler" and more basic questions on which it depends, we will eventually reach an intuition, presumably an intuition about simple natures. When the reductive stage is taken to this point, then we can begin the constructive stage. Having intuited the answer to the last question in the reductive series, we can turn the procedure on its head, and begin answering the questions that we have successively raised, in an order the reverse of the order in which we have raised them. What this should involve is starting with the intuition that we have attained through the reductive step, and deducing down from there, until we have answered the question originally raised. Should everything work out as Descartes hopes it will, when we are finished, it is evident that we will have certain knowledge as he understands it; the answer arrived at in this way will constitute a conclusion deduced ultimately from an initial intuition.

Descartes' strategy here is extremely ingenious. The stated goal of the method is certain knowledge, a science deduced from intuitively known premises. So, the principal goal of the reduction seems to be an intuition from which the original question posed can be answered deductively. But, he says, to simply launch off looking for such an intuition is folly; it is like searching the streets randomly for lost treasure. What the method gives us, at least as he understood it by the late 1620s, is a *workable procedure* for *discovering* an appropriate path between intuition and the answer to the question posed. This workable procedure is the reduction of a question to more and more basic questions, questions that we can identify as questions whose answers are presupposed for answering the question originally posed.

There are numerous problems with the method as Descartes seems to have understood it by the late 1620s. As the example shows, the intuitive step remains profoundly obscure; it is not at all clear how one is supposed to pass from a question about the nature of illumination and the nature of a natural power to an intuition about simple natures. Nor is it obvious that once we have such an intuition, we will be able actually to do the deductions necessary to solve the original problem posed, as Descartes seems to assume we can. It is, for example, a nontrivial problem to pass from the law of refraction to the actual shape of the anaclastic curve. But the most basic, the most substantive assumption that the method makes is that of the structure of knowledge itself, the idea that from any question (at least, any question we are capable of resolving) there is a string of questions that lead us back to an intuition. It is precisely because knowledge is ordered and interconnected in this way that the method is possible at all. Were knowledge to lack such order and connection, then method as Descartes conceives it would be impossible; and once we recognize such an order among things we seek to know, then the method for acquiring knowledge becomes relatively straightforward. It is here that the question of method links up with the question of the structure of Cartesian science in this period.

. . . .

Chapter 7. Motion and Its Laws: Part One

. . . .

Law 1

. . . .

[Descartes' first law of motion] as stated both in *The World* and in the *Principles,* is a *very* general law, a law that concerns *all* states of bodies, at least all states of bodies considered individually [*en particulier*] or considered as simple and undivided. The persistence of motion in a body, the principle that, no doubt, motivated this law, arises as a special case of a principle that mandates the persistence of at least shape and rest, in addition to motion.

The law Descartes presents has a number of features worth commenting on. Puzzling at first glance is a restriction that Descartes introduces into the law. In *The World,* the law is taken to govern bodies taken by themselves [*en particulier*], and in the *Principles,* the law is restricted to bodies that are simple and undivided [*simplex et indivisa*]. Descartes is by no means clear about this. But I suspect that this feature of the law is intended to restrict the scope of the law to genuine states of the genuine individuals in Descartes' physical world; that is, the states that persist are to be modes of extended substance and those alone. These are the states that pertain to the genuine particulars, to bodies insofar as they are simple and undivided. And these are precisely the examples Descartes gives in stating the law. In *The World* the examples of states that persist include size, shape, rest, and motion, and in the *Principles* it is shape, rest, and motion, all modes of extended substance. What Descartes means to exclude here are complex states like being hot or cold, which, it may have seemed to Descartes, can change without any *obvious* external cause. Insofar as heat and cold involve the greater or lesser motion of parts, for Descartes, those states fall outside the scope of this law; they are not states that one can talk about in connection with "simple and undivided" bodies. But while excluded from this law, they are by no means excluded from Descartes' physics. Writing to Mersenne on 26 April 1643, and discussing the apparent exclusion of certain physical states from this law, Descartes remarks: "and heat, sounds, or other such qualities give me no trouble, since they are only motions which are found in the air, where they find different obstacles that hinder them" (AT III 649–50 [K 136]). While such states are not governed directly by the laws of persistence, they are treated in Descartes' physics insofar as they are analyzable into genuine states of body which are governed by the basic laws.

Motion, of course, is one of the basic modes that the law is certainly intended to cover. But Descartes' statement of the law appears to leave open precisely what features of motion are meant to be included. At *very least* it is obvious that Descartes means to say that a body moving will continue to move with *some* speed or other and with *some* determination or other. But what of the speed and determinations, the two modes, varieties, or properties that, Descartes holds, characterized a given motion? (AT III 650 [K 136]; AT IV 185). Does Descartes intend to say that these persist as well?

Every indication is that the law is meant to hold that a body in motion will move with a constant speed, unless acted upon by an external cause. While this proviso is missing from the version of the law Descartes gives in Pr II 37, it seems to be what he has in mind when in *The World* he mandates that a body in motion will remain in motion "avec une egale force," "with the same force." Furthermore, when Descartes appeals to the principle of the persistence of motion in a number of contexts outside of the formal presentation he gives in *The World* and the *Principles,* it is clear that constancy of speed is included. Writing to Mersenne on 13 November 1629 about the problem of free-fall, Descartes states his principle as follows: "whatever once begins to move in a vacuum, always moves with the same speed [*aequali celeritate*]" (AT I 71–72). It is not unreasonable to suppose that Descartes had this proviso in mind in all of the passages we cited earlier where he appeals to the persistence of motion in connection with the problem of free-fall. His strategy in dealing with that problem is to imagine the falling body as receiving equal increments of speed at equal intervals, and to derive the law of free-fall by summing those equal increments. This, of course, presupposes that a body continues, at any instant, to move with the speeds previously imprinted on it. This is where Descartes appeals to the principle of the persistence of motion, and this principle would do him little good unless what persists is not only motion but also a particular speed. This assumption can also be found in later texts, texts written at roughly the time the *Principles* was being drafted. For example, writing to Huygens on 18 or 19 February 1643, Descartes notes that a motion, once begun, "will always continue with the same speed," unless interfered with (AT III 619).

The question of determination is somewhat more difficult. This question is not unconnected with the meaning of law 2 of the *Principles* and corresponding law C of *The World,* where Descartes argues that motion is, in its nature, rectilinear, and so a body tends to move in a straight line tangent to the circle and thus recede from a curvilinear path. Later in this chapter we shall discuss this law in detail, its meaning, and its relation to the law of persistence under discussion now. But setting these difficult questions aside, for the moment, I think that it is plausible to say that the law of the persistence of motion is intended to cover the determination of a body to move in some particular direction . . . [in the sense of] the component of a body's speed in some particular direction. Descartes is, if anything, less explicit about this than he is about the persistence of speed. The law of the persistence of motion is what explicitly stands behind the claim that a ball bouncing off of an immobile surface will retain its speed. Now, important to the derivation of the law of reflection from this model is the further assumption that in colliding with the surface, there is no change in the horizontal determination of the ball, the determination the ball has to move from left to right (AT VI 94–95 [Ols. 76]). While Descartes does not make any appeal to any general principle to justify this assumption, it is not implausible that he thought it followed in an unproblematic way from the principle of persistence of determination; if motion persists unless interfered with, then so should the determination of that motion in a given direction, a genuine *part* that composes that motion.

There is one last remark to make about the law itself. The law, of course, is conditional; it asserts that a state of body (mode of extension) like size, shape, motion, or rest will persist until something causes it to change. However, the law gives no *precise* account of the conditions under which change will occur. This is a condition that Descartes never really specifies in *The World,* as I shall later argue. In the *Principles,* though, it is dealt with in law 3, which, Descartes tells us, "contains all of the particular causes of change that can happen to bodies, . . . at least, those that are corporeal" (Pr II 40). This connection between law 3 and law 1 is signaled in the French version of law 1, where Descartes indicates that everything remains in the same state until changed not merely by "external causes," as the Latin version puts it, but "through collision with others" (Pr II 37F; cf. Pr II 37L).

Writing to Mersenne on 18 December 1629, Descartes brought up his principle of the persistence of motion, and told him that "in my treatise I shall try to demonstrate it" (AT I 90 note a). The treatise in question was, almost certainly, *The World,* and in *The World* Descartes does attempt to argue for the persistence of motion, now subsumed into a more general law concerning the persistence of all states of body. The argument comes after his statement of a law of impact, law B, the second of the three laws he presents in *The World.* Descartes begins with a general consideration that is supposed to underlie both the law of persistence of states, law A, and the collision law B: "now, these two rules follow in an obvious way from this alone, that God is immutable, and acting always in the same way, he always produces the same effect." He then goes on to give the special grounds for his first law: "thus, assuming that he had placed a certain quantity of motions in the totality of matter from the first instant that he had created it, we must admit that he always conserves in it just as much, or we would not believe that he always acts in the same way" (AT XI 43).

It is, then, what will later become Descartes' conservation principle in the *Principles* that is supposed to support the first law of *The World.* It is fairly obvious that this will not do. First of all, whatever this argument might tell us about the persistence of motion, it does not tell us anything about any of the other states of body that, Descartes holds, also persist; though it may support the persistence of the special case of motion, it does not address the more general persistence principle that Descartes frames in *The World.* But, perhaps more importantly, it does not even support the special case of the persistence of motion. As I pointed out earlier, the conservation principle to which Descartes appeals here is a very general principle that governs the world as a whole. But it says nothing at all about how motion is to be distributed among individuals in the world, whether it is to persist in individual bodies, or whether it is to redistribute itself promiscuously and arbitrarily from body to body. The two principles are linked in an obvious way. The conservation principle is a consequence of the fact that once God causes motion in the world, by his immutability he is committed to continue the motion he created. The law of the persistence of motion simply adds that the motion created persists in the body that has it, unless something causes it to change. But though linked, they are not linked deductively; one cannot derive the persistence of motion from the conservation principle, as Descartes suggests in *The World.*

However, the discussion of persistence in *The World* suggests another kind of argument, an argument suggested by the very generality with which the law is stated there. Descartes' real interest is in the special case of motion, in the persistence of that particular state of body. Now, there are many states of body, like size, shape, and rest, that *everyone* would admit persist unless caused to change. Descartes, I think, was attempting to argue for the law of persistence of motion, a law to which he had appealed many times in his earlier writings, by making that controversial law a trivial consequence of a broader but widely accepted principle, and, in *The World,* mocking those who would except motion from the general law of the persistence of states. Immediately after stating law A, he contrasts the treatment of motion he is developing for the new world he is creating with the way motion is treated in the "old world," the world of the schools:

> No one does not believe that this same rule isn't observed in the old world with regard to size, shape, rest, and a thousand other similar things. But the philosophers have excepted motion. . . . But don't think, on account of that that I desire to contradict them: the motion of which they speak is very different than that which I conceive. (AT XI 38–39)

Descartes goes on to poke fun at the scholastic conception of motion, their definition of motion, the varieties of motion that they recognize, and contrast it with his own geometrical conception of local motion. . . . But one comment he makes about the schools is especially relevant to the issue at hand:

> And finally, the motion they speak of has such a strange nature, that unlike other things, which have their perfection as a goal, and try only to conserve themselves, [their motion] has no goal but rest, and contrary to all of the laws of nature, it tries to destroy itself. But on the other hand, [the motion] I assume follows the same laws of nature that all of the dispositions and qualities found in matter in general follow. (AT XI 40)

All other properties of body tend to persist. Why, Descartes suggests, should we make an exception of motion? Why should motion alone tend toward its own destruction and reduction to its opposite, rest?

What is rhetoric in *The World* becomes the official argument in the period of the *Principles.* As in *The World* the persistence of motion is presented as a consequence of a more general principle, the persistence of all states in bodies (at least insofar as they are "simple and undivided"). And similar to his treatment in *The World* Descartes speaks against those who would except motion from this law, attributing the mistake now to the errors of the senses and the rash judgments of youth. As in *The World* Descartes points out that those who except motion from the general principle of the persistence of states hold that "[motions] cease of their own nature, or tend toward rest. But this is, indeed, greatly opposed to the laws of nature. For rest is contrary to motion, and nothing can, from its own nature, proceed toward its own contrary, or toward its own destruction" (Pr II 37). But when Descartes appeals to the immutability of God to prove the law, it is the *law as a whole,* the general principle of the persistence of all appropriate states, that he means to prove, and not just the special case of motion. This is not entirely clear in the sections of the *Principles* where law 1 is presented and defended. But it comes out quite explicitly in a letter to Mersenne from 26 April 1643. Descartes gives Mersenne the law in its most general form, "everything that is or exists always remains in the same state that it is, unless something external changes it." He then continues:

> I prove this through metaphysics. For since God, who is the creator of all things, is entirely perfect and immutable, it seems repugnant to me that any simple thing that exists and, consequently, of which God was the creator, has in itself the principle of its own destruction. (AT III 649 [K 136])

The consideration Descartes explicitly put forward with respect to motion both in *The World* and in the *Principles,* that a state should not tend toward its own opposite, toward its own destruction, is presented here as the general consideration on which the general law rests. Put positively, Descartes holds that God's immutability and constancy of operation entails that if an individual body is in a particular state, then God will sustain that body in that state, unless there is a reason external to that body for altering the state. . . .

For Descartes, in the end, the persistence of the state of motion would seem to depend on its *opposition* to a state of rest. This is closely connected with . . . the importance to Descartes of making motion a genuine mode of body, a mode of extended substance. If everything is to be explicable in terms of motion alone, then motion *must* be a genuine mode of extended substance. But if motion is a genuine mode, then it cannot be an arbitrary matter of point of view whether a body is in motion or not; there must be a genuine distinction between motion and rest. Earlier I emphasized the foundational aspects of the doctrine of motion, the extent to which it is required in order to ground Descartes' mature conception of the physical world. But here we see another motivation. In construing motion as a genuine mode of body, Descartes can then include it among the states of body that an immutable God sustains and, in that way, prove a central principle of his physics, a principle he first learned from Beeckman in autumn 1618. To be sure, it did not all come together at once. Descartes began with the special case, the persistence of motion, and it took him more than ten years to link the persistence of motion to the more general principle of the persistence of states of body, as he did in the early 1630s with the composition of *The World.* Though motion is not carefully defined there, it is clearly designated as a state, a state whose opposite is rest. But with the composition of the *Principles,* Descartes makes clearer the sense in which motion and rest are modes of body and distinct modes of body, the sense in which motion and rest are states and opposite states. With this the position is complete, and the proof of the persistence of motion in individual bodies rests solidly on its Cartesian foundations.

Law 2

The first law of persistence, law A in *The World* and law 1 in the *Principles,* asserts that motion persists along with the other states of a body. To this Descartes adds a second law of persistence, both in *The World* (law C) and in the *Principles* (law 2), a law that asserts that motion persists in a rectilinear path unless interfered with by an external cause. This law is not found in any surviving Cartesian texts before *The World.* But once stated there, it appears in an almost unaltered form in the *Principles;* though there are some differences of detail, the exposition of law 2 in the *Principles* is one of the few places where Descartes has virtually translated an earlier discussion into Latin. In both works, the development of the law takes place in two different contexts. The law is first announced and proved together with the other laws of nature, in chapter VII of *The World* and part II of the *Principles* (AT XI 43–47; Pr II 39). But Descartes makes some important later clarifications of the law in both works, when the law gets its principal application in grounding the theory of light, for light turns out to be the centrifugal pressure of celestial matter turning in vortices. This further discussion is found in chapter XIII of *The World* and part III of the *Principles* (AT XI 84–86; Pr III 55–60).

As with law 1, this law is taken to be grounded in divine immutability. . . . [Our] first problem in dealing with this law is determining what precisely it is meant to assert. The basic law is stated as follows in *The World:*

> When a body moves, even if its motion is most often on a curved path, . . . nevertheless, each of its parts, taken individually, always tends to continue its motion in a straight line. (AT XI 43–44)

The law is given in a similar form in the *Principles:*

> Each and every part of matter, regarded by itself, never tends to continue moving in any curved lines, but only along straight lines. (Pr II 39)

In the postil to the Latin edition of this section, the law is summarized by saying that "all motion, in and of itself, is straight."

But closely linked to the tendency of a body in motion to move in a straight path are some assertions about the behavior of bodies in circular motion, which are regarded sometimes as consequences of the law (Pr II 39 postil) and sometimes as part of the law itself (Pr III 55). Consider a body revolving around a center; for example, a stone in a sling. If we consider all of the causes that determine its motion, then the stone "tends" [*tendere, tendre*] circularly (AT XI 85; Pr III 57). But if we consider only "the force of motion it has in it" (Pr III 57), then, Descartes claims, it "is in action to move," or "is inclined" to go, or "is determined to move" or "tends" to move in a straight line, indeed, along the tangent to the circle at any given point (AT XI 45–46; Pr II 39; Pr III 57; AT XI 85). And, Descartes concludes, "from this it follows that everything which is moved circularly tends to recede from the center of the circle that it describes" (Pr II 39, postil).

In order to understand what the law asserts, we must understand what Descartes means when he asserts that a body has a tendency or an inclination. Descartes explains the notion as follows in a passage from *The World:*

> When I say that a body tends in some direction, I don't want anyone to imagine on account of that that it has in itself a thought or a volition that pushes it there, but only that it is disposed to move in that direction, whether it really moves or whether some other body prevents it from moving. (AT XI 84)

This same view is also found in the *Principles,* where Descartes also attempts to dementalize the term "tendency," latinized there as "conatus" (Pr III 56). But in the *Principles,* he is a bit clearer about what this tendency is. A body with a tendency or a conatus to move in a particular direction "is situated and incited to motion in such a way that it would really go [in that direction] if it were not hindered by some other cause" (Pr III 56). So, a body tends in a particular direction (a straight line, a path along the tangent, a motion away from the center of a rotation) insofar as it would actually go there unless otherwise impeded. And so, this law asserts the conditional persistence of a certain sort of motion in exactly the same way that the first law asserts the conditional persistence of the states of a body in general: the motion in question persists unless an external cause prevents it from persisting. Indeed, in stating one consequence of this law, Descartes uses exactly the same formula he uses in connection with the first law: "all bodies which move in circles recede from the center of their motion insofar as they can [*quantum in se est*]" (Pr III 55).

We should be very careful to distinguish the notion of tendency from the notion of determination, with which it might be confused. Determination is an aspect of the motion a body has. In its primary sense, it is a directional component of the motion a body has, a measure of the extent to which a body moving with a particular speed and with a particular direction advances in a direction that may or may not be identical with the direction of the body's motion. But Descartes does not use the word "tendency" in this way. For him a tendency is not a motion or an aspect of motion, but a property a body has by virtue of which it would move if it were unimpeded. It is not motion, but, as he suggests in one context, motion in potentiality, and in another, a "first preparation for motion [*prima praeparatio ad motum*]" (AT I 450–51 [K 42]; Pr III 63). Like motion, tendency has determination, a potential determination, as it were. A body in motion that tends in a particular direction is said to be "determined [*determinatum*] to continue its motion in a certain direction, following a straight line" (Pr II 39). And, as we shall see, Descartes will decompose the potential determinations to go in different directions that a body that tends to go in a particular direction has; as with bodies in motion, a body that tends in some direction as a whole will have tendencies in other directions as well, though Descartes has no clear technical terminology to express them.

It is obvious why Descartes chose to express this law in terms of tendencies rather than more straightforwardly in terms of a state of body that persists conditional on a lack of interference. In his plenum, this condition of noninterference can *never* be met. Since all is full, a body in motion must push another out of its way in order to move, and the space it leaves must be filled by another body that comes and takes its place. This, he argues, entails that all motion must be circular (Pr II 33). Descartes clearly has this in mind when framing this law (AT VIIIA 63, ll. 23–26). Though a body would go straight if it were not interfered with, other bodies are always interfering with it.

As noted earlier in this section, the rectilinear tendency of a body in motion is closely connected to two features of bodies in circular motion.

First of all, Descartes asserts, a body in circular motion tends away from that circle along the tangent. It is interesting to note that this is something that does not follow directly from the claim that bodies tend to move rectilinearly. Though a body moving in a circular path may, at any moment, tend to move in a rectilinear path, from that fact alone nothing follows about which rectilinear path it tends to move in at any given moment. Descartes takes his claim to be "confirmed by experience, since if [the stone] is then ejected from the sling, it will not continue to move toward B [on the circle] but toward C [on the tangent]" (Pr II 39). But he never offers any serious argument for the claim that the body tends to move along the tangent, and he seems to have regarded it as simply obvious.

But wherever the conclusion may come from, it leads Descartes to another important fact about circular motion, that a body in circular motion tends away from the center of rotation. Though this conclusion is taken to be obvious in the context of his initial presentation of the laws of nature, both in *The World* and in the *Principles* (AT XI 44; Pr II 39), in the later application of the law of rectilinear tendency to the theory of light he attempts to argue for the conclusion. The strategy for establishing this centrifugal tendency is ingenious.

Descartes throughout works under the methodological assumption that to reason about tendencies, we must reason about the motions that would result were those tendencies to be realized. So, he begins by decomposing the rectilinear motion of a body along a tangent to a circle into two different motions (see fig. 1):

> But to understand this last point distinctly, imagine this inclination that the stone has to move from A to C as if it were composed of two others, that is, one to turn around the circle AB, and the other to go straight along the line VXY, in such proportion that, finding itself at the point of the sling marked V when the sling is at the point of the circle marked A, it ought to find itself later at the point marked X when the sling would be at B, and at the point marked Y when it would be at F, and thus it ought always to remain on the straight line ACG. (AT XI 85)

. . . A rectilinear tendency along the tangent to a circle is divided into a tendency to revolve around a fixed point, together with a tendency to move in a straight line along a rotating radius of the circle. In presenting this argument in the *Principles* he makes this more graphic by imagining the rectilinear motion of an ant along the tangent to a circle as having been produced by the ant walking along a stick while that stick rotates around a center (Pr III 58; see fig. 2).

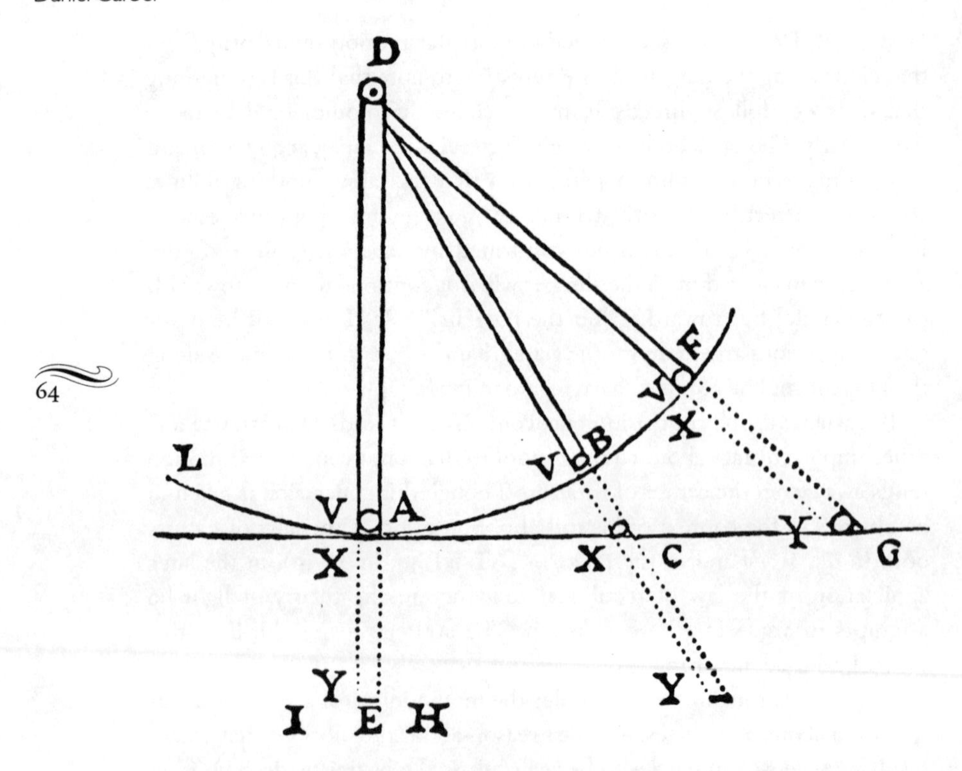

FIGURE 1 From [René Descartes,] *L'homme de René Descartes. . . . A quoy l'on a ajouté Le monde, ou Traité de la lumière, du mesme Autheur* (Paris, 1677), 441.1.

Descartes continues the passage from *The World* quoted above as follows:

> Knowing that one of the parts of its inclination, namely, that which carries it around the circle AB is in no way hindered by this sling, you can well see that the sling resists only the other part [of the inclination], namely, that which makes it move along the line DVXY, if it were not hindered. Consequently it tends (that is, it makes an effort) only to go directly away from the center D. (AT XI 85–86; see also Pr III 58–59)

The sling does not impede the circular component of the stone's motion, but it does impede the paracentric component, the tendency the ball has, on this analysis, to move away from the center. But in impeding the motion, the sling does not eliminate the tendency to motion (Pr II 57). If the sling were not impeding the stone, it would move out along the rotating radius. And so, Descartes concludes, the stone in the sling, indeed *any* body rotating around a center, has a tendency to move away from the center of

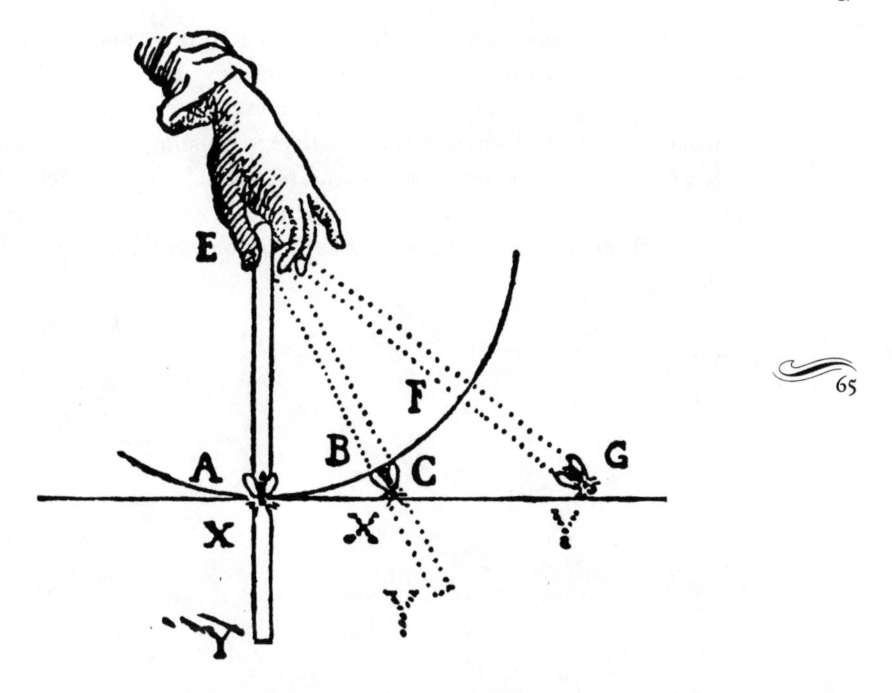

FIGURE 2 From *Renati Des-Cartes Principia Philosophiae* (Amsterdam, 1644), 99.

rotation, a tendency deriving from the tendency the ball has to move in a straight line.

The argument, though very ingenious, is very problematic. Now, Descartes certainly intends the circular motion in question in the argument to be uniform circular motion around a center, motion of a constant angular velocity, and by the first law he is committed to the view that the tangential motion toward which the body tends is uniform rectilinear motion. But what Descartes does not realize is that as he sets the case up, the uniformity of the rotational motion of DVXY around D is inconsistent with the uniformity of the motion of the stone along the tangent ACG. Suppose that the stone moving along the tangent ACG were to move with a uniform speed (see fig. 1). Then the rotating radius DVXY would have to move more and more slowly as the stone went further and further out the tangent; indeed, if the speed of the stone on the tangent were finite, DVXY would never become parallel with the tangent ACG. Suppose, now, that the radius DVXY were rotating uniformly around D, and that the stone, moving out along DVXY, were to remain on the tangent ACG. Then it is obvious that the motion of the stone along ACG would be accelerated as the angle ADF approached a right angle. (It is also obvious that in this circumstance, Descartes' ant in fig. 2 would have to run faster and faster to keep up, until the stick on which it was running became parallel to the original tangent, at which point it would have to run infinitely fast, no small feat even for an imaginary insect!) And so, I think, uniform rectilinear motion along the tangent cannot be broken down into components in the way Descartes tries to do; and without this decomposition, the argument collapses.

. . . .

PART II

Newton's Inductivism & the Law of Gravity

Isaac Newton, who revolutionized physics, astronomy, and mathematics during the seventeenth and eighteenth centuries, was born in Woolsthorpe, England, on December 25, 1642 (eight years before Descartes' death). He entered Trinity College, Cambridge University, as an undergraduate in 1661, where during the next few years he studied works in mathematics and physics (including Descartes' *Geometry* and his *Principles of Philosophy*), as well as scholastic philosophy, logic, and the classics. In 1669, at the age of twenty-six, he was appointed Lucasian Professor of Mathematics at Cambridge University, where he remained until 1696. During his years at Cambridge, Newton's research culminated in some of the most important discoveries, inventions, and theories in the history of science. These include, in mathematics, the differential and integral calculus (which he and the German mathematician and philosopher Leibniz invented separately); in optics, the design of a reflecting telescope and the experimental proof that white light is not simple but contains all the colors of the rainbow; in mechanics, the first modern formulation of the three laws of motion (still studied today in physics courses) and, most important, the law of universal gravitation, which he used to explain the motions of the planets and their satellites. Newton's most important book, a monument of scientific achievement, *Mathematical Principles of Natural Philosophy* (usually called the *Principia*), was first published in Latin in 1687 and was revised and reissued twice during his lifetime, in 1713 and 1726. It contains, among many things, the formulation and mathematical development of his three laws of motion and a derivation of his law of universal gravitation.

In 1696 Newton left Cambridge and became warden of the Mint in London, where he supervised revision in coinage and prosecuted counterfeiters. He became president of the Royal Society (the premier scientific society in England) in 1703. The next year he published his *Opticks,* a more popular scientific work detailing the many experiments on light that he had performed during the 1660s and later. He was knighted in 1705 and lived for another twenty-two years, performing duties as president of the Royal Society and making additions and revisions to the *Principia* and the *Opticks*. Newton died in 1727 at the age of eighty-four.

Newton's Methodological Rules

Newton's *Principia* contains three books. In book 3, which he calls "The System of the World," he derives his most famous law, that of universal gravitation. In modern terminology, the law asserts that every body in the universe exerts a force on every other body that is proportional to the product of the masses of these bodies and inversely proportional to the square of the distance between them. Before deriving the law he formulates four "rules for the study of natural philosophy" that he will use in the derivation. These are very general rules of scientific method that he believes apply to science (or "natural philosophy") generally.

Newton realized that his law of gravity would be controversial. For one thing, it postulates a single force of gravity, not the different ones that were commonly thought to exist for each planet, for example. Second, it makes an extremely general claim about *all* bodies in the universe, even bodies beyond the range of what can be observed. Third, and most important, it postulates a force that acts over a distance, without indicating any medium through which that force is propagated. Newton himself probably believed that there was such a medium, but in the *Principia* he provides no account of it. If there were no such medium, then there would be genuine "action at a distance," which was contrary both to everyday experience of contact forces acting to push or pull, and to the scientific views, held from Aristotle to Descartes, that forces are exerted only when one body touches another. By explicitly stating rules of reasoning he could use to arrive at his law, Newton thought that he could more readily satisfy any critics.

Newton's first rule of reasoning states that "no more causes of natural things should be admitted than are both true and sufficient to explain their phenomena." This is often referred to as the *vera causa* (true cause) rule. Several points are worthy of comment. First, Newton focuses on the idea of admitting causes in nature in order to explain phenomena. One of the examples he gives (in the discussion of rule 2) pertains to the falling of stones. In explaining why stones fall it does not suffice simply to give the manner of their fall—for example, by saying (following Galileo) that they fall with uniform acceleration. One needs to cite a cause—in this case, the force of gravity exerted on the stone by the earth. Second, Newton, like Descartes, emphasizes the idea of truth (he does so again in rule 4). It is not enough to postulate causes that, *if* they exist, will explain the phenomena. They must exist, they must be real. Third, if one cause suffices to explain a given phenomenon, then more than one should not be admitted. Newton adds, "Nature is simple and does not indulge in the luxury of superfluous causes" (or, in an older and grander translation, "Nature is pleased with simplicity and affects not the pomp of superfluous causes").[1]

These three points are all important to Newton in discussing gravity. A gravitational force is a cause of the motions of the planets and their satellites. This force really exists. It is not simply a "useful fiction" or mere device for calculating orbits.

And one universal force of gravity suffices to explain celestial motions; a different force for each pair of bodies is not needed.

Newton's second rule states that "the causes assigned to natural effects of the same kind must be, so far as possible, the same." One of his examples is the falling of stones in Europe and America: if stones fall in the same way in Europe and America, then we should infer that the cause of their falling is the same on both continents. Newton takes this second rule to follow from the first. His idea is that since, according to rule 1, causes should not be multiplied beyond necessity, from the fact that two effects are the same one should infer that there is one cause for both, not two separate causes.

This rule will prove crucial in Newton's derivation of universal gravitation. Newton establishes that the forces drawing the satellites of Jupiter toward Jupiter and the satellites of Saturn toward Saturn are directed to the center of these planets and are inversely proportional to the square of the distance from the centers of these planets. From the fact that these effects are of the same kind, using his rule 2 Newton infers that they have the same cause. There is one cause—the force of gravity—not several different forces, that causes these effects.

Newton's third rule states that "those qualities of bodies that cannot be intended and remitted [more about this phrase in a moment] and that belong to all bodies on which experiments can be made should be taken as qualities of all bodies universally." One of his examples is the extension of bodies in space. From the fact that all observed bodies are extended one may infer that all bodies are extended, even ones beyond the range of our senses.

This rule and the discussion that follows are crucial for understanding Newton's views about scientific method. Before Newton, Descartes had claimed that certain qualities ("simple natures") are known to the mind via intuition. Descartes asserted that the essence of matter is to be extended in space, and that although this idea might be suggested to us by experience, we can come to know the truth of the claim that bodies are extended only by an intuition of the mind. Newton emphatically rejects such a claim, writing that "the qualities of bodies can be known only through experiments," and that "the extension of bodies is known to us only through our senses." Newton, in contrast to Descartes, is an empiricist: knowledge of the physical world comes only from experience, not from pure thought.

Rule 3 is an *inductive* rule. It permits one to make a generalization, or induction, from the fact that all observed bodies have certain qualities to the claim that all bodies, whether or not observed or even observable, have those qualities. Newton adds the proviso that this applies to qualities that "cannot be intended and remitted." This phrase, which Newton appropriates from scholastic philosophy, is a puzzling one that different Newton scholars interpret in different ways. In the edition from which our selection is drawn, the translators add in brackets, "i.e., qualities that cannot be increased or diminished." But if a quality that can be "intended and remitted" is one that can be increased and diminished, as the translators suggest, then, contrary to what Newton wants, rule 3 cannot apply to gravity, since, according to Newton, the gravity of bodies "is diminished as [they] recede from the earth." For this reason some Newton commentators suggest that we ignore the "intension and remission" qualification, which was omitted by Newton himself in an earlier formulation of this rule.

Rule 3 is crucial for Newton's derivation of the law of universal gravitation. Rules 1 and 2 allow him to infer that the inverse-square force that is responsible for the gravitation of the planets and their satellites is the same force in all these cases. Rule 3 allows him to generalize this to all bodies in the universe: "It will have to be concluded by this third rule that all bodies gravitate toward one another."

Newton's rule 4 states that "in experimental philosophy [empirical science], propositions gathered from phenomena by induction should be considered either exactly true or very nearly true notwithstanding any contrary hypotheses, until yet other phenomena make such propositions either more exact or liable to exceptions." Suppose that, following rule 3, one has made an inductive generalization from the fact that all observed bodies are extended to the proposition that all bodies are extended. The fourth rule permits one to claim that the inferred proposition is "exactly or very nearly true." If we discover a type of body that is not extended in space, then we can say that all bodies except for those of this type are extended. Or, more positively, if we discover some further feature of bodies in virtue of which they are extended, we can make the inferred proposition more exact by saying that all bodies with this feature are extended.

Rule 4, then, includes the idea that propositions (including Newton's laws of motion and gravitation) that are inferred from phenomena using the other rules can in principle be shown to be incorrect or liable to exceptions. Following the rules does not guarantee truth. But for Newton the only way to demonstrate falsity or exceptions is to discover new phenomena that are incompatible with the inferred propositions.

In rule 4 itself, and in one sentence that follows its formulation, Newton introduces the concept of a *hypothesis*. He claims that the mere fact that one can formulate a hypothesis contrary to the inductively inferred proposition does not at all "nullify" the proposition inferred. At this point Newton does not define "hypothesis," but he does so at the end of the third book of the *Principia*. There he says that a hypothesis is any proposition that "is not deduced from the phenomena." Although Newton does not define "deduce," he seems to have in mind any inference of a sort permitted by his four rules of reasoning or by reasoning of a sort employed in mathematics (what today we would call "deductive").

What about the word "phenomenon," which appears in rules 1 and 3? Again, in the *Principia* Newton offers examples (which will be noted in a moment) but no definition. In an unpublished definition intended for the second edition of the *Principia* he writes that a phenomenon is "whatever can be seen and is perceptible . . . either things external which become known by the five senses, or things internal which we contemplate in our minds by thinking."[2] These examples follow: fire is hot, water is wet, gold is heavy, sun is light, I am, I think. In the *Principia* itself, following his four rules of reasoning, Newton lists six "phenomena," which are Kepler's laws of planetary motion applied to the motions of the planets and their satellites.[3] For example, the first phenomenon is that the satellites of Jupiter sweep out areas proportional to their times and that their periods of revolution are proportional to the $\frac{3}{2}$ths power of their distance from Jupiter; in a modern formulation, restricting the case to the period T of revolution, T^2 is proportional to r^3, where r is the satellite's distance from Jupiter. In his discussion of the phenomenon Newton makes it clear that the orbits in question "do not differ sensibly from circles" and that the motions are uniform.

Judging from the unpublished definition and the examples, "phenomena" for Newton are facts (e.g., the fact that fire is hot), rather than events or experiences. Moreover, they are facts that not only are establishable by observation but that have been established in this way by some member or members of the scientific community. They are not disputed facts, but ones the community in general accepts (or would readily accept once the results have been made known). Accordingly, Newton regards "phenomena" very differently from the way Descartes regards "intuitions." Although both are considered starting points for scientific investigations, and both are established as true, for Newton, phenomena are established by observation, while for Descartes, intuitions are established by pure thought. Moreover, for Descartes, intuitions are self-evident and not inferred from anything. For Newton, phenomena (particularly the six listed in the *Principia*) are inferred from observations and are usually not self-evident.

This leads to another important difference between Newtonian and Cartesian methodology. For Descartes, the ideal goal in a scientific investigation is to arrive at truths that are known with absolute certainty—that are beyond any possible doubt. One can achieve this certainty by raising questions that are more and more basic, leading to self-evident intuitions that answer the most basic questions, and then proceeding by Cartesian deductions to answer the others. For Newton, the goal of a scientific investigation is also to arrive at truths, but because these are empirically based, they do not carry absolute certainty. They can be refuted or revised if new phenomena are discovered. Nevertheless, if propositions are "deduced from the phenomena" using his rules, then, he says, they have the "highest evidence that a proposition can have in this [experimental] philosophy."[4] They are beyond reasonable doubt, but not all possible doubt. To achieve this end, says Newton, we start not with self-evident intuitions of pure thought that cannot be doubted, but with facts that have been empirically established within the scientific community.

Now we can return to Newton's remarks about "hypotheses" (which, we recall, are propositions not deduced from phenomena). Suppose that, as Newton claims, from the six phenomena regarding motions of the planets and their satellites, using his four rules of reasoning, including the inductive rule 3, Newton derives the universal law of gravity. Suppose, further, that in order to cast doubt on this law, a skeptic imagines a contrary possibility that is consistent with the observed phenomena (e.g., that the law of gravity operates *only* for the planets and their satellites but nowhere else in the universe). In accordance with rule 4, to such a skeptic Newton would reply that this contrary supposition is a hypothesis (i.e., something not deduced from the phenomena using one or more of his four rules plus mathematical deduction); and because of this it casts no doubt whatever on the universal law of gravity that is deduced from the phenomena. Only the discovery of new phenomena contrary to the law will cast doubt on the law.

At the end of book 3 of the *Principia,* where Newton defines "hypothesis," he claims indeed that "hypotheses . . . have no place in experimental philosophy" (General Scholium). And in various letters he writes that from the fact that one can derive a phenomenon from a hypothesis, one cannot infer that the hypothesis is true, since one may be able to derive the same phenomenon from contrary hypotheses. For Newton, the derivation must begin with the phenomena and, using his rules of reasoning, proceed to a general proposition. Only this order, not the reverse order, will justify a claim of truth for that proposition.

Despite Newton's claim that hypotheses have no place in experimental philosophy, he does introduce them from time to time. For example, in the very same book of the *Principia* in which he makes this claim he introduces what he himself calls "hypothesis 1," namely, that "the center of the system of the world is at rest." (This follows proposition 10 of book 3.) And in later editions of his *Opticks,* first published in 1704, Newton introduces a set of "queries" about the nature of light, the most important of which contain hypotheses in his sense (e.g., that light consists of particles, not waves). One way to understand Newton in such cases is this: If a proposition is a hypothesis in Newton's sense (i.e., if it has not been "deduced from the phenomena"), its truth is not to be inferred; in particular, it is not to be claimed to be true on the grounds that from it verified observational claims can be derived. Nevertheless, if, as in the *Opticks,* some weaker evidence can be offered in its support, one should proceed to do so, without claiming truth for the supported hypothesis. An inference to the truth of a proposition is possible only if that proposition either describes a phenomenon or can be deduced from phenomena.[5]

Newton's methodological rules have been interpreted and evaluated by many writers. The present volume reprints sections of the historical introduction and interpretation by the eminent Newton scholar I. Bernard Cohen, as well as a selection critical of Newton's rules by the nineteenth-century scientist and historian and philosopher of science William Whewell. In later parts of the book two twentieth-century philosophers of science, Karl Popper and Paul Feyerabend, develop general methodological views incompatible with Newton's. Rejecting Newton's rules as being much too conservative, Whewell and Feyerabend claim that Newton's emphasis on reasoning from effects to causes of a type known, and from particular instances to general laws via induction, will not permit scientists to introduce novel hypotheses. Popper considers any form of inductive generalization to be invalid.

The Law of Gravity

Following the "rules for the study of natural philosophy" and the six "phenomena," Newton derives forty-two "propositions," the first seven of which are included below. Proposition 1 is that the forces by which the moons of Jupiter are continually drawn away from rectilinear motion and maintained in their orbits are directed to the center of Jupiter (they are central or centripetal forces), and they vary inversely as the square of the distance from Jupiter. This proposition Newton derives from phenomenon 1 of book 3 and from proposition 2 and corollary 6 to proposition 4 derived earlier in book 1 of the *Principia*. Phenomenon 1, we recall, is that the satellites of Jupiter sweep out areas proportional to their times and that their periods of revolution are proportional to the 3/2ths power of the distance from Jupiter. Proposition 2 of book 1 states that every body that moves in a curved line around a point and that sweeps out areas around that point proportional to the times is urged by a centripetal force tending toward that point. Corollary 6 to proposition 4 in book 1 states that in a situation in which a body moves with uniform motion in a circle around a point, if the periodic times are proportional to the 3/2ths power of the radius, then the centripetal force will be inversely proportional to the square of the radius, and conversely. Since, according to phenomenon 1, the satellites of Jupiter do move in approximately circular paths (with uniform motion) around Jupiter and their periods are proportional to the 3/2ths power of their radii, proposition 1 of book 3 follows immediately.

Propositions 2–5 make the same assertions as proposition 1, except that they describe the forces acting on the moons of other planets circulating around those planets and about the forces acting on the planets themselves circulating around the sun. In each case there is a centripetal force acting on the body drawing it away from rectilinear motion, and this force varies inversely as the square of the distance from the center of the attracting body.

In his discussion of proposition 4 (which asserts that the moon gravitates toward the earth and is kept in its orbit by the force of gravity), Newton introduces an important claim. It is that the force by which the moon is kept in its orbit is the same force as the one causing heavy bodies to fall toward the center of the earth, which during Newton's time was commonly called "gravity." The argument should be of special interest to readers of the present volume, since Newton explicitly invokes three of his "rules for the study of natural philosophy."

Suppose, he says, that several moons revolved around the earth. By "the argument of induction" (i.e., rule 3), from similar cases with the moons of Jupiter and

Saturn, we could infer that these moons would obey Kepler's law concerning the areas swept out. Therefore, by proposition 1 of book 3 the centripetal forces keeping these moons in their orbits would vary inversely as the square of the distance of the moon from the center of the earth. Now, if the lowest of these moons nearly touched the tops of the highest mountains on the earth, the centripetal force keeping it in orbit would be approximately equal to the force of gravity acting on bodies at the tops of those mountains that is directed toward the center of the earth. (In the preceding discussion Newton presents an empirical argument and calculation showing that this is so.) But if the force acting to keep this moon in its orbit were different from the force causing bodies near the earth to fall toward the earth, then (since these forces have the same strength) the moon would be subject to two forces acting in the same direction and would fall away from a rectilinear path toward the earth twice as fast as it in fact does. Since both forces are directed toward the center of the earth and have equal magnitudes, by rule 1 (don't multiply causes beyond necessity) and rule 2 (from same type of effect infer the same cause), Newton concludes that they are the same force. That is, the force causing the moon to continue in its orbit is the same force as that causing bodies near the earth to fall toward the earth. Since the accelerative effects are the same, the cause is the same.

At the end of his discussion of proposition 5 (according to which the moons of Jupiter gravitate toward Jupiter, those of Saturn gravitate toward Saturn, and the planets gravitate toward the sun), Newton makes an important generalization. It is that the centripetal force by which the celestial bodies are kept in their orbits is the same force as gravity (i.e., the force that causes bodies to fall toward the earth and the moon to have the orbit it does). He claims that by his methodological rules 1, 2, and 4, this cause, which operates on our moon as well as on bodies on the earth, should be inferred to be the same cause operating to keep all the planets and their moons in the orbits they have.

At the very end of the third book of the *Principia* Newton includes a "General Scholium." In its final paragraphs he indicates that although he has "explained the phenomena of the heavens [the motions of the planets and their satellites] and of our sea [the tides] by the force of gravity," he has "not yet assigned a cause to gravity." He has not determined what causes bodies to exert a gravitational force on all other bodies. Newton indicates that he will not "feign" (or, in an earlier translation, "frame")[6] a hypothesis (i.e., a proposition not deduced from the phenomena) that assigns such a cause. It is here that he adds that hypotheses "have no place in experimental philosophy." For Newton in the *Principia* it suffices to establish by deduction from the phenomena, using his rules of reasoning, that gravity does exist universally and that it is subject to the general law he formulates. He considers the fact of universal gravitation established unless and until new phenomena are discovered that call it into question.

NOTES

1. Andrew Motte's 1729 translation.

2. From a manuscript sheet translated by J. E. McGuire, "Body and Void in Newton's De Mundi Systemate: Some New Sources," *Archive for History of Exact Sciences* 3 (1966):238–39.

3. Kepler's three laws are these: (1) each planet moves around the sun in an elliptical orbit with the sun at one focus of the ellipse; (2) a line drawn from the sun to each planet sweeps out equal areas in equal times; (3) the squares of the periods of the planets are proportional to the cubes of the mean distances from the sun (or, as Newton puts it, their periods are proportional to the $3/2$ths power of the distance).

4. Letter from Newton to Cotes in 1713. Reprinted in H. S. Thayer, ed., *Newton's Philosophy of Nature* (New York: Hafner, 1953), 6.

5. For a more extended discussion of Newton's views about hypotheses, and about his arguments for the particle theory of light, see Peter Achinstein, *Particles and Waves* (New York: Oxford University Press, 1991), chap. 2.

6. Again, Andrew Motte.

5

Newton's Methodological Rules

From The Principia, Book 3 / Isaac Newton

Rules for the Study of Natural Philosophy

Rule 1

No more causes of natural things should be admitted than are both true and sufficient to explain their phenomena.

As the philosophers say: Nature does nothing in vain, and more causes are in vain when fewer suffice. For nature is simple and does not indulge in the luxury of superfluous causes.

Rule 2

Therefore, the causes assigned to natural effects of the same kind must be, so far as possible, the same.

Examples are the cause of respiration in man and beast, or of the falling of stones in Europe and America, or of the light of a kitchen fire and the sun, or of the reflection of light on our earth and the planets.

Reprinted from *The Principia: Mathematical Principles of Natural Philosophy,* trans. I. Bernard Cohen and Anne Whitman (Berkeley: University of California Press, 1999), 794–96.

Translators' notes have been omitted. Bracketed interpolations are theirs.

Rule 3

Those qualities of bodies that cannot be intended and remitted [i.e., qualities that cannot be increased and diminished] and that belong to all bodies on which experiments can be made should be taken as qualities of all bodies universally.

For the qualities of bodies can be known only through experiments; and therefore qualities that square with experiments universally are to be regarded as universal qualities; and qualities that cannot be diminished cannot be taken away from bodies. Certainly idle fancies ought not to be fabricated recklessly against the evidence of experiments, nor should we depart from the analogy of nature, since nature is always simple and ever consonant with itself. The extension of bodies is known to us only through our senses, and yet there are bodies beyond the range of these senses; but because extension is found in all sensible bodies, it is ascribed to all bodies universally. We know by experience that some bodies are hard. Moreover, because the hardness of the whole arises from the hardness of its parts, we justly infer from this not only the hardness of the undivided particles of bodies that are accessible to our senses, but also of all other bodies. That all bodies are impenetrable we gather not by reason but by our senses. We find those bodies that we handle to be impenetrable, and hence we conclude that impenetrability is a property of all bodies universally. That all bodies are movable and persevere in motion or in rest by means of certain forces (which we call forces of inertia) we infer from finding these properties in the bodies that we have seen. The extension, hardness, impenetrability, mobility, and force of inertia of the whole arise from the extension, hardness, impenetrability, mobility, and force of inertia of each of the parts; and thus we conclude that every one of the least parts of all bodies is extended, hard, impenetrable, movable, and endowed with a force of inertia. And this is the foundation of all natural philosophy. Further, from phenomena we know that the divided, contiguous parts of bodies can be separated from one another, and from mathematics it is certain that the undivided parts can be distinguished into smaller parts by our reason. But it is uncertain whether those parts which have been distinguished in this way and not yet divided can actually be divided and separated from one another by the forces of nature. But if it were established by even a single experiment that in the breaking of a hard and solid body, any undivided particle underwent division, we should conclude by the force of this third rule not only that divided parts are separable but also that undivided parts can be divided indefinitely.

Finally, if it is universally established by experiments and astronomical observations that all bodies on or near the earth gravitate [*lit.* are heavy]

toward the earth, and do so in proportion to the quantity of matter in each body, and that the moon gravitates [is heavy] toward the earth in proportion to the quantity of its matter, and that our sea in turn gravitates [is heavy] toward the moon, and that all planets gravitate [are heavy] toward one another, and that there is a similar gravity [heaviness] of comets toward the sun, it will have to be concluded by this third rule that all bodies gravitate toward one another. Indeed, the argument from phenomena will be even stronger for universal gravity than for the impenetrability of bodies, for which, of course, we have not a single experiment, and not even an observation, in the case of the heavenly bodies. Yet I am by no means

80

affirming that gravity is essential to bodies. By inherent force I mean only the force of inertia. This is immutable. Gravity is diminished as bodies recede from the earth.

Rule 4

In experimental philosophy, propositions gathered from phenomena by induction should be considered either exactly or very nearly true notwithstanding any contrary hypotheses, until yet other phenomena make such propositions either more exact or liable to exceptions.

This rule should be followed so that arguments based on induction may not be nullified by hypotheses.

6

Newton's "Phenomena" and Derivation of the Law of Gravity

From The Principia, Book 3 / Isaac Newton

Phenomena

Phenomenon 1

The circumjovial planets [or satellites of Jupiter], by radii drawn to the center of Jupiter, describe areas proportional to the times, and their periodic times—the fixed stars being at rest—are as the $\frac{3}{2}$ powers of their distances from that center.

This is established from astronomical observations. The orbits of these planets do not differ sensibly from circles concentric with Jupiter, and their motions in these circles are found to be uniform. Astronomers agree that their periodic times are as the $\frac{3}{2}$ power of the semidiameters of their orbits, and this is manifest from the following table [see p. 82].

Using the best micrometers, Mr. Pound has determined the elongations of the satellites of Jupiter and the diameter of Jupiter in the following way. The greatest heliocentric elongation of the fourth satellite from the center of Jupiter was obtained with a micrometer in a telescope 15 feet long and came out roughly 8′16″ at the mean distance of Jupiter from the earth.

Reprinted from *The Principia: Mathematical Principles of Natural Philosophy*, trans. I. Bernard Cohen and Anne Whitman (Berkeley: University of California Press, 1999), 797–811.

Translators' notes have been omitted. Bracketed interpolations are theirs.

Periodic times of the satellites of Jupiter			
$1^{d}18^{h}27^{m}34^{s}$	$3^{d}13^{h}13^{m}42^{s}$	$7^{d}3^{h}42^{m}36^{s}$	$16^{d}16^{h}32^{m}9^{s}$

Distances of the satellites from the center of Jupiter, in semidiameters of Jupiter

	1	2	3	4
From the observations of				
Borelli	5⅔	8⅔	14	24⅔
Townly, by a micrometer	5.52	8.78	13.47	24.72
Cassini, by a telescope	5	8	13	23
Cassini, by eclipses of the satellites	5⅔	9	$14\frac{23}{60}$	$25\frac{3}{10}$
From the periodic times	5.667	9.017	14.384	25.299

82

That of the third satellite was obtained with a micrometer in a telescope 123 feet long and came out 4′42″ at the same distance of Jupiter from the earth. The greatest elongations of the other satellites, at the same distance of Jupiter from the earth, come out 2′56″47‴ and 1′51″6‴, on the basis of the periodic times.

The diameter of Jupiter was obtained a number of times with a micrometer in a telescope 123 feet long and, when reduced to the mean distance of Jupiter from the sun or the earth, always came out smaller than 40″, never smaller than 38″, and quite often 39″. In shorter telescopes this diameter is 40″ or 41″. For the light of Jupiter is somewhat dilated by its nonuniform refrangibility, and this dilation has a smaller ratio to the diameter of Jupiter in longer and more perfect telescopes than in shorter and less perfect ones. The times in which two satellites, the first and the third, crossed the disk of Jupiter, from the beginning of their entrance [i.e., from the moment of their beginning to cross the disk] to the beginning of their exit and from the completion of their entrance to the completion of their exit, were observed with the aid of the same longer telescope. And from the transit of the first satellite, the diameter of Jupiter at its mean distance from the earth came out 37⅛″ and, from the transit of the third satellite, 37⅜″. The time in which the shadow of the first satellite passed across the body of Jupiter was also observed, and from this observation the diameter of Jupiter at its mean distance from the earth came out roughly 37″. Let us assume that this diameter is very nearly 37¼″; then the greatest elongations of the first, second, third, and fourth satellites will be equal respectively to 5.965, 9.494, 15.141, and 26.63 semidiameters of Jupiter.

Phenomenon 2

The circumsaturnian planets [or satellites of Saturn], by radii drawn to the center of Saturn, describe areas proportional to the times, and their periodic times—the fixed stars being at rest—are as the 3⁄2 powers of their distances from that center.

Cassini, in fact, from his own observations has established their distances from the center of Saturn and their periodic times as follows.

Periodic times of the satellites of Saturn				
$1^d21^h18^m27^s$	$2^d17^h41^m22^s$	$4^d12^h25^m12^s$	$15^d22^h41^m14^s$	$79^d7^h48^m00^s$

Distances of the satellites from the center of Saturn, in semidiameters of the ring					
From the observations	$1\frac{19}{20}$	$2\frac{1}{2}$	$3\frac{1}{2}$	8	24
From the periodic times	1.93	2.47	3.45	8	23.35

Observations yield a value of the greatest elongation of the fourth satellite from the center of Saturn that is very near eight semidiameters. But the greatest elongation of this satellite from the center of Saturn, as determined by an excellent micrometer in Huygens's 123-foot telescope, came out $8\frac{7}{10}$ semidiameters. And from this observation and the periodic times, the distances of the satellites from the center of Saturn are, in semidiameters of the ring, 2.1, 2.69, 3.75, 8.7, and 25.35. The diameter of Saturn in the same telescope was to the diameter of the ring as 3 to 7, and the diameter of the ring on the 28th and 29th day of May 1719 came out 43″. And from this the diameter of the ring at the mean distance of Saturn from the earth is 42″, and the diameter of Saturn is 18″. These are the results obtained with the longest and best telescopes, because the apparent magnitudes of heavenly bodies, as seen in longer telescopes, have a greater proportion to the dilation of light at the edges of these bodies than when seen in shorter telescopes. If all erratic light [i.e., dilated light] is disregarded, the diameter of Saturn will not be greater than 16″.

Phenomenon 3

The orbits of the five primary planets—Mercury, Venus, Mars, Jupiter, and Saturn—encircle the sun.

That Mercury and Venus revolve about the sun is proved by their exhibiting phases like the moon's. When these planets are shining with a full face, they are situated beyond the sun; when half full, to one side of the sun; when horned, on this side of the sun; and they sometimes pass across the sun's disk like spots. Because Mars also shows a full face when near conjunction with the sun, and appears gibbous in the quadratures, it is certain that Mars goes around the sun. The same thing is proved also with respect to Jupiter and Saturn from their phases being always full; and in these two planets, it is manifest from the shadows that their satellites project upon them that they shine with light borrowed from the sun.

Phenomenon 4

84

The periodic times of the five primary planets and of either the sun about the earth or the earth about the sun—the fixed stars being at rest—are as the 3/2 powers of their mean distances from the sun.

This proportion, which was found by Kepler, is accepted by everyone. In fact, the periodic times are the same, and the dimensions of the orbits are the same, whether the sun revolves about the earth, or the earth about the sun. There is universal agreement among astronomers concerning the measure of the periodic times. But of all astronomers, Kepler and Boulliau have determined the magnitudes of the orbits from observations with the most diligence; and the mean distances that correspond to the periodic times as computed from the above proportion do not differ sensibly from the distances that these two astronomers found [from observations], and for the most part lie between their respective values, as may be seen in the following table.

Periodic times of the planets and of earth about the sun with respect to the fixed stars, in days and decimal parts of a day

♄	♃	♂	♁	♀	☿
10759.275	4332.514	686.9785	365.2565	224.6176	87.9692

Mean distances of the planets and of the earth from the sun

	♄	♃	♂	♁	♀	☿
According to Kepler	951000	519650	152350	100000	72400	38806
According to Boulliau	954198	522520	152350	100000	72398	38585
According to the periodic times	954006	520096	152369	100000	72333	38710

There is no ground for dispute about the distances of Mercury and Venus from the sun, since these distances are determined by the elongations of the planets from the sun. Furthermore, with respect to the distances of the superior planets from the sun, any ground for dispute is eliminated by the eclipses of the satellites of Jupiter. For by these eclipses the position of the shadow that Jupiter projects is determined, and this gives the heliocentric longitude of Jupiter. And from a comparison of the heliocentric and geocentric longitudes, the distance of Jupiter is determined.

Phenomenon 5

The primary planets, by radii drawn to the earth, describe areas in no way proportional to the times but, by radii drawn to the sun, traverse areas proportional to the times.

For with respect to the earth they sometimes have a progressive [direct or forward] motion, they sometimes are stationary, and sometimes they even have a retrograde motion; but with respect to the sun they move always forward, and they do so with a motion that is almost uniform—but, nevertheless, a little more swiftly in their perihelia and more slowly in their aphelia, in such a way that the description of areas is uniform. This is a proposition very well known to astronomers and is especially provable in the case of Jupiter by the eclipses of its satellites; by means of these eclipses we have said that the heliocentric longitudes of this planet and its distances from the sun are determined.

Phenomenon 6

The moon, by a radius drawn to the center of the earth, describes areas proportional to the times.

This is evident from a comparison of the apparent motion of the moon with its apparent diameter. Actually, the motion of the moon is somewhat perturbed by the force of the sun, but in these phenomena I pay no attention to minute errors that are negligible.

Propositions

Proposition 1
Theorem 1

The forces by which the circumjovial planets [or satellites of Jupiter] are continually drawn away from rectilinear motions and are maintained in their respective orbits are directed to the center of Jupiter and are inversely as the squares of the distances of their places from that center.

The first part of the proposition is evident from phen. 1 and from prop. 2 or prop. 3 of book 1, and the second part from phen. 1 and from corol. 6 to prop. 4 of book 1.

The same is to be understood for the planets that are Saturn's companions [or satellites] by phen. 2.

Proposition 2
Theorem 2

The forces by which the primary planets are continually drawn away from rectilinear motions and are maintained in their respective orbits are directed to the sun and are inversely as the squares of their distances from its center.

86 The first part of the proposition is evident from phen. 5 and from prop. 2 of book 1, and the latter part from phen. 4 and from prop. 4 of the same book. But this second part of the proposition is proved with the greatest exactness from the fact that the aphelia are at rest. For the slightest departure from the ratio of the square would (by book 1, prop. 45, corol. 1) necessarily result in a noticeable motion of the apsides in a single revolution and an immense such motion in many revolutions.

Proposition 3
Theorem 3

The force by which the moon is maintained in its orbit is directed toward the earth and is inversely as the square of the distance of its places from the center of the earth.

The first part of this statement is evident from phen. 6 and from prop. 2 or prop. 3 of book 1, and the second part from the very slow motion of the moon's apogee. For that motion, which in each revolution is only three degrees and three minutes forward [or in consequentia, i.e., in an easterly direction] can be ignored. For it is evident (by book 1, prop. 45, corol. 1) that if the distance of the moon from the center of the earth is to the semidiameter of the earth as D to 1, then the force from which such a motion may arise is inversely as $D^{2\,4/243}$, that is, inversely as that power of D of which the index is $2\frac{4}{243}$; that is, the proportion of the force to the distance is inversely as a little greater than the second power of the distance, but is $59\frac{3}{4}$ times closer to the square than to the cube. Now this motion of the apogee arises from the action of the sun (as will be pointed out below) and accordingly is to be ignored here. The action of the sun, insofar as it draws the moon away from the earth, is very nearly as the distance of the moon from the earth, and so (from what is said in book 1, prop. 45, corol. 2) is to the centripetal force of the moon as roughly 2 to 357.45, or 1 to $178\frac{29}{40}$. And if so small a force of the sun is ignored, the remaining force by which the moon is maintained in its orbit will be inversely as D^2. And this will be even more fully established by comparing this force with the force of gravity as is done in prop. 4 below.

COROLLARY. If the mean centripetal force by which the moon is maintained in its orbit is increased first in the ratio of $177\frac{29}{40}$ to $178\frac{29}{40}$, then also in the squared ratio of the semidiameter of the earth to the mean distance of the center of the moon from the center of the earth, the result will be the lunar centripetal force at the surface of the earth, supposing that that force, in descending to the surface of the earth, is continually increased in the ratio of the inverse square of the height.

Proposition 4
Theorem 4

The moon gravitates toward the earth and by the force of gravity is always drawn back from rectilinear motion and kept in its orbit.

The mean distance of the moon from the earth in the syzygies is, according to Ptolemy and most astronomers, 59 terrestrial semidiameters, 60 according to Vendelin and Huygens, $60\frac{1}{3}$ according to Copernicus, $60\frac{2}{5}$ according to Street, and $56\frac{1}{2}$ according to Tycho. But Tycho and all those who follow his tables of refractions, by making the refractions of the sun and moon (entirely contrary to the nature of light) be greater than those of the fixed stars—in fact greater by about four or five minutes—have increased the parallax of the moon by that many minutes, that is, by about a twelfth or fifteenth of the whole parallax. Let that error be corrected, and the distance will come to be roughly $60\frac{1}{2}$ terrestrial semidiameters, close to the value that has been assigned by others. Let us assume a mean distance of 60 semidiameters in the syzygies; and also let us assume that a revolution of the moon with respect to the fixed stars is completed in 27 days, 7 hours, 43 minutes, as has been established by astronomers; and that the circumference of the earth is 123,249,600 Paris feet, according to the measurements made by the French. If now the moon is imagined to be deprived of all its motion and to be let fall so that it will descend to the earth with all that force urging it by which (by prop. 3, corol.) it is [normally] kept in its orbit, then in the space of one minute, it will by falling describe $15\frac{1}{12}$ Paris feet. This is determined by a calculation carried out either by using prop. 36 of book 1 or (which comes to the same thing) by using corol. 9 to prop. 4 of book 1. For the versed sine of the arc which the moon would describe in one minute of time by its mean motion at a distance of 60 semidiameters of the earth is roughly $15\frac{1}{12}$ Paris feet, or more exactly 15 feet, 1 inch, and $1\frac{4}{9}$ lines [or twelfths of an inch]. Accordingly, since in approaching the earth that force is increased as the inverse square of the distance, and so at the surface of the earth is 60 × 60 times greater than at the moon, it follows that a body falling with that force, in our regions, ought in the space of one minute to describe 60 × 60 × 15 1/12 Paris feet, and in the space of one second 15 1/12 feet, or more exactly 15 feet, 1 inch, and 1 4/9 lines. And heavy bodies do actually descend to the earth with this very force. For a pendulum beating seconds in the latitude of Paris is 3 Paris feet and 8 1/2 lines in length, as Huygens observed. And the height that a heavy body describes by falling in the time of one second is to half the length of this pendulum as the square of the ratio of the circumference of a circle to its diameter (as Huygens also showed), and so is 15 Paris feet, 1 inch, 1 7/9 lines. And therefore that force by which the moon is kept in its orbit, in descending from the moon's orbit to the surface of the earth, comes out equal to the force of gravity here on earth, and so (by rules 1 and 2) is that very force which we generally call gravity. For if gravity were different from this force, then bodies making for the earth by both forces acting together would descend twice as fast, and in the space of one second would by falling describe 30 1/6 Paris feet, entirely contrary to experience.

This calculation is founded on the hypothesis that the earth is at rest. For if the earth and the moon move around the sun and in the meanwhile also revolve around their common center of gravity, then, the law of gravity remaining the same, the distance of the centers of the moon and earth from each other will be roughly 60 1/2 terrestrial semidiameters, as will be evident to anyone who computes it. And the computation can be undertaken by book 1, prop. 60.

Scholium

The proof of the proposition can be treated more fully as follows. If several moons were to revolve around the earth, as happens in the system of Saturn or of Jupiter, their periodic times (by the argument of induction) would observe the law which Kepler discovered for the planets, and therefore their centripetal forces would be inversely as the squares of the distances from the center of the earth, by prop. 1 of this book 3. And if the lowest of them were small and nearly touched the tops of the highest mountains, its centripetal force, by which it would be kept in its orbit, would (by the preceding computation) be very nearly equal to the gravities of bodies on the tops of those mountains. And this centripetal force would cause this little moon, if it were deprived of all the motion with which it proceeds in its orbit, to descend to the earth—as a result of the absence of the centrifugal force with which it had remained in its orbit—and to do so with the same velocity with which heavy bodies fall on the tops of those mountains, because the forces with which they descend are equal. And if the force by which the lowest little moon descends were different from gravity and that little moon also were heavy toward the earth in the manner of bodies on the tops of mountains, this little moon would descend twice as fast by both forces acting together. Therefore, since both forces—namely, those of heavy bodies and those of the moons—are directed toward the center of the earth and are similar to each other and equal, they will (by rules 1 and 2) have the same cause. And therefore that force by which the moon is kept in its orbit is the very one that we generally call gravity. For if this were not so, the little moon at the top of a mountain must either be lacking in gravity or else fall twice as fast as heavy bodies generally do.

Proposition 5
Theorem 5

The circumjovial planets [or satellites of Jupiter] gravitate toward Jupiter, the circumsaturnian planets [or satellites of Saturn] gravitate toward Saturn, and the circumsolar [or primary] planets gravitate toward the sun, and by the force of their gravity they are always drawn back from rectilinear motions and kept in curvilinear orbits.

For the revolutions of the circumjovial planets about Jupiter, of the circumsaturnian planets about Saturn, and of Mercury and Venus and the other circumsolar planets about the sun are phenomena of the same kind as the revolution of the moon about the earth, and therefore (by rule 2) depend on causes of the same kind, especially since it has been proved that the forces on which those revolutions depend are directed toward the centers of Jupiter, Saturn, and the sun, and decrease according to the same ratio and law (in receding from Jupiter, Saturn, and the sun) as the force of gravity (in receding from the earth).

COROLLARY 1. Therefore, there is gravity toward all planets universally. For no one doubts that Venus, Mercury, and the rest [of the planets, primary and secondary,] are bodies of the same kind as Jupiter and Saturn. And since, by the third law of motion, every attraction is mutual, Jupiter will gravitate toward all its satellites, Saturn toward its satellites, and the earth will gravitate toward the moon, and the sun toward all the primary planets.

COROLLARY 2. The gravity that is directed toward every planet is inversely as the square of the distance of places from the center of the planet.

COROLLARY 3. All the planets are heavy toward one another by corols. 1 and 2. And hence Jupiter and Saturn near conjunction, by attracting each other, sensibly perturb each other's motions, the sun perturbs the lunar motions, and the sun and moon perturb our sea, as will be explained in what follows.

Scholium

Hitherto we have called "centripetal" that force by which celestial bodies are kept in their orbits. It is now established that this force is gravity, and therefore we shall call it gravity from now on. For the cause of the centripetal force by which the moon is kept in its orbit ought to be extended to all the planets, by rules 1, 2, and 4.

Proposition 6
Theorem 6

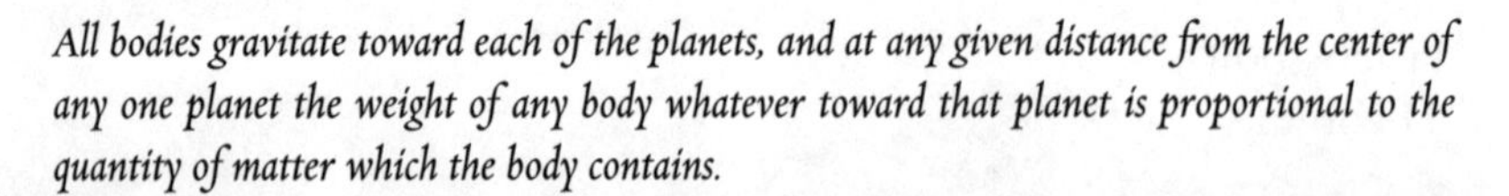

All bodies gravitate toward each of the planets, and at any given distance from the center of any one planet the weight of any body whatever toward that planet is proportional to the quantity of matter which the body contains.

Others have long since observed that the falling of all heavy bodies toward the earth (at least on making an adjustment for the inequality of the retardation that arises from the very slight resistance of the air) takes place in equal times, and it is possible to discern that equality of the times, to a very high degree of accuracy, by using pendulums. I have tested this with gold, silver, lead, glass, sand, common salt, wood, water, and wheat. I got two wooden boxes, round and equal. I filled one of them with wood, and I suspended the same weight of gold (as exactly as I could) in the center of oscillation of the other. The boxes, hanging by equal eleven-foot cords, made pendulums exactly like each other with respect to their weight, shape, and air resistance. Then, when placed close to each other [and set into vibration], they kept swinging back and forth together with equal oscillations for a very long time. Accordingly, the amount of matter in the gold (by book 2, prop. 24, corols. 1 and 6) was to the amount of matter in the wood as the action of the motive force upon all the gold to the action of the motive force upon all the [added] wood—that is, as the weight of one to the weight of the other. And it was so for the rest of the materials. In these experiments, in bodies of the same weight, a difference of matter that would be even less than a thousandth part of the whole could have been clearly noticed. Now, there is no doubt that the nature of gravity toward the planets is the same as toward the earth. For imagine our terrestrial bodies to be raised as far as the orbit of the moon and, together with the moon, deprived of all motion, to be released so as to fall to the earth simultaneously; and by what has already been shown, it is certain that in equal times these falling terrestrial bodies will describe the same spaces as the moon, and therefore that they are to the quantity of matter in the moon as their own weights are to its weight. Further, since the satellites of Jupiter revolve in times that are as the $3/2$ power of their distances from the center of Jupiter, their accelerative gravities toward Jupiter will be inversely as the squares of the distances from the center of Jupiter, and, therefore, at equal distances from Jupiter their accelerative gravities would come out equal. Accordingly, in equal times in falling from equal heights [toward Jupiter] they would describe equal spaces, just as happens with heavy bodies on this earth of ours. And by the same argument the circumsolar [or primary] planets, let fall from equal distances from the sun, would describe equal spaces in equal times in their descent to the sun. Moreover, the forces by which unequal bodies are equally accelerated are as the bodies; that is, the weights [of the primary planets toward the sun] are as the quantities of matter in the planets. Further, that the weights of Jupiter and its satellites toward the sun are proportional to the quantities of their matter is evident from the extremely regular motion of the satellites, according to book 1, prop. 65, corol. 3. For if some of these were more strongly attracted toward the sun in proportion to the quantity of their matter than the rest, the motions of the satellites (by book 1, prop. 65, corol. 2) would be perturbed by that inequality of attraction. If, at equal distances from the sun, some satellite were heavier [or gravitated more] toward the sun in proportion to the quantity of its matter than Jupiter in proportion to the quantity of its own matter, in any given ratio, say d to e, then the distance between the center of the sun and the center of the orbit of the satellite would always be greater than the distance between the center of the sun and the center of Jupiter and these distances would be to each other very nearly as the square root of d to the square root of e, as I found by making a certain calculation. And if the satellite were less heavy [or gravitated less] toward the sun in that ratio of d to e, the distance of the center of the orbit of the satellite from the sun would be less than the distance of the center of Jupiter from the sun in that same ratio of the square root of d to the square root of e. And so if, at equal distances from the sun, the accelerative gravity of any satellite toward the sun were greater or smaller than the accelerative gravity of Jupiter toward the sun, by only a thousandth of the whole gravity, the distance of the center of the orbit of the satellite from the sun would be greater or smaller than the distance of Jupiter from the sun by $1/2000$ of the total distance, that is, by a fifth of the distance of the outermost satellite from the center of Jupiter; and this eccentricity of the orbit would be very sensible indeed. But the orbits of the satellites are concentric with Jupiter, and therefore the accelerative gravities of Jupiter and of the satellites toward the sun are equal to one another. And by the same argument the weights [or gravities] of Saturn and its companions toward the sun, at equal distances from the sun, are as the quantities of matter in them; and the weights of the moon and earth toward the sun are either nil or exactly proportional

to their masses. But they do have some weight, according to prop. 5, corols. 1 and 3.

But further, the weights [or gravities] of the individual parts of each planet toward any other planet are to one another as the matter in the individual parts. For if some parts gravitated more, and others less, than in proportion to their quantity of matter, the whole planet, according to the kind of parts in which it most abounded, would gravitate more or gravitate less than in proportion to the quantity of matter of the whole. But it does not matter whether those parts are external or internal. For if, for example, it is imagined that bodies on our earth are raised to the orbit of the moon and compared with the body of the moon, then, if their weights were to the weights of the external parts of the moon as the quantities of matter in them, but were to the weights of the internal parts in a greater or lesser ratio, they would be to the weight of the whole moon in a greater or lesser ratio, contrary to what has been shown above.

COROLLARY 1. Hence, the weights of bodies do not depend on their forms and textures. For if the weights could be altered with the forms, they would be, in equal matter, greater or less according to the variety of forms, entirely contrary to experience.

COROLLARY 2. All bodies universally that are on or near the earth are heavy [or gravitate] toward the earth, and the weights of all bodies that are equally distant from the center of the earth are as the quantities of matter in them. This is a quality of all bodies on which experiments can be performed and therefore by rule 3 is to be affirmed of all bodies universally. If the aether or any other body whatever either were entirely devoid of gravity or gravitated less in proportion to the quantity of its matter, then, since (according to the opinion of Aristotle, Descartes, and others) it does not differ from other bodies except in the form of its matter, it could by a change of its form be transmuted by degrees into a body of the same condition as those that gravitate the most in proportion to the quantity of their matter; and, on the other hand, the heaviest bodies, through taking on by degrees the form of the other body, could by degrees lose their gravity. And accordingly the weights would depend on the forms of bodies and could be altered with the forms, contrary to what has been proved in corol. 1.

COROLLARY 3. All spaces are not equally full. For if all spaces were equally full, the specific gravity of the fluid with which the region of the air would be filled, because of the extreme density of its matter, would not be less than the specific gravity of quicksilver or of gold or of any other body with the greatest density, and therefore neither gold nor any other body could descend in air. For bodies do not ever descend in fluids unless they have a greater specific gravity. But if the quantity of matter in a given space could be diminished by any rarefaction, why should it not be capable of being diminished indefinitely?

COROLLARY 4. If all the solid particles of all bodies have the same density and cannot be rarefied without pores, there must be a vacuum. I say particles have the same density when their respective forces of inertia [or masses] are as their sizes.

COROLLARY 5. The force of gravity is of a different kind from the magnetic force. For magnetic attraction is not proportional to the [quantity of] matter attracted. Some bodies are attracted [by a magnet] more [than in proportion to their quantity of matter], and others less, while most bodies are not attracted [by a magnet at all]. And the magnetic force in one and the same body can be intended and remitted [i.e., increased and decreased] and is sometimes far greater in proportion to the quantity of matter than the force of gravity; and this force, in receding from the magnet, decreases not as the square but almost as the cube of the distance, as far as I have been able to tell from certain rough observations.

Proposition 7
Theorem 7

Gravity exists in all bodies universally and is proportional to the quantity of matter in each.

We have already proved that all planets are heavy [or gravitate] toward one another and also that the gravity toward any one planet, taken by itself, is inversely as the square of the distance of places from the center of the planet. And it follows (by book 1, prop. 69 and its corollaries) that the gravity toward all the planets is proportional to the matter in them.

Further, since all the parts of any planet A are heavy [or gravitate] toward any planet B, and since the gravity of each part is to the gravity of the whole as the matter of that part to the matter of the whole, and since to every action (by the third law of motion) there is an equal reaction, it follows that planet B will gravitate in turn toward all the parts of planet A, and its gravity toward any one part will be to its gravity toward the whole of the planet as the matter of that part to the matter of the whole. Q.E.D.

COROLLARY 1. Therefore the gravity toward the whole planet arises from and is compounded of the gravity toward the individual parts. We have examples of this in magnetic and electric attractions. For every attraction toward a whole arises from the attractions toward the individual parts. This will be understood in the case of gravity by thinking of several smaller planets coming together into one globe and composing a larger planet. For the force of the whole will have to arise from the forces of the component parts. If anyone objects that by this law all bodies on our earth would have to gravitate toward one another, even though gravity of this kind is by no means detected by our senses, my answer is that gravity toward these bodies is far smaller than what our senses could detect, since such gravity is to the gravity toward the whole earth as [the quantity of matter in each of] these bodies to the [quantity of matter in the] whole earth.

COROLLARY 2. The gravitation toward each of the individual equal particles of a body is inversely as the square of the distance of places from those particles. This is evident by book 1, prop. 74, corol. 3.

. . . .

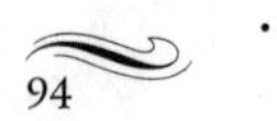

7

Newton on "Hypotheses," God, and Gravity

From The Principia, General Scholium / Isaac Newton

. . . .

The six primary planets revolve about the sun in circles concentric with the sun, with the same direction of motion, and very nearly in the same plane. Ten moons revolve about the earth, Jupiter, and Saturn in concentric circles, with the same direction of motion, very nearly in the planes of the orbits of the planets. And all these regular motions do not have their origin in mechanical causes, since comets go freely in very eccentric orbits and into all parts of the heavens. And with this kind of motion the comets pass very swiftly and very easily through the orbits of the planets; and in their aphelia, where they are slower and spend a longer time, they are at the greatest possible distance from one another, so as to attract one another as little as possible.

This most elegant system of the sun, planets, and comets could not have arisen without the design and dominion of an intelligent and powerful being. And if the fixed stars are the centers of similar systems, they will all be constructed according to a similar design and subject to the dominion of *One,* especially since the light of the fixed stars is of the same nature as the

---

Reprinted from *The Principia: Mathematical Principles of Natural Philosophy,* trans. I. Bernard Cohen and Anne Whitman (Berkeley: University of California Press, 1999), 939–43.

Translators' notes have been omitted. Bracketed interpolations are theirs.

light of the sun, and all the systems send light into all the others. And so that the systems of the fixed stars will not fall upon one another as a result of their gravity, he has placed them at immense distances from one another.

He rules all things, not as the world soul but as the lord of all. And because of his dominion he is called Lord God *Pantokrator.*[1] For "god" is a relative word and has reference to servants, and godhood is the lordship of God, not over his own body as is supposed by those for whom God is the world soul, but over servants. The supreme God is an eternal, infinite, and absolutely perfect being; but a being, however perfect, without dominion is not the Lord God. For we do say my God, your God, the God of Israel, the God of Gods, and Lord of Lords, but we do not say my eternal one, your eternal one, the eternal one of Israel, the eternal one of the gods; we do not say my infinite one, or my perfect one. These designations [i.e., eternal, infinite, perfect] do not have reference to servants. The word "god" is used far and wide to mean "lord," but every lord is not a god. The lordship of a spiritual being constitutes a god, a true lordship constitutes a true god, a supreme lordship a supreme god, an imaginary lordship an imaginary god. And from true lordship it follows that the true God is living, intelligent, and powerful; from the other perfections, that he is supreme, or supremely perfect. He is eternal and infinite, omnipotent and omniscient, that is, he endures from eternity to eternity, and he is present from infinity to infinity; he rules all things, and he knows all things that happen or can happen. He is not eternity and infinity, but eternal and infinite; he is not duration and space, but he endures and is present. He endures always and is present everywhere, and by existing always and everywhere he constitutes duration and space. Since each and every particle of space is *always,* and each and every indivisible moment of duration is *everywhere,* certainly the maker and lord of all things will not be *never* or *nowhere.*

Every sentient soul, at different times and in different organs of senses and motions, is the same indivisible person. There are parts that are successive in duration and coexistent in space, but neither of these exist in the person of man or in his thinking principle, and much less in the thinking substance of God. Every man, insofar as he is a thing that has senses, is one and the same man throughout his lifetime in each and every organ of his senses. God is one and the same God always and everywhere. He is omnipresent not only *virtually* but also *substantially;* for action requires substance [*lit.* for active power [virtus] cannot subsist without substance]. In him all things are contained and move, but he does not act on them nor they on him. God experiences nothing from the motions of bodies; the bodies feel no resistance from God's omnipresence.

It is agreed that the supreme God necessarily exists, and by the same necessity he is *always* and *everywhere.* It follows that all of him is like himself:

he is all eye, all ear, all brain, all arm, all force of sensing, of understanding, and of acting, but in a way not at all human, in a way not at all corporeal, in a way utterly unknown to us. As a blind man has no idea of colors, so we have no idea of the ways in which the most wise God senses and understands all things. He totally lacks any body and corporeal shape, and so he cannot be seen or heard or touched, nor ought he to be worshiped in the form of something corporeal. We have ideas of his attributes, but we certainly do not know what is the substance of any thing. We see only the shapes and colors of bodies, we hear only their sounds, we touch only their external surfaces, we smell only their odors, and we taste their flavors. But there is no direct sense and there are no indirect reflected actions by which we know innermost substances; much less do we have an idea of the substance of God. We know him only by his properties and attributes and by the wisest and best construction of things and their final causes, and we admire him because of his perfections; but we venerate and worship him because of his dominion. For we worship him as servants, and a god without dominion, providence, and final causes is nothing other than fate and nature. No variation in things arises from blind metaphysical necessity, which must be the same always and everywhere. All the diversity of created things, each in its place and time, could only have arisen from the ideas and the will of a necessarily existing being. But God is said allegorically to see, hear, speak, laugh, love, hate, desire, give, receive, rejoice, be angry, fight, build, form, construct. For all discourse about God is derived through a certain similitude from things human, which while not perfect is nevertheless a similitude of some kind. This concludes the discussion of God, and to treat of God from phenomena is certainly a part of natural philosophy.

Thus far I have explained the phenomena of the heavens and of our sea by the force of gravity, but I have not yet assigned a cause to gravity. Indeed, this force arises from some cause that penetrates as far as the centers of the sun and planets without any diminution of its power to act, and that acts not in proportion to the quantity of the *surfaces* of the particles on which it acts (as mechanical causes are wont to do) but in proportion to the quantity of *solid* matter, and whose action is extended everywhere to immense distances, always decreasing as the squares of the distances. Gravity toward the sun is compounded of the gravities toward the individual particles of the sun, and at increasing distances from the sun decreases exactly as the squares of the distances as far out as the orbit of Saturn, as is manifest from the fact that the aphelia of the planets are at rest, and even as far as the farthest aphelia of the comets, provided that those aphelia are at rest. I have not as yet been able to deduce from phenomena the reason for these properties of gravity, and I do not feign hypotheses. For whatever is not deduced from the phenomena must be called a hypothesis; and hypotheses, whether metaphysical or physical, or based on occult qualities, or mechanical, have no place in experimental philosophy. In this experimental philosophy, propositions are deduced from the phenomena and are made general by induction. The impenetrability, mobility, and impetus of bodies, and the laws of motion and the law of gravity have been found by this method. And it is enough that gravity really exists and acts according to the laws that we have set forth and is sufficient to explain all the motions of the heavenly bodies and of our sea.

. . . .

NOTE

1. [Newton's note a] That is, universal ruler.

 8

Cohen's Discussion of Newton's Methodology

From A Guide to Newton's *Principia* / I. Bernard Cohen

Chapter 8. The Structure of Book 3

. . . .

2. From Hypotheses to Rules and Phenomena

In the first edition of the *Principia,* book 3 opened with a set of "Hypotheses," nine in number. This designation made the *Principia* an easy target for criticism. Since Newton claimed that in the first two books he was concerned primarily with mathematics and since the third book appeared to be founded on "hypotheses," a critic could legitimately complain that the *Principia* did not disclose a system of natural philosophy. This was the harsh line taken by the author of a book review in the *Journal des Sçavans,* probably the Belgian Cartesian Pierre-Sylvain Régis,[1] who concluded that "in order to produce an *opus* as perfect as possible, Newton has only to give us a *Physics* [i.e., a *Natural Philosophy*] as exact as his *Mechanics.*" Newton will do so, he added, "when he substitutes true motions for those that he has supposed." Possibly it was in answer to this criticism that Newton altered the opening of book 3 in the second edition. He now divided the "hypotheses" into two

---

major groups. The former hyps. 1 and 2, which are methodological, became the first two rules in a new category, "Regulae Philosophandi," or "Rules for Natural Philosophy," and a new rule 3 was introduced. In the third edition, rule 4 was added.[2]

The second major group of original hypotheses were numbered 5, 6, 7, 8, and 9. In the second edition, these became "phenomena," bearing numbers 1, 3, 4, 5, and 6. A new one, phen. 2, was added at that time. These "phenomena" are indeed phenomenological, dealing with such matters as the observed periodic times of planets and satellites and the third law of Kepler, together with actual tables of data.

The original hyp. 3, that "Every body can be transformed into a body of any other kind and successively take on all the intermediate degrees of qualities," was eliminated in the second edition. Hyp. 4 ("The center of the system of the world is at rest") became "hypothesis 1" in the second edition but was moved from the beginning of book 3 to a position (between props. 10 and 11) where it is needed. In the second edition, Newton also converted lem. 3 of the first edition (following prop. 38) into a "hypothesis 2."

The "hypotheses" of the first edition constitute a curious grouping. They include procedural rules for natural philosophy, phenomena, and two hypotheses which would be hypotheses in any system of thought or of science. If all these "hypotheses" are to be taken as a group, it is difficult to think of any single neutral name for so motley a collection. In retrospect, this assemblage seems all the more odd in light of the later General Scholium, written in 1713 for the second edition. For there Newton not only will utter the celebrated slogan "Hypotheses non fingo" but will declare that "whatever is not deduced from the phenomena must be called a hypothesis," whereas in the opening part of book 3 in the first edition, a whole set of phenomena are designated as hypotheses.

3. Newton's "Rules for Natural Philosophy"

The second edition of the *Principia* contains three "Regulae Philosophandi," or "Rules for Natural Philosophy." We have seen that rules 1 and 2 had been, respectively, hyps. 1 and 2 in the first edition. They declare a traditional principle of parsimony, to admit no more causes of phenomena than are "both true and sufficient" (rule 1) and to assign the same causes, as far as possible, to "natural effects of the same kind" (rule 2).

Rule 3, appearing for the first time in the second edition, is of a different sort. Its message is that there is a certain set of "qualities" that (1) are found in all bodies within our range of direct experience on earth and (2) do not vary, and that these are to be considered qualities of all bodies universally, that is, of bodies everywhere in the universe. This feature of transferring qualities from those on which we can make experiments to those on which we cannot has been called "transdiction" and also "projection."[3] Thus extension (the property of occupying space) and inertia (i.e., force of inertia or mass) are properties of this sort and so may be considered, by rule 3, to be properties of the heavenly bodies: sun, planets, planetary satellites, comets, and stars. It is by this third rule that Newton can assign to the heavenly bodies the property of gravity or of exerting a force "in proportion to the quantity of matter," and it is "by this third rule that all bodies gravitate toward one another."[4]

In stating the third rule, Newton used the words "intendi" and "remitti," the present passive infinitives corresponding to the nouns "intensio" and "remissio," that is, "intension" and "remission," which go back to the late-medieval doctrine of the "latitude of forms." This medieval usage referred to any quality—motion, displacement, and even love and grace—which could be quantified and so undergo an intension or remission by degrees. Motte correctly rendered Newton's text, referring to "qualities of bodies, which admit neither intension nor remission of degrees," but the Cajori version altered "intension" to "intensification" without considering what the expression "intension . . . of degrees" might have meant to a reader in Newton's day. That is, the sense of "intensification" in today's usage is a process of increasing to an extreme degree or becoming more concentrated, whereas "intension" signified a process of becoming greater in magnitude.[5]

John Harris's *Lexicon Technicum* (London, 1704) defines "INTENSION in Natural Philosophy" as "the encrease of the Power or Energy of any Quality, such as *Heat, Cold,* &c. for of all Qualities they say, they are *Intended* and *Remitted;* that is, capable of Increase and Diminution." Harris adds that "the *Intension* of all Qualities, increases reciprocally as the Squares of the Distances from the Centre of the Radiating Quality decreases [*sic*]." A proof "that all Qualities are *Remitted,* or have their Power or Efficacy abated, in a *Duplicate Ratio* of the distance from the Centre of the *Radiation*" is given by Harris under "QUALITY." The proof is taken from John Keill's *Introductio ad Veram Physicam* (Oxford, 1702).

Rule 4 was introduced in the third edition. It is, in many ways, the most important of all. Its purpose is to validate the results of induction against any imagined (and unverified) hypotheses. Inductions, Newton declares, "should be considered either exactly or very nearly true" until new phenomena may make them "either more exact or liable to exceptions."

4. Newton's "Phenomena"

The "phenomena" which serve as the basis for Newton's system of the world are not phenomena in the sense in which that term is generally understood. Newton does not have in mind a single "occurrence, a circumstance, or a fact that is perceptible to the senses." We may take note that in philosophy, a phenomenon is usually taken to mean that "which appears real to the mind, regardless of whether its underlying existence is proved or its nature understood." In physics, however, a phenomenon is merely "an observable event." But Newton's phenomena are not simply individual observations, the raw data of sense experience or observable events, but are generalizations based upon such data or events and even can be theoretical conclusions that are phenomenologically based. An example is phen. 3, that the orbits of the "five primary planets—Mercury, Venus, Mars, Jupiter, and Saturn—encircle the sun." It is the same for Kepler's third or harmonic law of planetary motion in phen. 4, and the law of areas in phen. 5. While the evidence for phenomena 1–5 is quite convincing, the case for phen. 6, the area law for the moon's orbital motion (reckoned with respect to the center of the earth), is weak, being no more than a correlation of the "apparent motion" with the "apparent diameter."

A notable change was introduced in the content of phen. 1 (formerly hyp. 1) in the second edition. Newton now had better data concerning the satellites of Jupiter. Newton also (phen. 2) introduced the satellites of Saturn, whose existence he had not been willing to acknowledge at the time of the first edition.[6]

In the second edition a correction was made in phen. 4 of an error in the first edition that had escaped Halley's critical eye as well as Newton's. That is, in discussing Kepler's third law in the first edition (in hyp. 7), Newton had said that the periodic times of the planets and the size of their orbits are the same "whether the planets revolve about the earth or about the sun," which is pure nonsense. He obviously meant, as he said in the corrected version in the second edition, that the periodic times and the size of the orbits are the same "whether the sun revolves about the earth, or the earth about the sun"; that is, they are the same in the Copernican as in the Tychonic or Ricciolian systems, but definitely not the same in the Copernican as in the Ptolemaic system.

In stating phen. 4 (formerly hyp. 7), the third law of planetary motion, Newton finally does name Kepler as discoverer. He does not, however, similarly give credit to Kepler in phen. 5 (formerly hyp. 8) for the area law.

. . . .

6. Props. 1–5: The Principles of Motion of Planets and of Their Satellites; The First Moon Test

In prop. 1 Newton addresses the problem of the centripetal force acting on the satellites of Jupiter. Because (by phen. 1) the orbital motion of the satellites accords with the area law and the harmonic law, he can apply book 1, prop. 2 (or prop. 3) and prop. 4, corol. 6, to show, first, that there is a force directed toward the center of Jupiter and, second, that this force varies inversely as the square of the distance. In the second edition, Newton takes note of Cassini's discovery of satellites of Saturn in addition to the one discovered by Huygens; he extends the results of prop. 1 to these satellites. In prop. 2, he turns to the forces acting on the planets. The proof is the same as for prop. 1, differing only in the use of phen. 5 (the area law for planets) and phen. 4 (the harmonic law for planets). Newton additionally takes note that, by book 1, prop. 45, corol. 1, a small departure from the inverse second power of the distance would introduce a motion of the apsides sufficiently great to be observed in a single revolution and hence very noticeable in many revolutions.

Newton then turns in prop. 3 to the moon. By phen. 6 (the area law for the moon's orbital motion) and book 1, prop. 2 or 3, there must be a centrally directed force. Since there is but one moon encircling the earth, Newton cannot follow the simple path of the harmonic law to establish the inverse-square law. In prop. 3, he applies book 1, prop. 45, corol. 1, to the observed fact that the motion of the lunar apogee is very slow, being about $3°3'$ forward in each revolution or lunar month. According to that corollary, if the ratio of the moon's distance from the center of the earth to the earth's radius is D to 1, then the force producing this motion of the apogee must be as $1/D^n$, where n has the value of $2\frac{4}{243}$. Newton says in prop. 3 that "this motion of the apogee arises from the action of the sun" and "is to be ignored here." Drawing on book 1, prop. 45, corol. 2, he says that the action of the sun, "insofar as it draws the moon away from the earth," is to the force of the earth on the moon roughly as 2 to 357.45 or as 1 to $178\frac{29}{40}$.[7] Taking away that force leaves the net force of the earth on the moon to be inversely as the square of the distance from the center of the earth. That the law of force is as the inverse square of the distance will be confirmed by prop. 4.

In prop. 4, Newton states the first of the major results which indicate that the system of the world proposed in the *Principia* is radically different from any in existence. This proposition declares that the moon "gravitates" toward the earth and that it is the "force of gravity" which keeps the moon "in its orbit" by always drawing it "back from rectilinear motion." Here Newton is not making a supposition about the nature or mode of action of "gravity," merely assuming that every reader is familiar with "gravity," or weight. The purpose of prop. 4, therefore, is twofold. Newton wants to show, first, that gravity—terrestrial gravity, the gravity with which we on earth are all familiar—extends out to the moon and does so according to the law of the inverse square; and second, that it is this force of gravity which causes the moon to fall inward from its inertial linear path and so keep in its orbit.

In this moon test, Newton first computes how far the moon (or an object placed in the moon's orbit, 60 earth-radii from the earth's center) would actually fall in one minute if deprived of all forward motion. The result is 15 1/12 Paris feet.[8] If gravity diminishes by the inverse-square law and if the earth's gravity extends to the moon, then it follows that a heavy body on the earth's surface should fall freely in one minute through 60 × 60 or 3,600 of the above 15 1/12 Paris feet in one second (more accurately, 15 feet, 1 inch, 1 4/9 lines). This result agrees so closely with actual terrestrial experiments that Newton can conclude that the force keeping the moon in its orbit is that very force which "we generally call gravity." An alternative proof, given in the scholium, is based on considerations of a hypothetical terrestrial satellite. Later on, in book 3, prop. 37, corol. 7 (added in the second edition), Newton will present a second such moon test.

Using rule 2, Newton then proves (in prop. 5) that the satellites of Jupiter and of Saturn "gravitate" toward those planets, as do the primary planets toward the sun. Corol. 1 states expressly that "there is gravity toward all planets universally," a force that (prop. 5, corol. 2) is "inversely as the square of the distance." It follows (corol. 3) that "all the planets are heavy toward one another," the proof of which is found in the example of Saturn and Jupiter, near conjunction, where such interactions are said to "sensibly perturb each other's motions." A scholium explains that since the centripetal force producing the orbits of celestial bodies is gravity, he will now use that name.

. . . .

Chapter 9. The Concluding General Scholium

1. The General Scholium: "Hypotheses non fingo"

The first edition of the *Principia* had no proper conclusion, since Newton suppressed his draft "Conclusio." In the second edition, he planned to have a final discussion about "the attraction of the small particles of bodies," as he wrote to Cotes on 2 March 1712/13,[9] when most of the text had already been printed off. On further reflection, however, he abandoned the temptation to expose his theories of the forces, interactions, structure, and other aspects of particulate matter, and instead composed the concluding General Scholium, in which, he said, he had included a "short Paragraph about that part of Philosophy." One function of the General Scholium was to answer certain critics, notably the Cartesians and other strict adherents of the mechanical philosophy.

The opening paragraph of the General Scholium is a recapitulation of all the arguments in the main text to prove that the celestial phenomena are not compatible with "the hypothesis of vortices." It is followed by a discussion of the emptiness of the "celestial spaces . . . above the atmosphere of the earth," based on the long-term constancy of planetary and cometary motions.

The next topic is the argument from design, the proof of the existence of the creator from the perfection of his creation. Having established the presence of "an intelligent and powerful being," the architect of the universe, Newton proceeds to analyze the names and attributes of this creator. In the second edition, Newton concluded this paragraph by asserting that to discourse of "God" from phenomena is legitimate in "experimental philosophy." In the third edition, this was altered to "natural philosophy." This sentence was not in the original text of the General Scholium that Newton sent to Cotes. Newton wrote to Cotes shortly afterward, directing him to make this addition.

The next paragraph, the penultimate paragraph of the General Scholium, is probably the most discussed portion of all of Newton's writings. Briefly, he here declares that he has explained the phenomena of the heavens and the tides in the oceans "by the force of gravity," but he has "not yet assigned a cause to gravity." Summarizing some of the chief properties he has put forth in book 3, he takes note that this force does not operate "in proportion to the quantity of the *surfaces* of the particles on which it acts," but rather "in proportion to the quantity of *solid* matter." Accordingly, gravity does not act "as mechanical causes are wont to do."[10] Then Newton admits that he has been unable to discover the cause of the properties of gravity from phenomena, and he now declares: "Hypotheses non fingo."

Newton obviously did not mean by this phrase that he never "uses" or "makes" hypotheses, since this statement could easily be belied: for example, by the presence of one "hypothesis" in book 2 and two more in book 3. Alexandre Koyré suggested that by "fingo" Newton probably intended "I feign," in the sense of "inventing a fiction," since the Latin version (1706) of the *Opticks* (1704) used the cognate "confingere" for the English "feign." Thus Newton would be saying that he does not invent or contrive fictions (or "hypotheses") to be offered in place of sound explanations based on phenomena. In this category he included "metaphysical" and "physical" hypotheses, "mechanical" hypotheses, and hypotheses of "occult qualities."

In the *Opticks* and in a great number and variety of manuscripts, Newton used the English verb "feign" in the context of showing his disdain for hypotheses.[11]

From manuscript drafts we learn that Newton chose the verb "fingo" with care; it was not the first word to leap into his mind. He tried "fugio," as in "I flee from [or shun] hypotheses"; he tried "sequor," as in "I do not follow [or, perhaps, I am not a follower of] hypotheses. . . ." Most English-speaking readers know this phrase in Andrew Motte's translation as "I frame no hypotheses." There is no way of telling exactly what meaning Motte intended the verb "frame" to have. Certainly one use of "frame," in the context of theory, was "to fabricate." In any event, in Newton's day and in Motte's, one of the senses of "to frame" was decidedly pejorative, just as is the case today. In today's usage, however, the pejorative sense of "to frame" has a very different signification from what it did in Newton's day, since for us the verb "to frame" means "to concoct false evidence against" (a person). In Samuel Johnson's *Dictionary* (London, 1785), one of the definitions is "To invent, fabricate, in a bad sense: as, to *frame* a story or lie." Johnson gives, as an example, a quotation from Francis Bacon, "Astronomers, to solve the phaenomena, *framed* to their conceit eccentrics and epicycles." Surely, Motte was intelligent enough to recognize that for Newton "fingo" was a derogatory word. Accordingly, Motte would surely have intended that his translation "frame" would equally convey this sense to the reader.[12]

When Cotes received the text of the General Scholium, he was greatly concerned that in book 3, prop. 7, Newton seemed to him to "Hypothesim fingere," and he suggested that Newton write a further discussion in "an *Addendum* to be printed with the Errata."[13] Newton, however, proposed that the end of the penultimate paragraph be altered.

This paragraph, as originally sent to Cotes, concluded as follows:

> Indeed, I have not yet been able to deduce the reason [or cause] of these properties of gravity from phenomena, and I do not feign hypotheses. For whatever is not deduced from phenomena is to be called a hypothesis; and I do not follow *hypotheses,* whether metaphysical or physical, whether of occult qualities or mechanical. It is enough that gravity should really exist and act according to the laws expounded by us, and should suffice for all the motions of the celestial bodies and of our sea.

He now instructed Cotes to alter this and to make it more general. It now would read:

> I have not as yet been able to deduce from phenomena the reason for these properties of gravity, and I do not feign hypotheses. For whatever

> is not deduced from the phenomena must be called a hypothesis; and hypotheses, whether metaphysical or physical, or based on occult qualities, or mechanical, have no place in experimental philosophy. In this experimental philosophy, propositions are deduced from the phenomena and are made general by induction. The impenetrability, mobility, and impetus of bodies, and the laws of motion and the law of gravity have been found by this method. And it is enough that gravity really exists and acts according to the laws that we have set forth and is sufficient to explain all the motions of the heavenly bodies and of our sea.

This is the text as we know it. In a letter to Cotes, Newton wrote a gloss on the revision of the General Scholium, in which he explained that, "as in Geometry the word Hypothesis is not taken in so large a sense as to include the Axiomes & Postulates, so in Experimental Philosophy it is not to be taken in so large a sense as to include the first Principles or Axiomes which I call the laws of motion." These principles, he explained, "are deduced from Phaenomena & made general by Induction: which is the highest evidence that a Proposition can have in this philosophy." Furthermore, "the word Hypothesis is here used by me to signify only such a Proposition as is not a Phaenomenon nor deduced from any Phaenomena but assumed or supposed without any experimental proof." Newton believed that "the mutual & mutually equal attraction of bodies is a branch of the third Law of motion & . . . this branch is deduced from Phaenomena."[14]

"Satis est": Is It Enough?

Newton concludes the penultimate paragraph of the General Scholium in a series of assertions. First, in experimental philosophy, "propositions are deduced from the phenomena and are made general by induction." Second, this is the method by which "the impenetrability, mobility, and impetus of bodies" have been found, as also "the laws of motion and the law of gravity." Third, and in conclusion, "And it is enough [*satis est*] that gravity really exists [*revera existat*]" and that gravity (1) "acts according to the laws that we have set forth" and (2) "is sufficient to explain all the motions of the heavenly bodies and of our sea."

On the basis of these expressions, and without further inquiry, Newton was hailed by Ernst Mach as an early positivist, and others have followed Mach in thus attributing to Newton an apparent endorsement of a positivistic position. Anyone who is acquainted with Newton's writings, however, especially his correspondence and his further published and manuscript expressions on this subject, will know that Newton's position was altogether free of any taint of positivistic philosophy. We have seen the many attempts made by Newton, both before and after writing the General Scholium, to find some way of accounting for a force acting at a distance, an endeavor that occupied his attention to varying degrees from at least as early as the time of composing the *Principia* in the 1680s.[15] Not only did he explore possible modes of explanation in various manuscript essays and proposed revisions of the *Principia;* he introduced this topic in the second (revised) English edition of the *Opticks* in 1717/18, where it is mentioned specifically in the preface.[16]

One cannot know with any certainty exactly what Newton intended his readers to understand by "satis est." The evidence leaves no doubt, however, that for Newton, it was never "enough" merely that his explanation was based on a force that "really exists," whose laws he had set forth, and which serves to explain the observed phenomena of heaven and earth. Yet there can be no doubt that Newton firmly believed in the reality of universal gravity and its action in the solar system.

There is no way of telling whether the Newtonian style of initially dealing with the subject on a mathematical rather than a physical plane was only a subterfuge to avoid criticism or a sincere expression of methodological principle. But, from Newton's point of view, this style enabled him to develop the laws of the action of a gravity-like force in a mathematical analogue of the world of nature without having to be concerned with whether or not gravity exists.

The shift from the mathematical level of discourse of books 1 and 2 to the physical level of book 3, with constant reference to deductions or inductions[17] from phenomena, produced an elegant way of ordering or explaining some of the most basic phenomena of nature. In the General Scholium, Newton freely admitted that he had not been able to explain how a force can act over vast reaches of empty space. He had not succeeded in finding out how such a force acts. Nevertheless, he was able to assert that gravity really exists and that gravity does act to produce the principal phenomena of nature.

Mathematical deduction showed that if there are such long-range forces, then the planets move in orbits produced by an inverse-square force. Sound induction led him to the conclusion that the planetary and lunar forces are a variety of terrestrial gravity which, as he wrote, "really exists." It would seem, therefore, that what Newton is saying in the General Scholium is that gravity really exists, that it extends to the moon and beyond, and that it does provide a means of ordering or organizing the observed phenomena of nature. By using this force we gain an explanation of the phenomena of the heavenly bodies, the tides, and much more.

In the General Scholium, then, Newton is not telling his readers to abandon the search for the cause of gravity, nor is he denying the importance of finding out how gravity acts to produce its effects. Indeed, we know how hard Newton himself tried to solve these two problems. What he is saying, is that there are two jobs ahead. One is to find the cause and mode of operation of the universal force of gravity; the other is to apply the theory of gravity to yet new areas of phenomena, to use our mathematical skills to solve those vexing problems for which he himself had only found imperfect solutions. Among the latter was the complex of problems associated with the moon's motion, including the motion of the lunar apsis.

Newton was concerned by the fact that some scientists, notably those on the Continent, had accepted the inverse-square force for the sun-planet and planet-satellite forces (although not as attractions) but argued on philosophical grounds against accepting a force of universal gravity. To these critics Newton seems to be saying that even though the notion of attraction is not acceptable, that even though we cannot as of now understand the cause and mode of action of universal gravity, it is nevertheless fruitful to move ahead with the science of rational mechanics and celestial dynamics. For this purpose, "it is enough" that gravity does account for so many phenomena; therefore, we may legitimately use this concept to advance the subjects of rational mechanics and celestial dynamics and even to explore new fields of science relating to other forms and varieties of attractions. Such research, especially the discovery of new phenomena, might even produce an explanation of the cause of gravity or of the way in which gravity works. Newton was not a positivist; he did not say "satis est" in the sense that he had no concern for finding causes. But he did not believe that the search for a cause, or for a mode of producing effects, should cause a halt in the application of a useful concept.

The correctness of this reading of the message of the General Scholium seems to be confirmed by the final paragraph. For here, Newton himself is indicating a new path of research that might possibly illuminate the problem of the cause and mode of operation of universal gravity. This research topic that might possibly be fruitful involved the properties and laws of what he conceived to be "a certain very subtle spirit. . . ." In the paragraph about this "spirit" he mentioned many types of phenomena concerning it, but he did not include gravity in that list. Perhaps he was not then ready to do more than hint—by implication or by context—that this spirit might somehow be related to gravity. Or, perhaps, he was merely indicating that there are always new areas of research that may illuminate fundamental problems of cause and mode of operation. In various manuscripts, he indicated that perhaps the hoped-for way to account for the action of gravity might be found in some current research in the new field of electricity.[18]

Newton's suggestion of a field for fruitful research . . . is similar to his suggestions for research at the end of the *Opticks:* both declare a research program for the future.

. . . .

NOTES

1. Paul Mouy, *Le développement de la physique cartésienne* (Paris: Librairie Philosophique J. Vrin, 1934), p. 256.

2. I. B. Cohen, "Hypotheses in Newton's Philosophy," *Physics* 8 (1966): 63–184.

3. That is, a means of assigning to bodies beyond the immediate range of our sense experience the qualities or properties such as mass and inertia that we observe in bodies close at hand. Nelson Goodman has called this concept "projection," a name that has gained currency among philosophers and has replaced Mandelbaum's "transdiction." In the English-speaking philosophical community, a distinction is commonly made between hypotheses which are taken to be "projectable" and those which are not. See Maurice Mandelbaum, *Philosophy, Science, and Sense Perception* (Baltimore: Johns Hopkins Press, 1964). Philosophers have not generally adopted Mandelbaum's suggestion, preferring the term "projection."

4. Quoted from rule 3.

5. It may seem odd that this pair of technical terms, used by the medieval "calculatores," is not found in the early edition of the *Principia* (1687) but occurs for the first time in the second edition (1713). A half century or so later, the marquise du Châtelet gave these terms a modern sense, writing of "qualités des corps qui ne sont susceptibles ni d'augmentation ni de diminution."

6. In the first edition, Newton took account only of the single satellite of Saturn discovered by Huygens; in the later editions (in phen. 2, and in book 3, prop. 1) he referred to the satellites discovered by Cassini.

7. A careful reading of book 1, prop. 45, corol. 2, however, shows that Newton (while considering the action of an "extraneous force . . . added to or taken away from" the centripetal inverse-square force producing an elliptical orbit) writes, "Let us suppose the extraneous force to be 357.45 times less than the other force under the action of which the body revolves in the ellipse." Concerning Newton's claim in book 3, prop. 3, that (on the basis of book 1, prop. 45, corol. 2) the action of the sun in drawing the moon away from the earth is to the centripetal force of the earth on the moon "as roughly 2 to 357.45," see "Newt. Achievement" [Curtis Wilson, "The Newtonian Achievement in Astronomy," in *The General History of Astronomy,* vol. 2A, *Planetary Astronomy from the Renaissance to the Rise of Astrophysics: Tycho Brahe to Newton,* ed. René Taton and Curtis Wilson (Cambridge: Cambridge University Press, 1989)], p. 264.

8. As Curtis Wilson has pointed out to me, "in order to obtain the result of 15 1/12 Paris feet, Newton has to multiply the initial result of his calculation (which I find to equal 15.093 Paris feet/min.) by 178 29/40 divided by 177 29/40. In other words, Newton has, without mentioning the fact, subtracted out what he supposes to be the average subtractive force of the Sun, which counteracts some of the Earth's gravitational effect on the Moon. He has subtracted out too much; see Shinko Aoki, 'The Moon-Test in Newton's *Principia:* Accuracy of Inverse-Square Law of Universal Gravitation,' *Archive for History of Exact Sciences,* 1992, 44: 147–90. The exact agreement between Newton's calculation and the observation of length of fall per second at Paris is fudged. A good enough agreement

would have been attained without the fudging." See, further, *Never at Rest* [Richard S. Westfall, *Never at Rest: A Biography of Isaac Newton* (Cambridge: Cambridge University Press, 1980)], pp. 732–34.

9. *Corresp.* [*The Correspondence of Isaac Newton,* ed. H. W. Turnbull and A. R. Hall (Cambridge: Cambridge University Press, 1957–77)], vol. 5.

10. The force of mechanical pressure exerted on a body by a given blast of wind depends only on the area of surface exposed, whereas the effect of a given gravitational field would depend on the quantity of matter or mass.

11. We use the word "feign" today in a fictional sense for "pretend," as to "feign" sleep.

12. I. B. Cohen, "The First English Version of Newton's *Hypotheses non fingo,*" *Isis* 53 (1962): 379–88.

13. . . .*Corresp.,* vol. 3.

14. Newton to Cotes, 28 March 1713, *Corresp.,* vol. 3.

15. The most recent and thorough presentation of Newton's efforts is given in Dobbs, *Janus Faces* [Betty Jo Dobbs, *The Janus Faces of Genius: The Role of Alchemy in Newton's Thought* (Cambridge: Cambridge University Press, 1992)].

16. "In this Second Edition of these Opticks I . . . have added one Question," query 21, concerning the "Cause" of gravity, "chusing to propose it by way of a Question, because I am not yet satisfied about it for want of Experiments."

17. In the General Scholium and elsewhere, Newton refers to both deductions and inductions from phenomena.

18. I believe that my reconstruction has the virtue of explaining why Newton introduced the final paragraph, which, on first inspection, might seem to contravene his warning about hypotheses.

Whewell's Critique of Newton's Methodology

From

The Philosophy of the Inductive Sciences / William Whewell

Chapter 13. Newton

. . . .

4. With regard to the details of the process of discovery, Newton has given us some of his views, which are well worthy of notice, on account of their coming from him; and which are real additions to the philosophy of this subject. He speaks repeatedly of the *analysis* and *synthesis* of observed facts; and thus marks certain steps in scientific research, very important, and not, I think, clearly pointed out by his predecessors. Thus he says,[1] "As in Mathematics, so in Natural Philosophy, the investigation of difficult things by the method of analysis ought ever to precede the method of composition. This analysis consists in making experiments and observations, and in drawing general conclusions from them by induction, and admitting of no objections against the conclusions, but such as are taken

---

Reprinted from *The Philosophy of the Inductive Sciences, Founded upon Their History,* 2 vols. (London: John W. Parker, 1840; reprinted London: Routledge/Thoemmes Press, 1996), 2:439–56.

from experiments or other certain truths. And although the arguing from experiments and observations by induction be no demonstration of general conclusions; yet it is the best way of arguing which the nature of things admits of, and may be looked upon as so much the stronger, by how much the induction is more general." And he then observes . . . that by this way of analysis we proceed from compounds to ingredients, from motions to forces, from effects to causes, and from less to more general causes. The *analysis* here spoken of includes the steps which in this work we call the *decomposition* of facts, the exact *observation* and *measurement* of the phenomena, and the *colligation* of facts; the necessary intermediate step, the *selection* and *explication* of the appropriate conception, being passed over, in the fear of seeming to encourage the fabrication of hypotheses. The *synthesis* of which Newton here speaks consists of those steps of *deductive reasoning,* proceeding from the conception once assumed, which are requisite for the comparison of its consequences with the observed facts. This statement of the process of research, is, as far as it goes, perfectly exact.

5. In speaking of Newton's precepts on the subject, we are naturally led to the celebrated "Rules of Philosophizing," inserted in the second edition of the *Principia.* These rules have generally been quoted and commented on with an almost unquestioning reverence. Such Rules, coming from such an authority, cannot fail to be highly interesting to us; but at the same time, we cannot here evade the necessity of scrutinizing their truth and value, according to the principles which our survey of this subject has brought into view. The Rules stand at the beginning of that part of the *Principia* (the Third Book) in which he infers the mutual gravitation of the sun, moon, planets, and all parts of each. They are as follows:

"Rule I. We are not to admit other causes of natural things than such as both are true, and suffice for explaining their phenomena.

"Rule II. Natural effects of the same kind are to be referred to the same causes, as far as can be done.

"Rule III. The qualities of bodies which cannot be increased or diminished in intensity, and which belong to all bodies in which we can institute experiments, are to be held for qualities of all bodies whatever.

"Rule IV. In experimental philosophy, propositions collected from phenomena by induction, are to be held as true either accurately or approximately, notwithstanding contrary hypotheses; till other phenomena occur by which they may be rendered either more accurate or liable to exception."

In considering these Rules, we cannot help remarking, in the first place, that they are constructed with an intentional adaptation to the case with which Newton has to deal,—the induction of Universal Gravitation; and are intended to protect the reasonings before which they stand. Thus the first Rule is designed to strengthen the inference of gravitation from the celestial phenomena, by describing it as a *vera causa,* a true cause; the second countenances the doctrine that the planetary motions are governed by mechanical forces, as terrestrial motions are; the third rule appears intended to justify the assertion of gravitation, as a *universal* quality of bodies; and the fourth contains, along with a general declaration of the authority of induction, the author's usual protest against hypotheses, levelled at the Cartesian hypotheses especially.

6. *Of the First Rule.*—We, however, must consider these Rules in their general application, in which point of view they have often been referred to, and have had very great authority allowed them. One of the points which has been most discussed, is that maxim which requires that the causes of phenomena which we assign should be true causes, *verae causae.* Of course this does not mean that they should be *the* true or right cause; for although it is the philosopher's aim to discover such causes, he would be little aided in his search of truth, by being told that it is truth which he is to seek. The rule has generally been understood to prescribe that in attempting to account for any class of phenomena, we must assume such causes only, as *from other considerations,* we know to exist. Thus gravity, which was employed in explaining the motions of the moon and planets, was already known to exist and operate at the earth's surface.

114

Now the Rule thus interpreted is, I conceive, an injurious limitation of the field of induction. For it forbids us to look for a cause, except among the causes with which we are already familiar. But if we follow this rule, how shall we ever become acquainted with any new cause? Or how do we know that the phenomena which we contemplate do really arise from some cause which we already truly know? If they do not, must we still insist upon making them depend upon some of our known causes; or must we abandon the study of them altogether? Must we, for example, resolve to refer the action of radiant heat to the air, rather than to any peculiar fluid or ether, because the former is known to exist, the latter is merely assumed for the purpose of explanation? But why should we do this? Why should we not endeavour to learn the cause from the effects, even if it be not already known to us? We can infer causes, which are new when we first became acquainted with them. Chemical Forces, Optical Forces, Vital Forces, are known to us only by chemical and optical and vital phenomena; must we, therefore, reject their existence or abandon their study? They do not conform to the double condition, that they shall be sufficient and *also* real: they are true, only so far as they explain the facts, but are they, therefore, unintelligible or useless? Are they not highly important and instructive subjects of speculation? And if the gravitation which rules the motions of the planets had not existed at the earth's surface;—if it had been there masked and concealed by the superior effect of magnetism, or some other extraneous force, might not Newton still have inferred, from Kepler's laws, the tendency of the planets to the sun; and from their perturbations, their tendency to each other? His discoveries would still have been immense, if the cause which he assigned had not been a *vera causa* in the sense now contemplated.

7. But what do we mean by calling gravity a "true cause?" How do we learn its reality? Of course, by its effects, with which we are familiar;—by the weight and fall of bodies about us. These strike even the most careless observer. No one can fail to see that all bodies which we come in contact with are heavy;—that gravity acts in our neighbourhood here upon earth. Hence, it may be said, this cause is at any rate a true cause, whether it explains the celestial phenomena or not.

But if this be what is meant by a *vera causa,* it appears strange to require that in all cases we should find such a one to account for all classes of phenomena. Is it reasonable or prudent to demand that we shall reduce every set of phenomena, however minute, or abstruse, or complicated, to causes so obviously existing as to strike the most incurious, and to be familiar among men? How can we expect to find *such verae causae* for the delicate and recondite phenomena which an exact and skilful observer detects in chemical, or optical, or electrical experiments? The facts themselves are too fine for vulgar apprehension; their relations, their symmetries, their measures require a previous discipline to understand them. How then can their causes be found among those agencies with which the common unscientific herd of mankind are familiar? What likelihood is there that causes held for real by such persons, shall explain facts which such persons cannot see or cannot understand?

Again: if we give authority to such a rule, and require that the causes by which science explains the facts which she notes and measures and analyses, shall be causes which men, without any special study, have already come to believe in, from the effects which they casually see around them, what is this, except to make our first rude and unscientific persuasions the criterion and test of our most laborious and thoughtful inferences? What is it, but to give to ignorance and thoughtlessness the right of pronouncing upon the convictions of intense study and long disciplined thought? "Electrical atmospheres" surrounding electrized bodies, were at one time held to be a "true cause" of the effects which such bodies produce. These atmospheres, it was said, are obvious to the senses; we feel them like a spider's web on the hands and face. Aepinus had to answer such persons, by proving that there are no atmospheres, no effluvia, but only repulsion. He thus, for a *true cause* in the vulgar sense of the term, substituted an *hypothesis;* yet who doubts that what he did was an advance in the science of electricity?

8. Perhaps some persons may be disposed to say, that Newton's Rule does not enjoin us to take those causes only which we clearly know, or

suppose we know, to be really existing and operating, but only causes of *such kinds* as we have already satisfied ourselves do exist in nature. It may be urged that we are entitled to infer that the planets are governed in their motions by an attractive force, because we find, in the bodies immediately subject to observation and experiment, that such motions are produced by attractive forces, for example by that of the earth. It may be said that we might on similar grounds infer forces which unite particles of chemical compounds, or deflect particles of light, because we see adhesion and deflection produced by forces.

But it is easy to show that the Rule, thus laxly understood, loses all significance. It prohibits no hypothesis; for all hypotheses suppose causes *such as,* in some case or other, we have seen in action. No one would think of explaining phenomena by referring them to forces and agencies altogether different from any which are known; for on this supposition, how could he pretend to reason about the effects of the assumed causes, or undertake to prove that they would explain the facts? Some close similarity with some known kind of cause is requisite, in order that the hypothesis may have the appearance of an explanation. No forces, or virtues, or sympathies, or fluids, or ethers, would be excluded by *this* interpretation of *verae causae.* Least of all, would such an interpretation reject the Cartesian hypothesis of vortices; which undoubtedly, as I conceive, Newton intended to condemn by his Rule. For that *such* a case as a whirling fluid, carrying bodies round a centre in orbits, does occur, is too obvious to require proof. Every eddying stream, or blast that twirls the dust in the road, exhibits examples of such action, and would justify the assumption of the vortices which carry the planets in their courses; as indeed, without doubt, such facts suggested the Cartesian explanation of the solar system. The vortices, in this mode of considering the subject, are at the least as *real* a cause of motion as gravity itself.

9. Thus the Rule which enjoins "true causes," is nugatory, if we take *veræ causæ* in the extended sense of any causes of a real *kind,* and unphilosophical if we understand the term of *those very* causes which we familiarly suppose to exist. But it may be said that we are to designate as "true causes," not those which are collected in a loose, confused and precarious manner, by undisciplined minds, from obvious phenomena, but those which are justly and rigorously inferred. Such a cause, it may be added, gravity is; for the facts of the downward pressures and downward motions of bodies at the earth's surface lead us, by the plainest and strictest induction, to the assertion of such a force. Now to this interpretation of the Rule there is no objection; but then, it must be observed, that on this view, terrestrial gravity is inferred by the same process as celestial gravitation; and the cause is no more entitled to be called "true," because it is obtained from the former, than because it is obtained from the latter class of facts. We thus obtain an

intelligible and tenable explanation of a *vera causa;* but then, by this explanation its *verity* ceases to be distinguishable from its other condition, that it "suffices for the explanation of the phenomena." The assumption of universal gravitation accounts for the fall of a stone; it also accounts for the revolutions of the Moon or of Saturn; but since both these explanations are of the same kind, we cannot with justice make the one a criterion or condition of the admissibility of the other.

10. But still, the Rule, so understood, is so far from being unmeaning or frivolous, that it expresses one of the most important tests which can be given of a sound physical theory. It is true, the explanation of one set of facts may be of the same nature as the explanation of the other class: but then, that the cause explains *both* classes, gives it a very different claim upon our attention and assent from that which it would have if it explained one class only. The very circumstance that the two explanations coincide, is a most weighty presumption in their favour. It is the testimony of two witnesses in behalf of the hypothesis; and in proportion as these two witnesses are separate and independent, the conviction produced by their agreement is more and more complete. When the explanation of two kinds of phenomena, distinct and not apparently connected, leads us to the same cause, such a coincidence does give a reality to the cause, which it has not while it merely accounts for those appearances which suggested the supposition. This coincidence of propositions inferred from separate classes of facts, is exactly what we noticed in the last Book, as one of the most decisive characteristics of a true theory, under the name of the *Consilience of Inductions.*

That Newton's First Rule of Philosophizing, so understood, authorizes the inferences which he himself made, is really the ground on which they are so firmly believed by philosophers. Thus when the doctrine of a gravity varying inversely as the square of the distance from the body, accounted at the same time for the relations of times and distances in the planetary orbits and for the amount of the moon's deflection from the tangent of her orbit, such a doctrine became most convincing: or again, when the doctrine of the universal gravitation of all parts of matter, which explained so admirably the inequalities of the moon's motions, also gave a satisfactory account of a phenomenon utterly different, the precession of the equinoxes. And of the same kind is the evidence in favour of the undulatory theory of light, when the assumption of the length of an undulation, to which we are led by the colours of thin plates, is found to be identical with that length which explains the phenomena of diffraction; or when the hypothesis of transverse vibrations, suggested by the facts of polarization, explains also the laws of double refraction. When such a convergence of two trains of induction points to the same spot, we can no longer suspect that we are wrong. Such an accumulation of proof really persuades us that we have to do with a *vera causa.* And if this kind of proof be multiplied;—if we again find other facts of a sort uncontemplated in framing our hypothesis, but yet clearly accounted for when we have adopted the supposition;—we are still further confirmed in our belief; and by such accumulation of proof we may be so far satisfied, as to believe without conceiving it possible to doubt. In this case, when the validity of the opinion adopted by us has been repeatedly confirmed by its sufficiency in unforeseen cases, so that all doubt is removed and forgotten, the theoretical cause takes its place among the realities of the world, and becomes a *true cause.*

11. Newton's Rule then, to avoid mistakes, might be thus expressed; That "we may, provisorily, assume such hypothetical cause as will account for any given class of natural phenomena; but that when two different classes of facts lead us to the same hypothesis, we may hold it to be a *true cause.*" And this Rule will rarely or never mislead us. There are no instances, in which a doctrine recommended in this manner has afterwards been discovered to be false. There have been hypotheses which have explained many phenomena, and kept their ground long, and have afterwards been rejected. But these have been hypotheses which explained only one class of phenomena; and their fall took place when another kind of facts was examined and brought into conflict with the former. Thus the system of eccentrics and epicycles accounted for all the observed *motions* of the planets, and was the means of expressing and transmitting all astronomical knowledge for two thousand years. But then, how was it overthrown? By considering the *distances* as well as motions of the heavenly bodies. Here was a second class of facts; and when the system was adjusted so as to agree with the one class, it was at variance with the other. These cycles and epicycles could not be true, because they could not be made a just representation of the facts. But if the measures of distance as well as of position had conspired in pointing out the cycles and epicycles, as the paths of the planets, the paths so determined could not have been otherwise than their real paths; and the epicyclical theory would have been, at least geometrically, true.

12. *Of the Second Rule.*—Newton's Second Rule directs that "natural events of the *same kind* are to be referred to the *same causes,* so far as can be done." Such a precept at first appears to help us but little; for all systems, however little solid, profess to conform to such a rule. When any theorist undertakes to explain a class of facts, he assigns causes which according to him, will by their natural action, as seen in other cases, produce the effects in question. The events which he accounts for by his hypothetical cause, are, he holds, of the same kind as those which such a cause is known to produce. Kepler, in ascribing the planetary motions to magnetism, Descartes, in explaining them by means of vortices, held that they were referring celestial motions to the causes which give rise to terrestrial motions of the same kind. The

question is, *Are* the effects of the same kind? This once settled, there will be no question about the propriety of assigning them to the same cause. But the difficulty is, to determine *when* events are of the same kind. Are the motions of the planets of the same kind with the motion of a body moving freely in a curvilinear path, or do they not rather resemble the motion of a floating body swept round by a whirling current? The Newtonian and the Cartesian answered this question differently. How then can we apply this Rule with any advantage?

13. To this we reply, that there is no way of escaping this uncertainty and ambiguity, but by obtaining a clear possession of the ideas which our hypothesis involves, and by reasoning rigorously from them. Newton asserts that the planets move in free paths, acted on by certain forces. The most exact calculation gives the closest agreement of the results of this hypothesis with the facts. Descartes asserts that the planets are carried round by a fluid. The more rigorously the conceptions of force and the laws of motion are applied to this hypothesis, the more signal is its failure in reconciling the facts to one another. Without such calculation we can come to no decision between the two hypotheses. If the Newtonian hold that the motions of the planets are *evidently* of the *same kind* as those of a body describing a curve in free space, and therefore, like that, to be explained by a force acting upon the body; the Cartesian denies that the planets do move in free space. They are, he maintains, immersed in a plenum. It is only when it appears that comets pass through this plenum in all directions with no impediment, and that no possible form and motion of its whirlpools can explain the forces and motions which are observed in the solar system, that he is compelled to allow the Newtonian's classification of events of the *same kind.*

Thus it does not appear that this Rule of Newton can be interpreted in any distinct and positive manner, otherwise than as enjoining that, in the task of induction, we employ clear ideas, rigorous reasoning, and close and fair comparison of the results of the hypothesis with the facts. These are, no doubt, important and fundamental conditions of a just induction; but in this injunction we find no peculiar or technical criterion by which we may satisfy ourselves that we are right, or detect our errors. Still, of such general prudential rules, none can be more wise than one which thus, in the task of connecting facts by means of ideas, recommends that the ideas be clear, the facts correct, and the chain of reasoning which connects them without a flaw.

14. *Of the Third Rule.*—The Third Rule, that "qualities which are observed without exception be held to be universal," as I have already said, seems to be intended to authorize the assertion of gravitation as a universal attribute of matter. We formerly stated, in treating of Mechanical Ideas,[2] that this application of such a Rule appears to be a mode of reasoning far from

conclusive. The assertion of the universality of any property of bodies must be grounded upon the reason of the case, and not upon any arbitrary maxim. Is it intended by this Rule to prohibit any further examination how far gravity is an original property of matter, and how far it may be resolved into the result of other agencies? We know perfectly well that this was not Newton's intention; since the cause of gravity was a point which he proposed to himself as a subject of inquiry. It would certainly be very unphilosophical to pretend, by this Rule of Philosophizing, to prejudge the question of such hypotheses as that of Mosotti, That gravity is the excess of the electrical attraction over electrical repulsion: and yet to adopt this

hypothesis, would be to suppose electrical forces more truly universal than gravity; for according to the hypothesis, gravity, being the inequality of the attraction and repulsion, is only an accidental and partial relation of these forces. Nor would it be allowable to urge this Rule as a reason of assuming that double stars are attracted to each other by a force varying according to the inverse square of the distance; without examining, as Herschel and others have done, the orbits which they really describe. But if the Rule is not available in such cases, what is its real value and authority? and in what cases are they exemplified?

15. In a former part of this work,[3] it was shown that the fundamental laws of motion, and the properties of matter which these involve, are, after a full consideration of the subject, unavoidably assumed as universally true. It was further shown, that although our knowledge of these laws and properties be gathered from experience, we are strongly impelled, some philosophers think authorized, to look upon these as not only universally, but necessarily true. It was also stated, that the law of gravitation, though its universality may be deemed probable, does not apparently involve the same necessity as the fundamental laws of motion. But it was pointed out that these are some of the most abstruse and difficult questions of the whole of philosophy; involving the profound, perhaps insoluble, problem of the identity or diversity of ideas and things. It cannot, therefore, be deemed philosophical to cut these Gordian knots by peremptory maxims, which encourage us to decide without rendering a reason. Moreover, it appears clear that the reason which is rendered for this Rule by the Newtonians is quite untenable; namely, that we know extension, hardness, and inertia, to be universal qualities of bodies by experience alone, and that we have the same evidence of experience for the universality of gravitation. We have already observed that we cannot, with any propriety, say that we *find* by experience all bodies are extended. This could not be a just assertion, except we could conceive the possibility of our finding the contrary. But who can conceive our finding by experience some bodies which are not extended? It appears, then, that the reason given for the Third Rule of

Newton involves a mistake respecting the nature and authority of experience. And the Rule itself cannot be applied without attempting to decide by the casual limits of observation, questions which necessarily depend upon the relations of ideas.

16. *Of the Fourth Rule.*—Newton's Fourth Rule is, that "Propositions collected from phenomena by induction, shall be held to be true, notwithstanding contrary hypotheses; but shall be liable to be rendered more accurate, or to have their exceptions pointed out, by additional study of phenomena." This Rule contains little more than a general assertion of the authority of induction, accompanied by Newton's usual protest against hypotheses.

The really valuable part of the Fourth Rule is that which implies that a constant verification, and, if necessary, rectification, of truths discovered by induction, should go on in the scientific world. Even when the law is, or appears to be, most certainly exact and universal, it should be constantly exhibited to us afresh in the form of experience and observation. This is necessary, in order to discover exceptions and modifications if such exist; and if the law be rigorously true, the contemplation of it, as exemplified in the world of phenomena, will best give us that clear apprehension of its bearings which may lead us to see the ground of its truth.

The concluding clause of this Fourth Rule appears, at first, to imply that all inductive propositions are to be considered as merely provisional and limited, and never secure from exception. But to judge thus would be to underrate the stability and generality of scientific truths; for what man of science can suppose that we shall hereafter discover exceptions to the universal gravitation of all parts of the solar system? And it is plain that the author did not intend the restriction to be applied so rigorously; for in the Third Rule, as we have just seen, he authorizes us to infer universal properties of matter from observation, and carries the liberty of inductive inference to its full extent. The Third Rule appears to encourage us to assert a law to be universal, even in cases in which it has not been tried; the Fourth Rule seems to warn us that the law may be inaccurate, even in cases in which it has been tried. Nor is either of these suggestions erroneous; but both the universality and the rigorous accuracy of our laws are proved by reference to Ideas rather than to Experience; a truth which, perhaps, the philosophers of Newton's time were somewhat disposed to overlook.

17. The disposition to ascribe all our knowledge to Experience, appears in Newton and the Newtonians by other indications; for instance, it is seen in their extreme dislike to the ancient expressions by which the principles and causes of phenomena were described, as the *occult causes* of the Schoolmen, and the *forms* of the Aristotelians, which had been adopted by Bacon. Newton says,[4] that the particles of matter not only possess inertia, but also

active principles, as gravity, fermentation, cohesion; he adds, "These principles I consider not as Occult Qualities, supposed to result from the Specific Forms of things, but as General Laws of Nature, by which the things themselves are formed: their truth appearing to us by phenomena, though their causes be not yet discovered. For these are manifest qualities, and their causes only are occult. And the Aristotelians gave the name of *occult qualities,* not to manifest qualities, but to such qualities only as they supposed to lie hid in bodies, and to the unknown causes of manifest effects: such as would be the causes of gravity, and of magnetick and electrick attractions, and of fermentations, if we should suppose that these forces or actions arose from qualities unknown to us, and incapable of being discovered and made manifest. Such occult qualities put a stop to the improvement of Natural Philosophy, and therefore of late years have been rejected. To tell us that every species of things is endowed with an occult specific quality by which it acts and produces manifest effects, is to tell us nothing: but to derive two or three general principles of motion from phenomena, and afterwards to tell us how the properties and actions of all corporeal things follow from these manifest principles, would be a great step in philosophy, though the causes of those principles were not yet discovered: and therefore I scruple not to propose the principles of motion above maintained, they being of very general extent, and leave their causes to be found out."

18. All that is here said is highly philosophical and valuable; but we may observe that the investigation of *specific forms,* in the sense in which some writers had used the phrase, was no means a frivolous or unmeaning object of inquiry. Bacon and others had used *form* as equivalent to *law.*[5] If we could ascertain that arrangement of the particles of a crystal from which its external crystalline form and other properties arise, this arrangement would be the *internal form* of the crystal. If the undulatory theory be true, the *form* of light is transverse vibration: if the emission theory be maintained, the *form* of light is particles moving in straight lines, and deflected by various forces. Both the terms, *form* and *law,* imply an ideal connexion of sensible phenomena; form supposes matter which is moulded to the form; law supposes objects which are governed by the law. The former term refers more precisely to existences, the latter to occurrences. The latter term is now the more familiar, and is, perhaps, the better metaphor: but the former also contains the essential antithesis which belongs to the subject, and might be used in expressing the same conclusions.

But occult causes, employed in the way in which Newton describes, had certainly been very prejudicial to the progress of knowledge, by stopping inquiry with a mere word. The absurdity of such pretended explanations had not escaped ridicule. The pretended physician in the comedy gives an example of an occult cause or virtue.

Mihi demandatur
A doctissimo Doctore
Quare Opium facit dormire:
Et ego respondeo,
Quia est in eo
Virtus dormitiva,
Cujus natura est sensus assoupire.
[I am asked
by the most learned doctor
why opium puts us to sleep:
And I respond,
because it has
a dormitive virtue,
whose nature is to dull the senses.]

19. But the most valuable part of the view presented to us in the quotation just given from Newton is the distinct separation, already noticed as peculiarly brought into prominence by him, of the determination of the *laws* of phenomena, and the investigation of their *causes.* The maxim, that the former inquiry must precede the latter, and that if the general laws of facts be discovered, the result is highly valuable, although the causes remain unknown, is extremely important; and had not, I think, ever been so strongly and clearly stated, till Newton both repeatedly promulgated the precept, and added to it the weight of the most striking examples.

We have seen that Newton, along with views the most just and important concerning the nature and methods of science, had something of the tendency, prevalent in his time, to suspect or reject, at least speculatively, all elements of knowledge except observation. This tendency was, however, in him so corrected and restrained by his own wonderful sagacity and mathematical habits, that it scarcely led to any opinion which we might not safely adopt.

. . . .

NOTES

1. *Op[ticks]*, Qu. 31.
2. Book iii. c. 10.
3. Book iii. c. 9, 10, 11.
4. *Optic[k]s*, Qu. 31.
5. *Nov. Org.*, lib. ii. Aph. 2[,] . . . Aph. 17. . . .

PART III

Hypothetico-Deductivism, the Mill–Whewell Debate & the Wave Theory of Light

We turn now to a scientific method very different from the rules set out by Descartes and by Newton. It is called the "method of hypothesis," or "hypothetico-deductivism." Like Newton's methodology, and unlike Descartes', it is empirical. Observations, and not just pure thought, are required to establish a theory. However, its proponents reject Newton's idea that the only way to proceed is to argue from similar effects to the same cause and then generalize by induction.

Two different versions of the hypothetico-deductive (H-D) view will be presented in the readings. One, due to William Whewell, was developed in the middle of the nineteenth century. Whewell argued that the method he advocated is well illustrated by the development of the wave theory of light during the first part of the nineteenth century. John Stuart Mill, a contemporary of Whewell, who defended an inductivism similar to Newton's, sharply criticized Whewell's view as well as the reasoning of the wave theorists. The Mill–Whewell debate is valuable for focusing on the differences between two very prominent methodologies.

A different version of hypothetico-deductivism was formulated by Karl Popper in the middle of the twentieth century. Like Whewell, Popper rejects the idea, propounded by Newton and Mill, that one proceeds by making causal and inductive inferences to scientific propositions. Unlike Whewell, however, Popper believes that the only inferences scientists are allowed to make are deductive ones in which the conclusions must be true if the premises are; nondeductive (or "probabilistic") inferences are never justified.

In the introductory material that follows, we will begin with Thomas Young and the wave theory of light. Then we can see how Whewell analyzed the scientific method leading to the establishment of that theory, and how Mill argued for his own version of inductivism and criticized Whewellian hypothetico-deductivism.

Thomas Young and the Wave Theory of Light

Thomas Young (1773–1829), a British polymath, was a medical doctor whose most famous work was in optics, but who also was one of the first to translate Egyptian hieroglyphics.[1] He studied medicine in London, Edinburgh, and Göttingen (where he received an M.D. in 1796). From 1797 to 1803 he studied at Cambridge University, receiving another M.D. in 1808 in order to satisfy requirements of the Royal College of Physicians. Young's most important scientific contribution was to revive and defend the wave theory of light in 1800 and to formulate the principle of the interference of light.

In the seventeenth century the Dutch physicist Christian Huygens developed a theory according to which light consists of a series of irregular pulses in a medium called the "ether." These pulses are produced by motions of the particles of the body that is the source of light. The transmission of the pulses occurs in a finite time and "is propagated," Huygens writes, "as that of sound, by surfaces and spherical waves. I call these waves because of the resemblance to those which are formed when one throws a pebble into water and which represent gradual propagation in circles, although produced by a different cause and confined to a plane surface." Huygens introduced a principle, which came to bear his name, according to which "each particle of the medium through which the wave spreads does not communicate its motion only to that neighbor which lies in the straight line drawn from the luminous point, but shares it with all the particles which touch it and resist its motion. Each particle is thus to be considered the center of a wave."[2] Huygens used this idea in offering explanations of various observed optical phenomena, including reflection, refraction, and the fact that light travels in straight lines (rectilinear propagation).

Huygens' theory was not widely accepted, principally because of the influence of Newton, who in his *Opticks* raised many objections to the wave theory of light and suggested the superiority of a particle theory (although he did not claim that he had demonstrated its truth). According to the particle theory, light consists of material particles propagated from luminous bodies; these particles obey Newton's laws of motion, and they are subject to short-range forces of attraction and repulsion. Among Newton's objections to the wave theory is that if it were true, then light hitting an obstacle should bend into the shadow (or be "diffracted," as we now say). Newton observed that an analogous phenomenon occurs in the case of water waves and sound waves; but this had not been detected in the case of light. The particle theory that Newton had said was superior to the wave theory was widely accepted throughout the eighteenth century, despite various problems with it that its enthusiastic proponents attempted to solve.

In 1800 Thomas Young resuscitated the wave theory in a paper in which he suggested some new difficulties for the particle theory and showed how these are readily answered by the wave theory. In 1802 he published a second paper, "On the Theory of Light and Colours,"[3] in which he expressed the basic assumptions (or, as he called them, "hypotheses") of the wave theory: a rare and highly elastic luminiferous (light-bearing) ether pervades the universe; a luminous body excites undulations in this ether; the different color sensations depend on the frequency of vibrations excited by light in the retina; material bodies attract the ethereal medium so that the latter accumulates within them and around them for a small distance. Young claims that, contrary to what has usually been supposed, the first three hypotheses can be found in Newton's writings. And he defends these by appeal to passages from Newton. The fourth hypothesis, which was contrary to Newton's assumption that ordinary matter repels the ether, Young holds to be less fundamental than the others but "the simplest and best of any that have occurred" to him.[4]

Following this description of the four hypotheses are nine "propositions" that develop the wave theory in a qualitative way by indicating various properties of the undulations and how certain observed optical phenomena are to be explained by reference to them. These propositions are defended by invoking experiments and observations, analogies with sound, deductive argumentation, and appeals to authority. Historically the most important is the eighth proposition, which is an early formulation of the principle of interference, Young's most famous single contribution to optics. He writes, "When two undulations, from different origins, coincide, either perfectly or very nearly in direction, their joint effect is a combination of the motions belonging to each."[5] This is the case when undulations are in the same phase, and the effect is now called "constructive interference." In the discussion that follows this proposition, Young also speaks of the case when the undulations are 180 degrees out of phase, as a result of which they are destroyed ("destructive interference").

In 1807 Young published a two-volume work entitled *A Course of Lectures on Natural Philosophy and the Mechanical Arts*. Lecture 39, "On the Nature of Light and Colours," parts of which are included below, is a defense of the wave theory. Young begins by noting that "it is allowed on all sides, that light either consists in the emission of very minute particles from luminous substances, . . . or in the excitation of an undulatory motion, analogous to that which constitutes sound, in a highly light and elastic medium pervading the universe." He shows how both theories attempt to explain various optical phenomena. With regard to a significant number of them, including diffraction, uniform velocity of light, partial reflection, and double refraction, he claims that the particle theory, but not the wave theory, introduces improbable auxiliary explanatory hypotheses. Of major interest in this lecture is Young's first description of the double-slit experiment. The experiment involves a beam of homogeneous light falling on a screen in which there are two small slits that produce diffraction of the light. This results in an image consisting of light or dark stripes (the interference pattern) on a surface placed beyond the screen.

Young's work did not succeed in persuading particle theorists to mend their ways. Some regarded it simply as a defense, albeit with some new wrinkles, of an already discredited theory. Other particularly empirically minded scientists rejected it primarily on the grounds that it postulated a luminiferous ether—a hypothesis that was entirely speculative. The contributor to the wave theory who turned the tide in its favor was the French engineer Augustin Jean Fresnel. Instead of relying heavily on qualitative accounts often appealing to analogies between light and sound, as Young had done, Fresnel introduced quantitative mathematical analyses into wave-theoretical explanations of various optical phenomena. His most important contribution was a mathematical analysis of diffraction, for which he received a prize from the Paris Academy in 1819.

Following Young's revival of the wave theory at the start of the nineteenth century, and Fresnel's work beginning in 1818, by the 1830s and 1840s most physicists concerned with optics had come to support the wave theory. When William Whewell and John Stuart Mill conducted their debate about scientific method in the 1840s, both of them invoking the wave theory, that theory was clearly dominant over the Newtonian particle theory.

Hypothetico-Deductivism

Several versions of the H-D view exist. The simplest one, which neither Whewell nor Popper proposed, is a view that both Newton and Descartes were rejecting when they criticized the method of hypothesis.

Simple H-D View

The basic idea of the simple H-D view is that the scientist begins with one or more hypotheses. From these various conclusions are derived using deductive reasoning. The conclusions should include ones that can be established or refuted by observation and experiment. If they are established, then one may infer that the hypotheses are true or at least probable. If they are refuted, then the hypotheses are false and need to be revised or replaced.

The initial hypotheses may come from a variety of sources, including experience, analogies with other cases, authority, imagination, or even dreams. But their truth or probability is not *inferred* from any of these sources. Rather, these sources suggest or cause the ideas behind the hypotheses. Sometimes the hypotheses are spoken of as "guesses" or "conjectures," which are produced by scientists trying to solve certain problems. But from the sources producing these guesses the scientists do not infer their truth. The only inferences are these: (1) deductive inferences from the hypotheses to conclusions that can be tested observationally; (2) a nondeductive inference from the fact that the observational conclusions are true to the truth or probability of the hypotheses; or (3) if the observational conclusions are false, a deductive inference from this fact to the falsity of the hypotheses that entail them.

Whewell's Hypothetico-Deductivism

William Whewell (1794–1866) was a British scientist, philosopher, and historian of science. He produced scientific works on mineralogy and the tides and books on moral philosophy, as well as a volume entitled *The History of the Inductive Sciences* (1840) and one entitled *The Philosophy of the Inductive Sciences*, from which the selection below is taken. From 1841 until his death he was master of Trinity College, Cambridge University, where he made important changes in the teaching of science.

Whewell's hypothetico-deductivism is more complex and sophisticated than the simple H-D view rejected by Descartes and Newton. Whewell, like Newton, uses the term "induction," but unlike Newton he distinguishes two stages. First, there is the discovery of the hypothesis. In this stage "science begins with common observation of facts" (chap. 4, sec. 2). But in order to proceed from these observed facts to a hypothesis in the discovery stage, Whewell writes, "there is introduced some general conception, which is given, not by the phenomena, but by the mind" (chap. 5, sec. 2). Whewell speaks of this as a "colligation" of the facts. He cites the example of the astronomer Kepler, who begins with observations (made by Tycho Brahe) of the position of the planet Mars at various times of the year. What Kepler introduces, which is not given by the phenomena, is an idea that connects these observations, namely, that the observed points lie on an ellipse. This Whewell calls an "act of *invention*" (sec. 3), and he claims that it is a "conjecture." Its truth is not inferred from the facts that it connects, since frequently the mind invents various conflicting hypotheses to colligate the data, before an inference to truth or falsity is made in the second stage of the induction, which Whewell calls the testing of hypotheses.

In this second stage Whewell imposes three conditions the satisfaction of which will justify an inference to the truth of the hypothesis; or, to speak comparatively, the better satisfied these conditions are, the stronger that inference will be. First, the hypothesis should not only explain phenomena that have been observed but also predict ones not yet observed. Second, it should explain and predict phenomena "of a *kind different* from those which were contemplated in the formation of our hypothesis" (sec. 11). Whewell speaks of this as the "consilience of inductions." He cites Newton's law of gravity as an example, because it explains all three of Kepler's laws, even though no connection between these laws had been contemplated before. Third, Whewell introduces a condition pertaining to the development of a scientific theory over time—a novel idea not discussed by previous scientists or philosophers. Whewell notes that a hypothesis is usually part of a set of hypotheses, or system, or theory, and that the members of this set are not framed all at once but are introduced, added to, and altered over time. Some theories become simpler, more unified, and more coherent in this process. These are the ones that turn out to be true. In false theories the reverse is the case: they become more complex, less unified, less coherent. They introduce ad hoc hypotheses that do not fit in well with previous ones but are invoked simply to solve a particular, isolated problem. The idea that as theories change over time they should become more coherent we might call Whewell's "coherence" condition.

The wave theory of light, as it had been developed by the 1840s, is one of Whewell's principal examples. "Colligation" is present, since the mind of the physicist imposes on the observed optical phenomena the idea of waves in an ether. In the testing stage Whewell claims that the wave theory satisfies his three conditions for inferring truth much better than does its rival, the particle theory. The wave theory explains and predicts optical phenomena; it does so not just for phenomena (such as double refraction) on the basis of which the theory was first proposed but for ones that were not contemplated in its initial formulation (such as polarization). Whewell cites various examples of the coherence and progression to simplicity of the wave theory, by contrast with the lack of these characteristics in the development of the particle theory.[6]

Does the satisfaction of Whewell's three conditions justify a claim of truth for the proposition inferred, as Whewell insisted? As will be noted in a moment, a negative answer is given by John Stuart Mill, who engaged in an important debate with Whewell in the mid-nineteenth century. First, however, let us turn to one final version of hypothetico-deductivism.

Popper's Falsificationism

Karl Popper (1902–94), one of the most influential philosophers of science of the twentieth century, defended a form of hypothetico-deductivism different from the two above. There are some similarities. Like the other hypothetico-deductivists, Popper claims that science begins with observations—or, perhaps better, with a problem suggested by observations. Also, like the others, Popper claims that to solve the problem a hypothesis or set of hypotheses is introduced. Such hypotheses are not inferred from observations or anything else but are guesses or conjectures. Finally, like the simple hypothetico-deductivist, Popper claims that once hypotheses have been conjectured, scientists proceed to derive conclusions from them, particularly observational ones, by deductive reasoning. If these can be shown to be false by observation and experiment, then the set of hypotheses contains ones that are false—in which case the set must be revised or replaced. If no conclusions are testable, then, says Popper, the theory is nonscientific, since it is incapable of being falsified by observation. (Popper takes falsifiability to be a necessary condition for a theory to be *scientific*.)

Suppose that all of the observational conclusions that have been derived from the theory turn out to be true, as demonstrated by observation and experiment. It is at this point that Popper's version of hypothetico-deductivism differs substantially from others. The simple H-D view is that if all the observational conclusions derived from the theory are true, then one can infer that the theory is true or at least probable. On Whewell's view, if the observational conclusions are true, and if these include new predictions and if consilience and coherence are satisfied, then one can infer that the theory is true. Popper rejects all such inferences to the truth or probability of the theory from its observational success. He does so because he regards all such inferences as inductive, in a broad sense. They infer the truth of a general theory or hypothesis from the truth of particular instances. Following the eighteenth-century Scottish philosopher David Hume, Popper regards all such reasoning as unjustified. He claims that it cannot be shown to be valid a priori by deductive reasoning. Nor can it be justified empirically by its past success, since that would be using induction to justify itself.

If all the observational conclusions so far derived from a theory using deduction are true, all we can say is, "So far so good; the theory has not been shown to be false." We cannot infer that it is true or probable. Popper calls the idea of a failed attempt to falsify a theory "corroboration of a theory." The best procedure for the scientist is to propose the most general hypothesis that has not yet been refuted by observation, and try to refute it by testing observational conclusions derived from it. If all such conclusions turn out to be true, the theory is "corroborated": despite the scientist's best efforts, it has not been refuted. If such a theory is more general than another one that is also "corroborated," then the former has a higher degree of corroboration than the latter and is the better theory.

On Popper's view, some of Whewell's claims about the wave theory of light in the 1840s were correct, and some incorrect. Popper would agree with Whewell that wave theorists (in the discovery stage) made a conjecture that was not inferred from observed optical phenomena. He would agree that from the conjecture various optical phenomena could be derived and new ones predicted. And he might even consider that the wave theory in the mid-nineteenth century satisfied Whewell's criteria of consilience and coherence. But he would have rejected Whewell's inference (in his second stage of induction) from these facts to the truth or probability of the wave theory. Such an inference, involving as it does an inductive generalization, is unjustified. All that wave theorists could properly say in the mid-nineteenth century was that the theory, which is very general, remained unrefuted despite attempts to show that it is false.

Mill's Inductivism and His Debate with Whewell

John Stuart Mill (1806–73), an enormously influential British philosopher, made important contributions in various fields of philosophy, including logic, philosophy of science, philosophy of language, and ethics. His book *A System of Logic,* from which the selection below is taken, was published in 1843. It was highly successful both as a university textbook and as a volume read by nonacademics. It was revised various times and went into eight editions.

From the very first edition onward Mill is critical of Whewell. He rejects Whewell's claims about "colligation," particularly the claim that the mind introduces a conception not given by the observed phenomena. He cites an example used by Whewell, Kepler's conception of an ellipse as the path of the planet Mars. Mill claims that, contrary to Whewell, "Kepler did not *put* what he had conceived into the facts, but *saw* it in them" (bk. 3, chap. 1, sec. 4). From the observations of Tycho Brahe on the various positions of Mars, Kepler, using mathematical calculations, determined that all the observed positions lie on an ellipse. This Mill describes as a summary of the observed facts. It is not a Whewellian conjecture in which the mind produces a hypothesis. Nor, says Mill, is it an inductive inference, which Mill (unlike Whewell) regards as an inference from the fact that all the observed members of a class have a certain property to the conclusion that all members of the class have that property. According to Mill, the only inductive inference Kepler made regarding Mars was from the fact that all observed positions of Mars lie on an ellipse to the claim that all positions of Mars, past and present, observed and unobserved, do too.

Among inductively inferred propositions, Mill distinguishes those that describe conditional sequences of events from those describing unconditional ones. In both cases some type of event A is always followed by a type of event B, but in the conditional case this is dependent on some contingent fact that could change. For example, the fact that summers in New Orleans are always hot is a conditional sequence that depends on conditions on the earth that could change, producing another ice age. By contrast, Newton's law of gravity expresses an unconditional relationship: the law holds no matter what. Mill calls generalizations that express unconditional relationships between types of events "causal laws." In one of the most famous sections in his book (bk. 3, chap. 8) he presents methods ("Mill's methods of experimental inquiry") for determining whether a generalization is truly causal (i.e., whether one type of event or phenomenon causes another). This, he claims, involves experimentation; observation is not enough: observation without experiment "can ascertain sequences and co-existences, but cannot prove causation" (chap. 7, sec. 4).

Suppose we observe that events or phenomena of one type have been associated with those of another. We want to determine whether a causal relationship exists. Two of Mill's "methods" will be noted: the method of agreement, and the method of difference. Although Mill presents these in an idealized form, we can understand them as follows, with respect to a simple example.

Suppose that bodies on which forces have been exerted always accelerate. To determine whether this is a causal relationship (i.e., whether this association is unconditional), the "method of agreement" tells us that we can exert forces on a variety of different types of bodies (bodies with different masses, sizes, shapes, positions, etc.) such that the only relevant property in common is that all such bodies were subjected to a force. If there is an acceleration in all such cases, then we can infer that forces cause accelerations.

Another method, the "method of difference," tells us that we can test for causation by having two groups of bodies, where the only relevant difference is that in one group forces are exerted while in the other they are not. If bodies in the first group accelerate while those in the second do not, we can infer that forces cause accelerations.

Mill considers these, and several other methods that he describes, as an essential part of inductive reasoning to causal laws in science. Although both Mill and Newton are inductivists, and both stress the idea of reasoning to causes, as well as inductive generalization, Mill goes well beyond Newton in showing how reasoning to causes is supposed to operate.

Mill rejects hypothetico-deductivism in both its simple version and its Whewellian form. He considers a central hypothesis of the wave theory of light, namely, the claim that a luminiferous ether exists. On the simple H-D view, it is legitimate to infer that the ether exists if from this supposition together with other assumptions of the wave theory one can derive known phenomena of light, such as rectilinear propagation, reflection, refraction, and diffraction. On Whewell's version, one can make such an inference, provided that (1) the theory yields phenomena of various kinds that are different from ones that suggested the theory in the first place (consilience), and (2) the theory as it has developed over time is coherent. Mill's objection is simple and powerful. Suppose there is a different, conflicting theory from which the known phenomena can be derived that is at least as consilient and coherent as the wave theory. Then one would not be justified in inferring the truth or probability of the wave theory. On Mill's view, to make the latter inference one must show not only that the wave theory has the features required by Whewell but that no other conflicting theory has such features. The fact that physicists in the mid-nineteenth century could not think of such a theory, Mill urges, is no argument that there is none. (In fact Mill turned out to be right: the quantum theory of light developed in the early twentieth century was just such a theory, even though, to use Mill's expression [bk. 3, chap. 14, sec. 6], the minds of nineteenth-century physicists were "unfitted to conceive" such a theory.)

By contrast with the H-D method, Mill advocates the "deductive method" for cases involving several hypotheses and various computations (bk. 3, chap. 11). Here is a simple example. Suppose we want to determine at what angle to fire a cannon to achieve maximum range. The first part of Mill's "deductive method" consists in using induction (including his causal methods noted above) to determine the causes and laws operating in such cases (e.g., causes involving the force exerted on the cannonball, the force of gravity, and Newton's laws of motion). The second part of the "deductive method" consists in what Mill calls "ratiocination"—a process of deductive (including mathematical) reasoning to conclusions that can be tested empirically. So, for example, from the inductively derived forces and laws one deduces that the maximum range is produced at an angle of 45 degrees. The third part Mill calls "verification," in which empirical tests and observations are made to determine the truth of the derived claims. (We fire the cannon at various angles and determine whether the angle at which the maximum range is achieved is 45 degrees.)

Now, says Mill, what Whewell and other hypothetico-deductivists omit is the first inductive step. They start with a hypothesis that is not inductively arrived at, deduce consequences, and, if these are true, claim that the hypothesis is true or probable. This, says Mill, is fallacious reasoning, since conflicting hypotheses may yield the same results. A fundamental difference, then, between Mill and Whewell over inferences to hypotheses is this: Whewell claims that if the hypotheses generate known phenomena, and if, in addition, they satisfy consilience and coherence, then one can infer their truth. Mill denies this, since there is the possibility that other, conflicting hypotheses, even ones we do not or cannot know about, will do the same. Whewell will respond by saying that the satisfaction of his conditions makes it very improbable that such conflicting hypotheses exist. Mill will reply with a demand to be shown why this is so. Whewell simply asserts that it is so.[7] Mill and Whewell do agree on one important point, namely, that inductive inferences can be made to the truth or probability of general hypotheses. Both would reject Popper's claim that all such inferences are fallacious.

In the final selection in this part of the volume the editor reconstructs the scientific reasoning of the mid-nineteenth-century wave theorists to determine to what extent it corresponds to, or violates, the hypothetico-deductivism of Whewell and the inductivism of Mill.

NOTES

1. Parts of the material in this section are from the editor's book *Particles and Waves* (New York: Oxford University Press, 1991).

2. Christian Huygens, *Treatise on Light,* reprinted in part in *The Wave Theory of Light: Memoirs of Huygens, Young, and Fresnel,* ed. Henry Crew (New York: American Book Co., 1900), quotations on 11 and 21.

3. *Philosophical Transactions of the Royal Society* 92 (1802): 12–48. Reprinted in part in Crew, *Wave Theory of Light,* 47–61.

4. Quoted from Crew, *Wave Theory of Light,* 53.

5. *Ibid.,* 60.

6. Many commentators on Whewell, including Mill, consider him to be a hypothetico-deductivist. For an alternative interpretation, the reader is invited to consult Laura J. Snyder, "The Mill–Whewell Debate: Much Ado about Induction," *Perspectives on Science* 5 (1997): 159–98.

7. For a more extensive, formal discussion of this debate, see the editor's *Particles and Waves.*

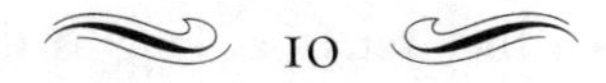

10

Young's Wave Theory of Light

From

A Course of Lectures on Natural Philosophy / Thomas Young

Lecture 39. On the Nature of Light and Colours

The nature of light is a subject of no material importance to the concerns of life or to the practice of the arts, but it is in many other respects extremely interesting, especially as it tends to assist our views both of the nature of our sensations, and of the constitution of the universe at large. The examination of the production of colours, in a variety of circumstances, is intimately connected with the theory of their essential properties, and their causes; and we shall find that many of these phenomena will afford us considerable assistance in forming our opinion respecting the nature and origin of light in general.

It is allowed on all sides, that light either consists in the emission of very minute particles from luminous substances, which are actually projected, and continue to move with the velocity commonly attributed to light, or in the excitation of an undulatory motion, analogous to that which constitutes

---

Reprinted from *A Course of Lectures on Natural Philosophy and the Mechanical Arts,* new ed. with references and notes by the Rev. P. Kelland, 2 vols. (London: Printed for Taylor and Walton, 1845), 1:359–70.

References to illustrations, as well as Kelland's notes and "additional remarks," have been omitted.

sound, in a highly light and elastic medium pervading the universe; but the judgments of philosophers of all ages have been much divided with respect to the preference of one or the other of these opinions. There are also some circumstances which induce those, who entertain the first hypothesis, either to believe, with Newton, that the emanation of the particles of light is always attended by the undulations of an ethereal medium, accompanying it in its passage, or to suppose, with Boscovich, that the minute particles of light themselves receive, at the time of their emission, certain rotatory and vibratory motions, which they retain as long as their projectile motion continues. These additional suppositions, however necessary they may have been thought for explaining some particular phenomena, have never been very generally understood or admitted, although no attempt has been made to accommodate the theory in any other manner to those phenomena.

We shall proceed to examine in detail the manner in which the two principal hypotheses respecting light may be applied to its various properties and affections; and in the first place to the simple propagation of light in right lines through a vacuum, or a very rare homogeneous medium. In this circumstance there is nothing inconsistent with either hypothesis; but it undergoes some modifications, which require to be noticed, when a portion of light is admitted through an aperture, and spreads itself in a slight degree in every direction. In this case it is maintained by Newton that the margin of the aperture possesses an attractive force, which is capable of inflecting the rays: but there is some improbability in supposing that bodies of different forms and of various refractive powers should possess an equal force of inflection, as they appear to do in the production of these effects; and there is reason to conclude from experiments, that such a force, if it existed, must extend to a very considerable distance from the surfaces concerned, at least a quarter of an inch, and perhaps much more, which is a condition not easily reconciled with other phenomena. In the Huygenian system of undulation, this divergence or diffraction is illustrated by a comparison with the motions of waves of water and of sound, both of which diverge when they are admitted into a wide space through an aperture, so much indeed that it has usually been considered as an objection to this opinion, that the rays of light do not diverge in the degree that would be expected if they were analogous to the waves of water. But as it has been remarked by Newton, that the pulses of sound diverge less than the waves of water, so it may fairly be inferred, that in a still more highly elastic medium, the undulations, constituting light, must diverge much less considerably than either.

With respect, however, to the transmission of light through perfectly transparent mediums of considerable density, the system of emanation labours under some difficulties. It is not to be supposed that the particles of light can perforate with freedom the ultimate atoms of matter, which compose a substance of any kind; they must, therefore, be admitted in all directions through the pores or interstices of those atoms; for if we allow such suppositions as Boscovich's, that matter itself is penetrable, that is, immaterial, it is almost useless to argue the question further. It is certain that some substances retain all their properties when they are reduced to the thickness of the ten millionth of an inch at most, and we cannot therefore suppose the distances of the atoms of matter in general to be so great as the hundred millionth of an inch. Now if ten feet of the most transparent water transmits, without interruption, one half of the light that enters it, each section or stratum of the thickness of one of these pores of matter must intercept only about one twenty thousand millionth, and so much must the space or area occupied by the particles be smaller than the interstices between them, and the diameter of each atom must be less than the hundred and forty thousandth part of its distance from the neighbouring particles; so that the whole space occupied by the substance must be as little filled as the whole of England would be filled by a hundred men, placed at the distance of about thirty miles from each other. This astonishing degree of porosity is not indeed absolutely inadmissible, and there are many reasons for believing the statement to agree in some measure with the actual constitution of material substances; but the Huygenian hypothesis does not require the disproportion to be by any means so great, since the general direction and even the intensity of an undulation would be very little affected by the interposition of the atoms of matter, while these atoms may at the same time be supposed to assist in the transmission of the impulse, by propagating it through their own substance. Euler indeed imagined that the undulations of light might be transmitted through the gross substance of material bodies alone, precisely in the same manner as sound is propagated; but this supposition is for many reasons inadmissible.

A very striking circumstance, respecting the propagation of light, is the uniformity of its velocity in the same medium. According to the projectile hypothesis, the force employed in the free emission of light must be about a million million times as great as the force of gravity at the earth's surface; and it must either act with equal intensity on all the particles of light, or must impel some of them through a greater space than others, if its action be less powerful, since the velocity is the same in all cases; for example, if the projectile force is weaker with respect to red light than with respect to violet light, it must continue its action on the red rays to a greater distance than on the violet rays. There is no instance in nature besides of a simple projectile moving with a velocity uniform in all cases, whatever may be its cause, and it is extremely difficult to imagine that so immense a force of repulsion can reside in all substances capable of becoming luminous, so that the light of decaying wood, or of two pebbles rubbed together, may be projected precisely with the same velocity as the light emitted by iron burning in oxygen gas, or by the reservoir of liquid fire on the surface of the sun. Another cause would also naturally interfere with the uniformity of the velocity of light, if it consisted merely in the motion of projected corpuscles of matter; Mr. Laplace has calculated, that if any of the stars were 250 times as great in diameter as the sun, its attraction would be so strong as to destroy the whole momentum of the corpuscles of light proceeding from it, and to render the star invisible at a great distance; and although there is no reason to imagine that any of the stars are actually of this magnitude, yet some of them are probably many times greater than our sun, and therefore large enough to produce such a retardation in the motion of their light as would materially alter its effects. It is almost unnecessary to observe that the uniformity of the velocity of light, in those spaces which are free from all material substances, is a necessary consequence of the Huygenian hypothesis, since the undulations of every homogeneous elastic medium are always propagated, like those of sound, with the same velocity, as long as the medium remains unaltered.

On either supposition, there is no difficulty in explaining the equality of the angles of incidence and reflection; for these angles are equal as well in the collision of common elastic bodies with others incomparably larger, as in the reflections of the waves of water and of the undulations of sound. And it is equally easy to demonstrate, that the sines of the angles of incidence and refraction must be always in the same proportion at the same surface, whether it be supposed to possess an attractive force, capable of acting on the particles of light, or to be the limit of a medium through which the undulations are propagated with a diminished velocity. There are, however, some cases of the production of colours, which lead us to suppose that the velocity of light must be smaller in a denser than in a rarer medium; and supposing this fact to be fully established, the existence of such an attractive force could no longer be allowed, nor could the system of emanation be maintained by any one.

The partial reflection from all refracting surfaces is supposed by Newton to arise from certain periodical retardations of the particles of light, caused by undulations, propagated in all cases through an ethereal medium. The mechanism of these supposed undulations is so complicated, and attended by so many difficulties, that the few who have examined them have been in general entirely dissatisfied with them; and the internal vibrations of the particles of light themselves, which Boscovich has imagined, appear scarcely to require a serious discussion. It may, therefore, safely be asserted, that in the projectile hypothesis this separation of the rays of light of the same kind by a partial reflection at every refracting surface, remains wholly unexplained. In the undulatory system, on the contrary, this separation follows as a necessary consequence. It is simplest to consider the ethereal medium which pervades any transparent substance, together with the material atoms of the substance, as constituting together a compound medium denser than the pure ether, but not more elastic; and by comparing the contiguous particles of the rarer and the denser medium with common elastic bodies of different dimensions, we may easily determine not only in what manner, but almost in what degree, this reflection must take place in different circumstances. Thus, if one of two equal bodies strikes the other, it communicates to it its whole motion without any reflection; but a smaller body striking a larger one is reflected, with the more force as the difference of their magnitude is greater; and a larger body, striking a smaller one, still proceeds with a diminished velocity; the remaining motion constituting, in the case of an undulation falling on a rarer medium, a part of a new series of motions which necessarily returns backwards with the appropriate velocity; and we may observe a circumstance nearly similar to this last in a portion of mercury spread out on a horizontal table; if a wave be excited at any part, it will be reflected from the termination of the mercury almost in the same manner as from a solid obstacle.

The total reflection of light, falling, with a certain obliquity, on the surface of a rarer medium, becomes, on both suppositions, a particular case of refraction. In the undulatory system, it is convenient to suppose the two mediums to be separated by a short space in which their densities approach by degrees to each other, in order that the undulation may be turned gradually round, so as to be reflected in an equal angle; but this supposition is not absolutely necessary, and the same effects may be expected at the surface of two mediums separated by an abrupt termination.

The chemical process of combustion may easily be imagined either to disengage the particles of light from their various combinations, or to agitate the elastic medium by the intestine motions attending it: but the operation of friction upon substances incapable of undergoing chemical changes, as well as the motions of the electric fluid through imperfect conductors, afford instances of the production of light in which there seems to be no easy way of supposing a decomposition of any kind. The phenomena of solar phosphori appear to resemble greatly the sympathetic sounds of musical instruments, which are agitated by other sounds conveyed to them through the air: it is difficult to understand in what state the corpuscles of light could be retained by these substances so as to be re-emitted after a short space of time; and if it is true that diamonds are often found, which exhibit a red light after having received a violet light only, it seems impossible to explain this property, on the supposition of the retention and subsequent emission of the same corpuscles.

The phenomena of the aberration of light agree perfectly well with the system of emanation; and if the ethereal medium, supposed to pervade the earth and its atmosphere, were carried along before it, and partook materially in its motions, these phenomena could not easily be reconciled with the theory of undulation. But there is no kind of necessity for such a supposition: it will not be denied by the advocates of the Newtonian opinion that all material bodies are sufficiently porous to leave a medium pervading them almost absolutely at rest; and if this be granted, the effects of aberration will appear to be precisely the same in either hypothesis.

142

The unusual refraction of the Iceland spar has been most accurately and satisfactorily explained by Huygens, on the simple supposition that this crystal possesses the property of transmitting an impulse more rapidly in one direction than in another; whence he infers that the undulations constituting light must assume a spheroidical instead of a spherical form, and lays down such laws for the direction of its motion, as are incomparably more consistent with experiment than any attempts which have been made to accommodate the phenomena to other principles. It is true that nothing has yet been done to assist us in understanding the effects of a subsequent refraction by a second crystal, unless any person can be satisfied with the name of polarity assigned by Newton to a property which he attributes to the particles of light, and which he supposes to direct them in the species of refraction which they are to undergo: but on any hypothesis, until we discover the reason why a part of the light is at first refracted in the usual manner, and another part in the unusual manner, we have no right to expect that we should understand how these dispositions are continued or modified, when the process is repeated.

In order to explain, in the system of emanation, the dispersion of the rays of different colours by means of refraction, it is necessary to suppose that all refractive mediums have an elective attraction, acting more powerfully on the violet rays, in proportion to their mass, than on the red. But an elective attraction of this kind is a property foreign to mechanical philosophy, and when we use the term in chemistry, we only confess our incapacity to assign a mechanical cause for the effect, and refer to an analogy with other facts, of which the intimate nature is perfectly unknown to us. It is not indeed very easy to give a demonstrative theory of the dispersion of coloured light upon the supposition of undulatory motion; but we may derive a very satisfactory illustration from the well known effects of waves of different breadths. The simple calculation of the velocity of waves, propagated in a liquid perfectly elastic, or incompressible, and free from friction, assigns to them all precisely the same velocity, whatever their breadth may be: the compressibility of the fluids actually existing introduces, however, a necessity for a correction according to the breadth of the wave, and it is very easy to observe, in a river or a pond of considerable depth, that the wider waves proceed much more rapidly than the narrower. We may, therefore, consider the pure ethereal medium as analogous to an infinitely elastic fluid, in which undulations of all kinds move with equal velocity, and material transparent substances, on the contrary, as resembling those fluids, in which we see the large waves advance beyond the smaller; and by supposing the red light to consist of larger or wider undulations and the violet of smaller, we may sufficiently elucidate the greater refrangibility of the red than of the violet light.

It is not, however, merely on the ground of this analogy that we may be induced to suppose the undulations constituting red light to be larger than those of violet light: a very extensive class of phenomena leads us still more directly to the same conclusion; they consist chiefly of the production of colours by means of transparent plates, and by diffraction or inflection, none of which have been explained upon the supposition of emanation, in a manner sufficiently minute or comprehensive to satisfy the most candid even of the advocates for the projectile system; while on the other hand all of them may be at once understood, from the effect of the interference of double lights, in a manner nearly similar to that which constitutes in sound the sensation of a beat, when two strings forming an imperfect unison, are heard to vibrate together.

Supposing the light of any given colour to consist of undulations of a given breadth, or of a given frequency, it follows that these undulations must be liable to those effects which we have already examined in the case of the waves of water and the pulses of sound. It has been shown that two equal series of waves, proceeding from centres near each other, may be seen to destroy each other's effects at certain points, and at other points to redouble them; and the beating of two sounds has been explained from a similar interference. We are now to apply the same principles to the alternate union and extinction of colours.

In order that the effects of two portions of light may be thus combined, it is necessary that they be derived from the same origin, and that they arrive at the same point by different paths, in directions not much deviating from each other. This deviation may be produced in one or both of the portions by diffraction, by reflection, by refraction, or by any of these effects combined; but the simplest case appears to be, when a beam of homogeneous light falls on a screen in which there are two very small holes or slits, which may be considered as centres of divergence, from whence the light is diffracted in every direction. In this case, when the two newly formed beams are received on a surface placed so as to intercept them, their light is divided by dark stripes into portions nearly equal, but becoming wider as the surface is more remote from the apertures, so as to subtend very nearly equal angles from the apertures at all distances, and wider also in the same proportion as the apertures are closer to each other. The middle of the two portions is always light, and the bright stripes on each side are at such distances, that the light coming to them from one of the apertures, must have passed through a longer space than that which comes from the other, by an interval which is equal to the breadth of one, two, three, or more of the supposed undulations, while the intervening dark spaces correspond to a difference of half a supposed undulation, of one and a half, of two and a half, or more.

144

From a comparison of various experiments, it appears that the breadth of the undulations constituting the extreme red light must be supposed to be, in air, about one 36 thousandth of an inch, and those of the extreme violet about one 60 thousandth; the mean of the whole spectrum, with respect to the intensity of light, being about one 45 thousandth. From these dimensions it follows, calculating upon the known velocity of light, that almost 500 millions of millions of the slowest of such undulations must enter the eye in a single second. The combination of two portions of white or mixed light, when viewed at a great distance, exhibits a few white and black stripes, corresponding to this interval: although, upon closer inspection, the distinct effects of an infinite number of stripes of different breadths appear to be compounded together, so as to produce a beautiful diversity of tints, passing by degrees into each other. The central whiteness is first changed to a yellowish, and then to a tawny colour, succeeded by crimson, and by violet and blue, which together appear, when seen at a distance, as a dark stripe; after this a green light appears, and the dark space beyond it has a crimson hue; the subsequent lights are all more or less green, the dark spaces purple and reddish; and the red light appears so far to predominate in all these effects, that the red or purple stripes occupy nearly the same place in the mixed fringes as if their light were received separately.

The comparison of the results of this theory with experiments fully establishes their general coincidence; it indicates, however, a slight correction in some of the measures, on account of some unknown cause, perhaps connected with the intimate nature of diffraction, which uniformly occasions the portions of light proceeding in a direction very nearly rectilinear, to be divided into stripes or fringes a little wider than the external stripes, formed by the light which is more bent.

When the parallel slits are enlarged, and leave only the intervening substance to cast its shadow, the divergence from its opposite margins still continues to produce the same fringes as before, but they are not easily visible, except within the extent of its shadow, being overpowered in other parts by a stronger light; but if the light thus diffracted be allowed to fall on the eye, either within the shadow or in its neighbourhood, the stripes will still appear; and in this manner the colours of small fibres are probably formed. Hence if a collection of equal fibres, for example a lock of wool, be held before the eye when we look at a luminous object, the series of stripes belonging to each fibre combine their effects, in such a manner, as to be converted into circular fringes or coronae. This is probably the origin of the coloured circles or coronae sometimes seen round the sun and moon, two or three of them appearing together, nearly at equal distances from each other and from the luminary, the internal ones being, however, like the stripes, a little dilated. It is only necessary that the air should be loaded with globules of moisture, nearly of equal size among themselves, not much exceeding one two thousandth of an inch in diameter, in order that a series of such coronae, at the distance of two or three degrees from each other, may be exhibited.

If, on the other hand, we remove the portion of the screen which separates the parallel slits from each other, their external margins will still continue to exhibit the effects of diffracted light in the shadow on each side; and the experiment will assume the form of those which were made by Newton on the light passing between the edges of two knives, brought very nearly into contact; although some of these experiments appear to show the influence of a portion of light reflected by a remoter part of the polished edge of the knives, which indeed must unavoidably constitute a part of the light concerned in the appearance of fringes, wherever their whole breadth exceeds that of the aperture, or of the shadow of the fibre.

The edges of two knives, placed very near each other, may represent the opposite margins of a minute furrow, cut in the surface of a polished substance of any kind, which, when viewed with different degrees of obliquity, present a series of colours nearly resembling those which are exhibited within the shadows of the knives: in this case, however, the paths of the two portions of light before their incidence are also to be considered, and the whole difference of these paths will be found to determine the appearance of colour in the usual manner: thus when the surface is so situated, that the image of the luminous point would be seen in it by regular reflection, the difference will vanish, and the light will remain perfectly white, but in other cases various colours will appear, according to the degree of obliquity. These colours may easily be seen, in an irregular form, by looking at any metal, coarsely polished, in the sunshine; but they become more distinct and conspicuous, when a number of fine lines of equal strength are drawn parallel to each other, so as to conspire in their effects.

It sometimes happens that an object, of which a shadow is formed in a beam of light, admitted through a small aperture, is not terminated by parallel sides; thus the two portions of light, which are diffracted from two sides of an object, at right angles with each other, frequently form a short series of curved fringes within the shadow, situated on each side of the diagonal, which were first observed by Grimaldi, and which are completely explicable from the general principle, of the interference of the two portions encroaching perpendicularly on the shadow.

But the most obvious of all the appearances of this kind is that of the fringes which are usually seen beyond the termination of any shadow, formed in a beam of light, admitted through a small aperture: in white light three of these fringes are usually visible, and sometimes four; but in light of one colour only, their number is greater; and they are always much narrower as they are remoter from the shadow. Their origin is easily deduced from the interference of the direct light with a portion of light reflected from the margin of the object which produces them, the obliquity of its incidence causing a reflection so copious as to exhibit a visible effect, however narrow that margin may be; the fringes are, however, rendered more obvious as the quantity of this reflected light is greater. Upon this theory it follows that the distance of the first dark fringe from the shadow should be half as great as that of the fourth, the difference of the lengths of the different paths of the light being as the squares of those distances; and the experiment precisely confirms this calculation, with the same slight correction only as is required in all other cases; the distances of the first fringes being always a little increased. It may also be observed, that the extent of the shadow itself is always augmented, and nearly in an equal degree with that of the fringes: the reason of this circumstance appears to be the gradual loss of light at the edges of every separate beam, which is so strongly analogous to the phenomena visible in waves of water. The same cause may also perhaps have some effect in producing the general modification or correction of the place of the first fringes, although it appears to be scarcely sufficient for explaining the whole of it.

146

A still more common and convenient method of exhibiting the effects of the mutual interference of light, is afforded us by the colours of the thin plates of transparent substances. The lights are here derived from the successive partial reflections produced by the upper and under surface of the plate, or when the plate is viewed by transmitted light, from the direct beam which is simply refracted, and that portion of it which is twice [or more times] reflected within the plate. The appearance in the latter case is much less striking than in the former, because the light thus affected is only a small portion of the whole beam, with which it is mixed; while in the former the two reflected portions are nearly of equal intensity, and may be separated from all other light tending to overpower them. In both cases, when the plate is gradually reduced in thickness to an extremely thin edge, the order of colours may be precisely the same as in the stripes and coronae already described; their distance only varying when the surfaces of the plate, instead of being plane, are concave, as it frequently happens in such experiments. The scale of an oxid, which is often formed by the effect of heat on the surface of a metal, in particular of iron, affords us an example of such a series formed in reflected light: this scale is at first inconceivably thin, and destroys none of the light reflected; it soon, however, begins to be of a dull yellow, which changes to red, and then to crimson and blue, after which the effect is destroyed by the opacity which the oxid acquires. Usually, however, the series of colours produced in reflected light follows an order somewhat different: the scale of oxid is denser than the air, and the iron below than the oxid; but where the mediums above and below the plate are either both rarer or both denser than itself, the different natures of the reflections at its different surfaces appear to produce a modification in the state of the undulations, and the infinitely thin edge of the plate becomes black instead of white, one of the portions of light at once destroying the other, instead of cooperating with it. Thus when a film of soapy water is stretched over a wine glass, and placed in a vertical position, its upper edge becomes extremely thin, and appears nearly black, while the parts below are divided by horizontal lines into a series of coloured bands; and when two glasses, one of which is slightly convex, are pressed together with some force, the plate of air between them exhibits the appearance of coloured rings, beginning from a black spot at the centre, and becoming narrower and narrower, as the curved figure of the glass causes the thickness of the plate of air to increase more and more rapidly. The black is succeeded by a violet, so faint as to be scarcely perceptible; next to this is an orange yellow, and then crimson and blue. When water or any other fluid, is substituted for the air between the glasses, the rings appear where the thickness is as much less than that of the plate of air, as the refractive density of the fluid is greater; a circumstance which necessarily follows from the proportion of the velocities with which light must, upon the Huygenian hypothesis, be supposed to move in different mediums. It is also a consequence equally necessary in this theory, and equally inconsistent with all others, that when the direction of the light is oblique, the effect of a thicker plate must be the same as that of a thinner plate, when the light falls perpendicularly upon it; the difference of the paths described by the different portions of light precisely corresponding with the observed phenomena.

Sir Isaac Newton supposes the colours of natural bodies in general to be similar to these colours of thin plates, and to be governed by the magnitude of their particles. If this opinion were universally true, we might always separate the colours of natural bodies by refraction into a number of different portions, with dark spaces intervening; for every part of a thin plate which exhibits the appearance of colour, affords such a divided spectrum, when viewed through a prism. There are accordingly many natural colours in which such a separation may be observed; one of the most remarkable of them is that of blue glass, probably coloured with cobalt, which becomes divided into seven distinct portions. It seems, however, impossible to suppose the production of natural colours perfectly identical with those of thin plates, on account of the known minuteness of the particles of colouring bodies, unless the refractive density of these particles be at least 20 or 30 times as great as that of glass or water; which is indeed not at all improbable with respect to the ultimate atoms of bodies, but difficult to believe with respect to any of their arrangements constituting the diversities of material substances.

148 The colours of mixed plates constitute a distinct variety of the colours of thin plates, which has not been commonly observed. They appear when the interstice between two glasses nearly in contact, is filled with a great number of minute portions of two different substances, as water and air, oil and air, or oil and water; the light which passes through one of the mediums, moving with a greater velocity, anticipates the light passing through the other; and their effects on the eye being confounded and combined, their interference produces an appearance of colours nearly similar to those of the colours of simple thin plates, seen by transmission; but at much greater thicknesses, depending on the difference of the refractive densities of the substances employed. The effect is observed by holding the glasses between the eye and the termination of a bright object, and it is most conspicuous in the portion which is seen on the dark part beyond the object, being produced by the light scattered irregularly from the surfaces of the fluid. Here, however, the effects are inverted, the colours resembling those of the common thin plates seen by reflection; and the same considerations on the nature of the reflections are applicable to both cases.

The production of the supernumerary rainbows, which are sometimes seen within the primary and without the secondary bow, appears to be intimately connected with that of the colours of thin plates. We have already seen that the light producing the ordinary rainbow is double, its intensity being only greatest at its termination, where the common bow appears, while the whole light is extended much more widely. The two portions concerned in its production must divide this light into fringes; but unless almost all the drops of a shower happen to be of the same magnitude, the effects of these fringes must be confounded and destroyed; in general, however, they must at least cooperate more or less in producing one dark fringe, which must cut off the common rainbow much more abruptly than it would otherwise have been terminated, and consequently assist the distinctness of its colours. The magnitude of the drops of rain, required for producing such of these rainbows as are usually observed, is between the 50th and the 100th of an inch; they become gradually narrower as they are more remote from the common rainbows, nearly in the same proportions as the external fringes of a shadow, or the rings seen in a concave plate.

The last species of the colours of double lights, which it will be necessary to notice, constitutes those which have been denominated, from Newton's experiments, the colours of thick plates, but which may be called, with more propriety, the colours of concave mirrors. The anterior surface of a mirror of glass, or any other transparent surface placed before a speculum of metal, dissipates irregularly in every direction two portions of light, one before and the other after its reflection. When the light falls obliquely on the mirror, being admitted through an aperture near the centre of its curvature, it is easy to show, from the laws of reflection, that the two portions, thus dissipated, will conspire in their effects, throughout the circumference of a circle, passing through the aperture; this circle will consequently be white, and it will be surrounded with circles of colours very nearly at equal distances, resembling the stripes produced by diffraction. The analogy between these colours and those of thin plates is by no means so close as Newton supposed it; since the effect of a plate of any considerable thickness must be absolutely lost in white light, after ten or twelve alternations of colours at most, while these effects would require the whole process to remain unaltered, or rather to be renewed, after many thousands or millions of changes.

It is presumed, that the accuracy, with which the general law of the interference of light has been shown to be applicable to so great a variety of facts, in circumstances the most dissimilar, will be allowed to establish its validity in the most satisfactory manner. The full confirmation or decided rejection of the theory, by which this law was first suggested, can be expected from time and experience alone; if it be confuted, our prospects will again be confined within their ancient limits, but if it be fully established, we may expect an ample extension of our views of the operations of nature, by means of our acquaintance with a medium, so powerful and so universal, as that to which the propagation of light must be attributed.

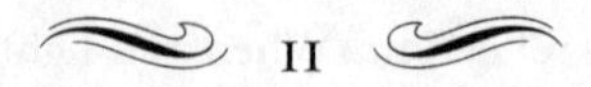 11

Whewell's Hypothetico-Deductivism

From

The Philosophy of the Inductive Sciences / William Whewell

Chapter 5. Of Certain Characteristics of Scientific Induction

. . . .

2. Induction is familiarly spoken of as the process by which we collect a *general proposition* from a number of *particular cases:* and it appears to be frequently imagined that the general proposition results from a mere juxtaposition of the cases, or at most, from merely conjoining and extending them. But if we consider the process more closely, as exhibited in the cases lately spoken of, we shall perceive that this is an inadequate account of the matter. The particular facts are not merely brought together, but there is a new element added to the combination by the very act of thought by which they are combined. There is a conception of the mind introduced in the general proposition, which did not exist in any of the observed facts. When the Greeks, after long observing the motions of the planets, saw that these motions might be rightly considered as produced by the motion of one

Reprinted from *The Philosophy of the Inductive Sciences, Founded upon Their History,* 2 vols. (London: John W. Parker, 1840; reprinted London: Routledge/Thoemmes Press, 1996), 2:213–39.

wheel revolving in the inside of another wheel, these wheels were creations of their minds, added to the facts which they perceived by sense. And even if the wheels were no longer supposed to be material, but were reduced to mere geometrical spheres or circles, they were not the less products of the mind alone,—something additional to the facts observed. The same is the case in all other discoveries. The facts are known, but they are insulated and unconnected, till the discoverer supplies from his own stores a principle of connexion. The pearls are there, but they will not hang together till some one provides the string. The distances and periods of the planets were all so many separate facts; by Kepler's Third Law they are connected into a single truth: but the conceptions which this law involves were supplied by Kepler's mind, and without these, the facts were of no avail. The planets described ellipses round the sun, in the contemplation of others as well as of Newton; but Newton conceived the deflection from the tangent in these elliptical motions in a new light,—as the effect of a central force following a certain law; and then it was that such a force was discovered truly to exist.

Thus[1] in each inference made by Induction, there is introduced some general conception, which is given, not by the phenomena, but by the mind. The conclusion is not contained in the premises, but includes them by the introduction of a new generality. In order to obtain our inference, we travel beyond the cases which we have before us; we consider them as mere exemplifications of some ideal case in which the relations are complete and intelligible. We take a standard, and measure the facts by it; and this standard is constructed by us, not offered by Nature. We assert, for example, that a body left to itself will move on with unaltered velocity; not because our senses ever disclosed to us a body doing this, but because (taking this as our ideal case) we find that all actual cases are intelligible and explicable by means of the Conception of *Forces,* causing change and motion, and exerted by surrounding bodies. In like manner, we see bodies striking each other, and thus moving and stopping, accelerating and retarding each other: but in all this, we do not perceive by our senses that abstract quantity *Momentum,* which is always lost by one body as it is gained by another. This Momentum is a creation of the mind, brought in among the facts, in order to convert their apparent confusion into order, their seeming chance into certainty, their perplexing variety into simplicity. This the Conception of *Momentum gained and lost* does: and in like manner, in any other case in which a truth is established by Induction, some Conception is introduced, some Idea is applied, as the means of binding together the facts, and thus producing the truth.

3. Hence in every inference by Induction there is some Conception *superinduced* upon the Facts: and we may henceforth conceive this to be the peculiar import of the term *Induction.* I am not to be understood as asserting

that the term was originally or anciently employed with this notion of its meaning; for the peculiar feature just pointed out in Induction has generally been overlooked. This appears by the accounts generally given of Induction. "Induction," says Aristotle,[2] "is when by means of one extreme term[3] we infer the other extreme term to be true of the middle term." Thus, (to take such exemplifications as belong to our subject,) from knowing that Mercury, Venus, Mars, describe ellipses about the Sun, we infer that all Planets describe ellipses about the Sun. In making this inference syllogistically, we assume that the evident proposition, "Mercury, Venus, Mars, do what all Planets do," may be taken *conversely,* "All Planets do what Mercury, Venus, Mars, do." But we remark that, in this passage, Aristotle (as was natural in his line of discussion) turns his attention entirely to the *evidence* of the inference; and overlooks a step which is of far more importance to our knowledge, namely, the *invention* of the second extreme term. In the above instance, the particular luminaries, Mercury, Venus, Mars, are one logical *extreme;* the general designation Planets is the *middle term;* but having these before us, how do we come to think of *description of ellipses,* which is the other extreme of the syllogism? When we have once invented this "second extreme term," we may, or may not, be satisfied with the evidence of the syllogism; we may, or may not, be convinced that, so far as this property goes, the extremes are co-extensive with the middle term[4]; but the *statement* of the syllogism is the important step in science. We know how long Kepler laboured, how hard he fought, how many devices he tried, before he hit upon this *term,* the elliptical motion. He rejected, as we know, many other "second extreme terms," for example, various combinations of epicyclical constructions, because they did not represent with sufficient accuracy the special facts of observation. When he had established his premiss, that "Mars does describe an ellipse about the Sun," he does not hesitate to *guess* at least that, in this respect, he might *convert* the other premiss, and assert that "All the Planets do what Mars does." But the main business was, the inventing and verifying the proposition respecting the ellipse. The Invention of the Conception was the great step in the *discovery;* the Verification of the Proposition was the great step in the *proof* of the discovery. If Logic consists in pointing out the conditions of proof, the Logic of Induction must consist in showing what are the conditions of proof in such inferences as this: but this subject must be pursued in the next chapter; I now speak principally of the act of *invention* which is requisite in every inductive inference.

4. Although in every inductive inference an act of invention is requisite, the act soon slips out of notice. Although we bind together facts by superinducing upon them a new conception, this conception, once introduced and applied, is looked upon as inseparably connected with the facts, and necessarily implied in them. Having once had the phenomena bound to-

gether in their minds in virtue of the conception, men can no longer easily restore them back to the detached and incoherent condition in which they were before they were thus combined. The pearls once strung, they seem to form a chain by their nature. Induction has given them a unity which it is so far from costing us an effort to preserve, that it requires an effort to imagine it dissolved. For instance, we usually represent to ourselves the earth as round, the earth and the planets as revolving about the sun, and as drawn to the sun by a central force; we can hardly understand how it could cost the Greeks, and Copernicus, and Newton so much pains and trouble to arrive at a view which is to us so familiar. These are no longer to us conceptions caught hold of and kept hold of by a severe struggle; they are the simplest modes of conceiving the facts: they are really facts. We are willing to *own* our obligation to those discoverers, but we hardly *feel* it: for in what other manner (we ask in our thoughts), could we represent the facts to ourselves?

Thus we see why it is that this step of which we now speak, the invention of a new Conception in every inductive inference, is so generally overlooked that it has hardly been noticed by preceding philosophers. When once performed by the discoverer, it takes a fixed and permanent place in the understanding of every one. It is a thought which, once breathed forth, permeates all men's minds. All fancy they nearly or quite knew it before. It oft was thought, or almost thought, though never till now expressed. Men accept it and retain it, and know it cannot be taken from them, and look upon it as their own. They will not and cannot part with it, even though they may deem it trivial and obvious. It is a secret, which once uttered, cannot be recalled, even though it be despised by those to whom it is imparted. As soon as the leading term of a new theory has been pronounced and understood, all the phenomena change their aspect. There is a standard to which we cannot help referring them. We cannot fall back into the helpless and bewildered state in which we gazed at them when we possessed no principle which gave them unity. Eclipses arrive in mysterious confusion: the notion of a *Cycle* dispels the mystery. The Planets perform a tangled and mazy dance; but *Epicycles* reduce the maze to order. The Epicycles themselves run into confusion; the conception of an *Ellipse* makes all clear and simple. And thus from stage to stage, new elements of intelligible order are introduced. But this intelligible order is so completely adopted by the human understanding, as to seem part of its texture. Men ask whether Eclipses follow a Cycle; whether the Planets describe Ellipses; and they imagine that so long as they do not *answer* such questions rashly, they take nothing for granted. They do not recollect how much they assume in *asking* the question:—how far the conceptions of Cycles and of Ellipses are beyond the visible surface of the celestial phenomena:—how many ages elapsed, how much thought, how much observation, were needed, before men's thoughts were fashioned into the words which they now so familiarly use. And thus they treat the subject, as we have seen Aristotle treating it; as if it were a question, not of invention, but of proof; not of substance, but of form: as if the main thing were not *what* we assert, but *how* we assert it. But for our purpose it is requisite to bear in mind the feature which we have thus attempted to mark; and to recollect that in every inference by induction, there is a Conception supplied by the mind and superinduced upon the Facts.

5. In collecting scientific truths by Induction we often find (as has already been observed,) a Definition and a Proposition established at the same time,—introduced together and mutually dependent on each other. The combination of the two constitutes the Inductive act; and we may consider the Definition as representing the superinduced Conception, and the Proposition as exhibiting the Colligation of Facts.

6. To discover a conception of the mind which will justly represent a train of observed facts is, in some measure, a process of conjecture, as I have stated already; and as I then observed, the business of conjecture is commonly conducted by calling up before our minds several suppositions, and selecting that one which most agrees with what we know of the observed facts. Hence he who has to discover the laws of nature may have to invent many suppositions before he hits upon the right one; and among the endowments which lead to his success, we must reckon that fertility of invention which ministers to him such imaginary schemes, till at last he finds the one which conforms to the true order of nature. A facility in devising hypotheses, therefore, is so far from being a fault in the intellectual character of a discoverer, that it is, in truth, a faculty indispensable to his task. It is, for his purposes, much better that he should be too ready in contriving, too eager in pursuing systems which promise to introduce law and order among a mass of unarranged facts, than that he should be barren of such inventions and hopeless of such success. Accordingly, as we have already noticed, great discoverers have often invented hypotheses which would not answer to all the facts, as well as those which would; and have fancied themselves to have discovered laws, which a more careful examination of the facts overturned.

The tendencies of our speculative nature,[5] carrying us onwards in pursuit of symmetry and rule, and thus producing all true theories, perpetually show their vigour by overshooting the mark. They obtain something, by aiming at much more. They detect the order and connexion which exist, by conceiving imaginary relations of order and connexion which have no existence. Real discoveries are thus mixed with baseless assumptions; profound sagacity is combined with fanciful conjecture; not rarely, or in peculiar instances, but commonly, and in most cases; probably in all, if we could read the thoughts of discoverers as we read the books of Kepler. To try wrong guesses is, with most persons, the only way to hit upon right ones. The character of the true philosopher is, not that he never conjectures hazardously, but that his conjectures are clearly conceived, and brought into rigid contact with facts. He sees and compares distinctly the ideas and the things;—the relations of his notions to each other and to phenomena. Under these conditions it is not only excusable, but necessary for him, to snatch at every semblance of general rule,—to try all promising forms of simplicity and symmetry.

Hence advances in knowledge[6] are not commonly made without the previous exercise of some boldness and license in guessing. The discovery of new truths requires, undoubtedly, minds careful and scrupulous in examining what is suggested; but it requires, no less, such as are quick and fertile in suggesting. What is invention, except the talent of rapidly calling before us the many possibilities, and selecting the appropriate one? It is true that when we have rejected all the inadmissible suppositions, they are often quickly forgotten; and few think it necessary to dwell on these discarded hypotheses, and on the process by which they were condemned. But all who discover truths must have reasoned upon many errors to obtain each truth; every accepted doctrine must have been one chosen out of many candidates. If many of the guesses of philosophers of bygone times now appear fanciful and absurd because time and observation have refuted them, others, which were at the time equally gratuitous, have been confirmed in a manner which makes them appear marvellously sagacious. To form hypotheses, and then to employ much labour and skill in refuting, if they do not succeed in establishing them, is a part of the usual process of inventive minds. Such a proceeding belongs to the *rule* of the genius of discovery, rather than (as has often been taught in modern time) to the *exception*.

7. But if it be an advantage for the discoverer of truth that he be ingenious and fertile in inventing hypotheses which may connect the phenomena of nature, it is indispensably requisite that he be diligent and careful in comparing his hypotheses with the facts, and ready to abandon his invention as soon as it appears that it does not agree with the course of actual occurrences. This constant comparison of his own conceptions and supposition with observed facts under all aspects, forms the leading employment of the discoverer: this candid and simple love of truth, which makes him willing to suppress the most favourite production of his own ingenuity as soon as it appears to be at variance with realities, constitutes the first characteristic of his temper. He must have neither the blindness which cannot, nor the obstinacy which will not, perceive the discrepancy of his fancies and his facts. He must allow no indolence, or partial views, or self-complacency, or delight in seeming demonstration, to make him tena-

cious of the schemes which he devises, any further than they are confirmed by their accordance with nature. The framing of hypotheses is, for the inquirer after truth, not the end, but the beginning of his work. Each of his systems is invented, not that he may admire it and follow it into all its consistent consequences, but that he may make it the occasion of a course of active experiment and observation. And if the results of this process contradict his fundamental assumptions, however ingenious, however symmetrical, however elegant his system may be, he rejects it without hesitation. He allows no natural yearning for the offspring of his own mind to draw him aside from the higher duty of loyalty to his sovereign, Truth: to her he not only gives his affections and his wishes, but strenuous labour and scrupulous minuteness of attention.

We may refer to what we have said of Kepler, Newton, and other eminent philosophers, for illustrations of this character. In Kepler we have remarked[7] the courage and perseverance with which he undertook and executed the task of computing his own hypotheses: and, as a still more admirable characteristic, that he never allowed the labour he had spent upon any conjecture to produce any reluctance in abandoning the hypothesis, as soon as he had evidence of its inaccuracy. And in the history of Newton's discovery that the moon is retained in her orbit by the force of gravity, we have noticed the same moderation in maintaining the hypothesis, after it had once occurred to the author's mind. The hypothesis required that the moon should fall from the tangent of her orbit every second through a space of sixteen feet; but according to his first calculations it appeared that in fact she only fell through a space of thirteen feet in that time. The difference seems small, the approximation encouraging, the theory plausible; a man in love with his own fancies would readily have discovered or invented some probable cause of the difference. But Newton acquiesced in it as a disproof of his conjecture, and "laid aside at that time any further thoughts of this matter."[8]

8. It has often happened that those who have undertaken to instruct mankind have not possessed this pure love of truth and comparative indifference to the maintenance of their own inventions. Men have frequently adhered with great tenacity and vehemence to the hypotheses which they have once framed; and in their affection for these, have been prone to overlook, to distort, and to misinterpret facts. In this manner hypotheses have so often been prejudicial to the genuine pursuit of truth, that they have fallen into a kind of obloquy; and have been considered as dangerous temptations and fallacious guides. Many warnings have been uttered against the fabrication of hypotheses by those who profess to teach philosophy; many disclaimers of such a course by those who cultivate science.

Thus we shall find Bacon frequently discommending this habit, under the name of "anticipation of the mind," and Newton thinks it necessary to say emphatically "hypotheses non fingo." It has been constantly urged that the inductions by which sciences are formed must be *cautious* and *rigorous;* and the various imaginations which passed through Kepler's brain, and to which he has given utterance, have been blamed or pitied as lamentable instances of an unphilosophical frame of mind. Yet it has appeared in the preceding remarks that hypotheses rightly used are among the helps, far more than the dangers, of science;—that scientific induction is not a "cautious" or a "rigorous" process in the sense of *abstaining from* such suppositions, but in *not adhering to* them till they are confirmed by fact, and in carefully seeking from facts confirmation or refutation. Kepler's character was, not that he was peculiarly given to the construction of hypotheses, but that he narrated with extraordinary copiousness and candour the course of his thoughts, his labours, and his feelings. In the minds of most persons, as we have said, the inadmissible suppositions, when rejected, are soon forgotten: and thus the trace of them vanishes from the thoughts, and the successful hypothesis alone holds its place in our memory. But in reality, many other transient suppositions must have been made by all discoverers;—hypotheses which are not afterwards asserted as true systems, but entertained for an instant;—"tentative hypotheses," as they have been called. Each of these hypotheses is followed by its corresponding train of observations, from which it derives its power of leading to truth. The hypothesis is like the captain, and the observations like the soldiers of an army: while he appears to command them, and in this way to work his own will, he does in fact derive all his power of conquest from their obedience, and becomes helpless and useless if they mutiny.

Since the discoverer has thus constantly to work his way onwards by means of hypotheses, false and true, it is highly important for him to possess talents and means for rapidly *testing* each supposition as it offers itself. In this as in other parts of the work of discovery, success has in general been mainly owing to the native ingenuity and sagacity of the discoverer's mind. Yet some rules tending to further this object have been delivered by eminent philosophers, and some others may perhaps be suggested. Of these we shall here notice only some of the most general, leaving for a future chapter the consideration of some more limited and detailed processes by which, in certain cases, the discovery of the laws of nature may be materially assisted.

9. A maxim which it may be useful to recollect is this;—that hypotheses may often be of service to science, when they involve a certain portion of incompleteness, and even of error. The object of such inventions is to bind together facts which without them are loose and detached; and if they do this, they may lead the way to a perception of the true rule by which the

phenomena are associated together, even if they themselves somewhat misstate the matter. The imagined arrangement enables us to contemplate as a whole a collection of special cases which perplex and overload our minds when they are considered in succession; and if our scheme has so much of truth in it as to conjoin what is really connected, we may afterwards duly correct or limit the mechanism of this connexion. If our hypothesis renders a reason for the agreement of cases really similar, we may afterwards find this reason to be false, but we shall be able to translate it into the language of truth.

A conspicuous example of such an hypothesis, one which was of the highest value to science, though very incomplete, and as a representation of nature altogether false, is seen in the *doctrine of epicycles* by which the ancient astronomers explained the motions of the sun, moon, and planets. This doctrine connected the places and velocities of these bodies at particular times in a manner which was, in its general features, agreeable to nature. Yet this doctrine was erroneous in its assertion of the circular nature of all the celestial motions, and in making the heavenly bodies revolve round the earth. It was, however, of immense value to the progress of astronomical science; for it enabled men to express and reason upon many important truths which they discovered respecting the motion of the stars, up to the time of Kepler. Indeed we can hardly imagine that astronomy could, in its outset, have made so great a progress under any other form, as it did in consequence of being cultivated in this shape of the incomplete and false epicyclical hypothesis.

We may notice another instance of an exploded hypothesis, which is generally mentioned only to be ridiculed, and which undoubtedly is both false in the extent of its assertion, and unphilosophical in its expression; but which still, in its day, was not without merit. I mean the doctrine of *Nature's horror of a vacuum (fuga vacui,)* by which the action of siphons and pumps and many other phenomena were explained, till Mersenne and Pascal taught a truer doctrine. This hypothesis was of real service; for it brought together many facts which really belong to the same class, although they are very different in their first aspect. A scientific writer of modern times[9] appears to wonder that men did not at once divine the weight of the air from which the phenomena formerly ascribed to the *fuga vacui* really result. "Loaded, compressed by the atmosphere," he says, "they did not recognize its action. In vain all nature testified that air was elastic and heavy; they shut their eyes to her testimony. The water rose in pumps and flowed in siphons at that time as it does at this day. They could not separate the boards of a pair of bellows of which the holes were stopped; and they could not bring together the same boards without difficulty if they were at first separated. Infants sucked the milk of their mothers; air entered rapidly into the lungs of animals at every inspiration; cupping-glasses produced tumours on the skin; and in spite of all these striking proofs of the weight and elasticity of the air, the ancient philosophers maintained resolutely that air was light, and explained all these phenomena by the horror which they said nature had for a vacuum." It is curious that it should not have occurred to the author while writing this, that if these facts, so numerous and various, can all be accounted for by *one* principle, there is a strong presumption that the principle is not altogether baseless. And in reality is it not true that nature *does* abhor a vacuum, and do all she can to avoid it? No doubt this power is not unlimited; and we can trace it to a mechanical cause, the pressure of the circumambient air. But the tendency, arising from this pressure, which the bodies surrounding a space void of air have to rush into it, may be expressed, in no extravagant or unintelligible manner, by saying that nature has a repugnance to a vacuum.

That imperfect and false hypotheses, though they may thus explain *some* phenomena, and may be useful in the progress of science, cannot explain *all* phenomena;—and that we are never to rest in our labours or acquiesce in our results, till we have found some view of the subject which *is* consistent with *all* the observed facts:—will of course be understood. We shall afterwards have to speak of the other steps of such a progress.

10. The hypotheses which we accept ought to explain phenomena which we have observed. But they ought to do more than this: they ought to *foretel* phenomena which have not yet been observed;—at least all of the same kind as those which the hypothesis was invented to explain. For our assent to the hypothesis implies that it is held to be true of all particular instances. That these cases belong to past or to future times, that they have or have not already occurred, makes no difference in the applicability of the rule to them. Because the rule prevails, it includes all cases; and will determine them all, if we can only calculate its real consequences. Hence it will predict the results of new combinations, as well as explain the appearances which have occurred in old ones. And that it does this with certainty and correctness, is one mode in which the hypothesis is to be verified as right and useful.

The scientific doctrines which have at various periods been established have been verified in this manner. For example, the Epicyclical Theory of the heavens was confirmed by its *predicting* truly eclipses of the sun and moon, configurations of the planets, and other celestial phenomena; and by its leading to the construction of Tables by which the places of the heavenly bodies were given at every moment of time. The truth and accuracy of these predictions were a proof that the hypothesis was valuable and, at least to a great extent, true; although, as was afterwards found, it involved a false representation of the structure of the heavens. In like manner, the discovery of the Laws of Refraction enabled mathematicians to *predict,* by calculation, what would be the effect of any new form or combination of transparent lenses. Newton's hypothesis of Fits of Easy Transmission and Easy Reflection in the particles of light, although not confirmed by other kinds of facts, involved a true statement of the law of the phenomena which it was framed to include and served to *predict* the forms and colours of thin plates for a wide range of given cases. The hypothesis that Light operates by Undulations and Interferences, afforded the means of *predicting* results under a still larger extent of conditions. In like manner in the progress of chemical knowledge, the doctrine of Phlogiston supplied the means of *foreseeing* the consequence of many combinations of elements, even before they were tried; but the Oxygen Theory, besides affording predictions, at least equally exact, with regard to the general results of chemical operations, included all the facts concerning the relations of weight of the elements and their compounds, and enabled chemists to *foresee* such facts in untried cases. And the Theory of Electromagnetic Forces, as soon as it was rightly understood, enabled those who had mastered it to *predict* motions such as had not been before observed, which were accordingly found to take place.

Men cannot help believing that the laws laid down by discoverers must be in a great measure identical with the real laws of nature, when the discoverers thus determine effects beforehand in the same manner in which nature herself determines them when the occasion occurs. Those who can do this, must, to a considerable extent, have detected nature's secret;—must have fixed upon the conditions to which she attends, and must have seized the rules by which she applies them. Such a coincidence of untried facts with speculative assertions cannot be the work of chance, but implies some large portion of truth in the principles on which the reasoning is founded. To trace order and law in that which has been observed, may be considered as interpreting what nature has written down for us, and will commonly prove that we understand her alphabet. But to predict what has not been observed, is to attempt ourselves to use the legislative phrases of nature; and when she responds plainly and precisely to that which we thus utter, we cannot but suppose that we have in a great measure made ourselves masters of the meaning and structure of her language. The prediction of results, even of the same kind as those which have been observed, in new cases, is a proof of real success in our inductive processes.

11. We have here spoken of the prediction of facts *of the same kind* as those from which our rule was collected. But the evidence in favour of our induction is of a much higher and more forcible character when it enables us to explain and determine cases of a *kind different* from those which were contemplated in the formation of our hypothesis. The instances in which this has occurred, indeed, impress us with a conviction that the truth of our

hypothesis is certain. No accident could give rise to such an extraordinary coincidence. No false supposition could, after being adjusted to one class of phenomena, so exactly represent a different class, when the agreement was unforeseen and uncontemplated. That rules springing from remote and unconnected quarters should thus leap to the same point, can only arise from *that* being the point where truth resides.

Accordingly the cases in which inductions from classes of facts altogether different have thus *jumped together,* belong only to the best established theories which the history of science contains. And as I shall have occasion to refer to this peculiar feature in their evidence, I will take the liberty of describing it by a particular phrase; and will term it the *Consilience of Inductions.*

It is exemplified principally in some of the greatest discoveries. Thus it was found by Newton that the doctrine of the attraction of the sun varying according to the inverse square of this distance, which explained Kepler's *third law* of the proportionality of the cubes of the distances to the squares of the periodic times of the planets, explained also his *first* and *second laws* of the elliptical motion of each planet; although no connexion of these laws had been visible before. Again, it appeared that the force of universal gravitation, which had been inferred from the *perturbations* of the moon and planets by the sun and by each other, also accounted for the fact, apparently altogether dissimilar and remote, of the *precession of the equinoxes.* Here was a most striking and surprising coincidence, which gave to the theory a stamp of truth beyond the power of ingenuity to counterfeit. In like manner in optics; the hypothesis of alternate fits of easy transmission and reflection would explain the colours of thin plates, and indeed was devised and adjusted for that very purpose; but it could give no account of the phenomena of the fringes of shadows. But the doctrine of interferences, constructed at first with reference to phenomena of the nature of the *fringes,* explained also the *colours of thin plates* better than the supposition of the fits invented for that very purpose. And we have in physical optics another example of the same kind, which is quite as striking as the explanation of precession by inferences from the facts of perturbation. The doctrine of undulations propagated in a spheroidal form was contrived at first by Huygens, with a view to explain the laws of *double refraction* in calc-spar; and was pursued with the same view by Fresnel. But in the course of the investigation it appeared, in a most unexpected and wonderful manner, that this same doctrine of spheroidal undulations, when it was so modified as to account for the directions of the two refracted rays, accounted also for the positions of their *planes of polarization*[10]; a phenomenon which, taken by itself, it had perplexed previous mathematicians, even to represent.

The theory of universal gravitation, and of the undulatory theory of light, are, indeed, full of examples of this Consilience of Inductions. With regard to the latter, it has been justly asserted by Herschel, that the history of the undulatory theory was a succession of *felicities.*[11] And it is precisely the unexpected coincidences of results drawn from distant parts of the subject which are properly thus described. Thus the laws of the *modification of polarization* to which Fresnel was led by his general views, accounted for the rule respecting the *angle at which light is polarized,* discovered by Brewster.[12] The conceptions of the theory pointed out peculiar *modifications* of the phenomena when *Newton's rings* were produced by polarized light, which were ascertained to take place in fact, by Arago and Airy.[13] When the beautiful phenomena of *dipolarized light* were discovered by Arago and Biot, Young was able to declare that they were reducible to the general laws of *interference* which he had already established.[14] And what was no less striking a confirmation of the truth of the theory, *measures* of the same element deduced from various classes of facts were found to coincide. Thus the *length* of a luminiferous undulation, calculated by Young from the measurement of *fringes* of shadows, was found to agree very nearly with the previous calculation from the colours of *thin plates.*[15]

No example can be pointed out, in the whole history of science, so far as I am aware, in which this Consilience of Inductions has given testimony in favour of an hypothesis afterwards discovered to be false. If we take one class of facts only, knowing the law which they follow, we may construct an hypothesis, or perhaps several, which may represent them: and as new circumstances are discovered, we may often adjust the hypothesis so as to correspond to these also. But when the hypothesis, of itself and without adjustment for the purpose, gives us the rule and reason of a class of facts not contemplated in its construction, we have a criterion of its reality, which has never yet been produced in favour of falsehood.

12. In the preceding section I have spoken of the hypothesis with which we compare our facts as being framed *all at once,* each of its parts being included in the original scheme. In reality, however, it often happens that the various suppositions which our system contains are *added* upon occasion of different researches. Thus in the Ptolemaic doctrine of the heavens, new epicycles and eccentrics were added as new inequalities of the motions of the heavenly bodies were discovered; and in the Newtonian doctrine of material rays of light, the supposition that these rays had "fits," was added to explain the colours of thin plates; and the supposition that they had "sides" was introduced on occasion of the phenomena of polarization. In like manner other theories have been built up of parts devised at different times.

This being the mode in which theories are often framed, we have to notice a distinction which is found to prevail in the progress of true and of false theories. In the former class all the additional suppositions *tend to simplicity* and harmony; the new suppositions resolve themselves into the old ones, or at least require only some easy modification of the hypothesis first assumed: the system becomes more coherent as it is further extended. The elements which we require for explaining a new class of facts are already contained in our system. Different members of the theory run together, and we have thus a constant convergence to unity. In false theories, the contrary is the case. The new suppositions are something altogether additional;—not suggested by the original scheme; perhaps difficult to reconcile with it. Every such addition adds to the complexity of the hypothetical system, which at last becomes unmanageable, and is compelled to surrender its place to some simpler explanation.

Such a false theory, for example, was the ancient doctrine of eccentrics and epicycles. It explained the general succession of the places of the Sun, Moon, and Planets; it would not have explained the proportion of their magnitudes at different times, if these could have been accurately observed; but this the ancient astronomers were unable to do. When, however, Tycho and other astronomers came to be able to observe the planets accurately in all positions, it was found that *no* combination of *equable* circular motions would exactly represent all the observations. We may see, in Kepler's works, the many new modifications of the epicyclical hypothesis which offered themselves to him; some of which would have agreed with the phenomena with a certain degree of accuracy, but not so great a degree as Kepler, fortunately for the progress of science, insisted upon obtaining. After these epicycles had been thus accumulated, they all disappeared and gave way to the simpler conception of an *elliptical* motion. In like manner, the discovery of new inequalities in the moon's motions encumbered her system more and more with new machinery, which was at last rejected all at once in favour of the *elliptical* theory. Astronomers could not but suppose themselves in a wrong path when the prospect grew darker and more entangled at every step.

Again; the Cartesian system of vortices might be said to explain the primary phenomena of the revolutions of planets about the sun, and satellites about planets. But the elliptical form of the orbits required new suppositions. Bernoulli ascribed this curve to the shape of the planet, operating on the stream of the vortex in a manner similar to the rudder of a boat. But then the motions of the aphelia, and of the nodes,—the perturbations,—even the action of gravity to the earth,—could not be accounted for without new and independent suppositions. Here was none of the simplicity of truth. The theory of gravitation on the other hand became more simple as the facts to be explained became more numerous. The attraction of the sun accounted for the motions of the planets; the attraction of the planets was the cause of the motion of the satellites. But this being assumed, the perturbations, the motions of the nodes and aphelia, only made it requisite to extend the attraction of the sun to the satellites, and that of the planets to each other:—the tides, the spheroidal form of the earth, the precession, still required nothing more than that the moon and sun should attract the parts of the earth, and that these should attract each other;—so that all the suppositions resolved themselves into the single one, of the universal gravitation of all matter. It is difficult to imagine a more convincing manifestation of simplicity and unity.

Again, to take an example from another science;—the doctrine of phlogiston brought together many facts in a very plausible manner,—combustion, acidification, and others,—and very naturally prevailed for a while. But the balance came to be used in chemical operations, and the facts of weight as well as of combination were to be accounted for. On the phlogistic theory, it appeared that this could not be done without a new supposition, and *that* a very strange one;—that phlogiston was an element not only not heavy, but absolutely light, so that it diminished the weight of the compounds into which it entered. Some chemists for a time adopted this extravagant view; but the wiser of them saw, in the necessity of such a supposition to the defence of the theory, an evidence that the hypothesis of an element *phlogiston* was erroneous. And the opposite hypothesis, which taught that oxygen was subtracted, and not phlogiston added, was accepted because it required no such novel and inadmissible assumption.

Again, we find the same evidence of truth in the progress of the undulatory theory of light, in the course of its application from one class of facts to another. Thus we explain reflection and refraction by undulations; when we come to thin plates, the requisite "fits" are already involved in our fundamental hypothesis, for they are the length of an undulation: the phenomena of diffraction also require such intervals; and the intervals thus required agree exactly with the others in magnitude, so that no new property is needed. Polarization for a moment appears to require some new hypothesis; yet this is hardly the case; for the direction of our vibrations is hitherto arbitrary:—we allow polarization to decide it, and we suppose the undulations to be transverse. Having done this for the sake of polarization, we turn to the phenomena of double refraction, and inquire what new hypothesis they require. But the answer is, that they require none: the supposition of transverse vibrations, which we have made in order to explain polarization, gives us also the law of double refraction. Truth may give rise to such a coincidence; falsehood cannot. Again, the facts of dipolarization come into view. But they hardly require any new assumption; for the difference of optical elasticity of crystals in different directions, which is already assumed in uniaxal crystals,[16] is extended to biaxal exactly according to the law of symmetry; and this being done, the laws of the phenomena, curious and complex as they are, are fully explained. The phenomena of circular polarization by internal reflection, instead of requiring a new hypothesis, are found to be given by an interpretation of an apparently inexplicable result of an old hypothesis. The circular polarization of quartz and its double refraction does indeed appear to require a new assumption, but still not one which at all disturbs the form of the theory; and in short, the whole history of this theory is a progress, constant and steady, often striking and startling, from one degree of evidence and consistence to another of higher order.

In the emission theory, on the other hand, as in the theory of solid epicycles, we see what we may consider as the natural course of things in the career of a false theory. Such a theory may, to a certain extent, explain the phenomena which it was at first contrived to meet; but every new class of facts requires a new supposition—an addition to the machinery: and as observation goes on, these incoherent appendages accumulate, till they overwhelm and upset the original frame-work. Such has been the hypothesis of the material emission of light. In its original form it explained reflection and refraction: but the colours of thin plates added to it the fits of easy transmission and reflection; the phenomena of diffraction further invested the emitted particles with complex laws of attraction and repulsion; polarization gave them sides: double refraction subjected them to peculiar forces emanating from the axes of the crystal: finally, dipolarization loaded them with the complex and unconnected contrivance of moveable polarization: and even when all this had been done, additional mechanism was wanting. There is here no unexpected success, no happy coincidence, no convergence of principles from remote quarters. The philosopher builds the machine, but its parts do not fit. They hold together only while he presses them. This is not the character of truth.

As another example of the application of the maxim now under consideration, I may perhaps be allowed to refer to the judgment which, in the History of Thermotics, I have ventured to give respecting Laplace's Theory of Gases. I have stated,[17] that we cannot help forming an unfavourable judgment of this theory, by looking for that great characteristic of true theory; namely, that the hypotheses which were assumed to account for *one class* of facts are found to explain *another class* of a different nature. Thus Laplace's first suppositions explain the connexion of compression with density, (the law of Boyle and Mariotte,) and the connexion of elasticity with heat, (the law of Dalton and Gay Lussac.) But the theory requires other assumptions when we come to latent heat; and yet these new assumptions produce no effect upon the calculations in any application of the theory. When the hypothesis, constructed with reference to the elasticity and temperature, is applied to another class of facts, those of latent heat, we have no Simplification of the Hypothesis, and therefore no evidence of the truth of the theory.

13. The two last sections of this chapter direct our attention to two circumstances, which tend to prove, in a manner which we may term irresistible, the truth of the theories which they characterize:—the *Consilience of Inductions* from different and separate classes of facts;—and the progressive *Simplification of the Theory* as it is extended to new cases. These two Characters are, in fact, hardly different; they are exemplified by the same cases. For if these Inductions, collected from one class of facts, supply an unexpected explanation of a new class, which is the case first spoken of, there will be no need for new machinery in the hypothesis to apply it to the newly-contemplated facts; and thus we have a case in which the system does not become more complex when its application is extended to a wider field, which was the character of true theory in its second aspect. The Consiliences of our Inductions give rise to a constant Convergence of our Theory towards Simplicity and Unity.

But, moreover, both these cases of the extension of the theory, without difficulty or new suppositions, to a wider range and to new classes of phenomena, may be conveniently considered in yet another point of view; namely, as successive steps by which we gradually ascend in our speculative views to a higher and higher point of generality. For when the theory, either by the concurrence of two indications, or by an extension without complication, has included a new range of phenomena, we have, in fact, a new induction of a more general kind, to which the inductions formerly obtained are subordinate, as particular cases to a general proposition. We have in such examples, in short, an instance of *successive generalization.*

NOTES

1. I repeat here remarks made at the end of the *Mechanical Euclid,* p. 178.

2. *Analyt*[*ica*] *Prior*[*a*], lib. ii., c. 23. . . .

3. The syllogism here alluded to would be this:—
Mercury, Venus, Mars, describe ellipses about the Sun;
All Planets do what Mercury, Venus, Mars, do;
Therefore all Planets describe ellipses about the Sun.

4. . . . Aristot. [*Analytica Priora,* lib. ii., c. 23].

5. I here take the liberty of characterizing inventive minds in general in the same phraseology which, in the History of Science, I have employed in reference to particular examples. These expressions are what I have used in speaking of the discoveries of Copernicus.—*Hist. Ind. Sci.* [*History of the Inductive Sciences, from the Earliest to the Present Time,* 3 vols. (London, 1837)], vol. i. p. 373.

6. These observations are made on occasion of Kepler's speculations, and are illustrated by reference to his discoveries.—[*Ib.*], 411–414.

7. [*Ib.*], 414.

8. *Ib.*, ii., 160.
9. Deluc, *Modifications de l'Atmosphere,* partie i.
10. *Hist. Ind. Sci.*, ii. 420.
11. [*Ib.*], 435.
12. *Ib.*, 423.
13. *Ib.*, 450.
14. *Ib.*, 426.
15. *Ib.*, 406.
16. [*Ib.*], 427.
17. [*Ib.*], 530.

12

Popper's Falsificationism

From

The Logic of Scientific Discovery / Karl R. Popper

Chapter I. A Survey of Some Fundamental Problems

A scientist, whether theorist or experimenter, puts forward statements, or systems of statements, and tests them step by step. In the field of the empirical sciences, more particularly, he constructs hypotheses, or systems of theories, and tests them against experience by observation and experiment.

I suggest that it is the task of the logic of scientific discovery, or the logic of knowledge, to give a logical analysis of this procedure; that is, to analyse the method of the empirical sciences.

But what are these 'methods of the empirical sciences'? And what do we call 'empirical science'?

1. The Problem of Induction

According to a widely accepted view—to be opposed in this book—the empirical sciences can be characterized by the fact that they use '*inductive methods*', as they are called. According to this view, the logic of scientific discovery would be identical with inductive logic, *i.e.* with the logical analysis of these inductive methods.

Reprinted from *The Logic of Scientific Discovery* (New York: Basic Books, 1959), 27–30, 32–33.

It is usual to call an inference 'inductive' if it passes from *singular statements* (sometimes also called 'particular' statements), such as accounts of the results of observations or experiments, to *universal statements,* such as hypotheses or theories.

Now it is far from obvious, from a logical point of view, that we are justified in inferring universal statements from singular ones, no matter how numerous; for any conclusion drawn in this way may always turn out to be false: no matter how many instances of white swans we may have observed, this does not justify the conclusion that *all* swans are white.

The question whether inductive inferences are justified, or under what conditions, is known as *the problem of induction.*

The problem of induction may also be formulated as the question of how to establish the truth of universal statements which are based on experience, such as the hypotheses and theoretical systems of the empirical sciences. For many people believe that the truth of these universal statements is '*known by experience*'; yet it is clear that an account of an experience—of an observation or the result of an experiment—can in the first place be only a singular statement and not a universal one. Accordingly, people who say of a universal statement that we know its truth from experience usually mean that the truth of this universal statement can somehow be reduced to the truth of singular ones, and that these singular ones are known by experience to be true; which amounts to saying that the universal statement is based on inductive inference. Thus to ask whether there are natural laws known to be true appears to be only another way of asking whether inductive inferences are logically justified.

Yet if we want to find a way of justifying inductive inferences, we must first of all try to establish a *principle of induction.* A principle of induction would be a statement with the help of which we could put inductive inferences into a logically acceptable form. In the eyes of the upholders of inductive logic, a principle of induction is of supreme importance for scientific method: '. . . this principle', says Reichenbach, 'determines the truth of scientific theories. To eliminate it from science would mean nothing less than to deprive science of the power to decide the truth or falsity of its theories. Without it, clearly, science would no longer have the right to distinguish its theories from the fanciful and arbitrary creations of the poet's mind.'[1]

Now this principle of induction cannot be a purely logical truth like a tautology or an analytic statement. Indeed, if there were such a thing as a purely logical principle of induction, there would be no problem of induction; for in this case, all inductive inferences would have to be regarded as purely logical or tautological transformations, just like inferences in deductive logic. Thus the principle of induction must be a synthetic statement;

that is, a statement whose negation is not self-contradictory but logically possible. So the question arises why such a principle should be accepted at all, and how we can justify its acceptance on rational grounds.

Some who believe in inductive logic are anxious to point out, with Reichenbach, that 'the principle of induction is unreservedly accepted by the whole of science and that no man can seriously doubt this principle in everyday life either'.[2] Yet even supposing this were the case—for after all, 'the whole of science' might err—I should still contend that a principle of induction is superfluous, and that it must lead to logical inconsistencies.

That inconsistencies may easily arise in connection with the principle of induction should have been clear from the work of Hume; also, that they can be avoided, if at all, only with difficulty. For the principle of induction must be a universal statement in its turn. Thus if we try to regard its truth as known from experience, then the very same problems which occasioned its introduction will arise all over again. To justify it, we should have to employ inductive inferences; and to justify these we should have to assume an inductive principle of a higher order; and so on. Thus the attempt to base the principle of induction on experience breaks down, since it must lead to an infinite regress.

Kant tried to force his way out of this difficulty by taking the principle of induction (which he formulated as the 'principle of universal causation') to be '*a priori* valid'. But I do not think that his ingenious attempt to provide an *a priori* justification for synthetic statements was successful.

My own view is that the various difficulties of inductive logic here sketched are insurmountable. So also, I fear, are those inherent in the doctrine, so widely current today, that inductive inference, although not 'strictly valid', *can attain some degree of 'reliability' or of 'probability'*. According to this doctrine, inductive inferences are 'probable inferences'.[3] 'We have described', says Reichenbach, 'the principle of induction as the means whereby science decides upon truth. To be more exact, we should say that it serves to decide upon probability. For it is not given to science to reach either truth or falsity . . . but scientific statements can only attain continuous degrees of probability whose unattainable upper and lower limits are truth and falsity.'[4]

At this stage I can disregard the fact that the believers in inductive logic entertain an idea of probability that I shall later reject as highly unsuitable for their own purposes. . . . I can do so because the difficulties mentioned are not even touched by an appeal to probability. For if a certain degree of probability is to be assigned to statements based on inductive inference, then this will have to be justified by invoking a new principle of induction, appropriately modified. And this new principle in its turn will have to be justified, and so on. Nothing is gained, moreover, if the principle of induction, in its turn, is taken not as 'true' but only as 'probable'. In short, like every other form of inductive logic, the logic of probable inference, or 'probability logic', leads either to an infinite regress, or to the doctrine of *apriorism*. . . .

The theory to be developed in the following pages stands directly opposed to all attempts to operate with the ideas of inductive logic. It might be described as the theory of *the deductive method of testing*, or as the view that a hypothesis can only be empirically *tested*—and only *after* it has been advanced.

. . . .

3. Deductive Testing of Theories

According to the view that will be put forward here, the method of critically testing theories, and selecting them according to the results of tests, always proceeds on the following lines. From a new idea, put up tentatively, and not yet justified in any way—an anticipation, a hypothesis, a theoretical system, or what you will—conclusions are drawn by means of logical deduction. These conclusions are then compared with one another and with other relevant statements, so as to find what logical relations (such as equivalence, derivability, compatibility, or incompatibility) exist between them.

We may if we like distinguish four different lines along which the testing of a theory could be carried out. First there is the logical comparison of the conclusions among themselves, by which the internal consistency of the system is tested. Secondly, there is the investigation of the logical form of the theory, with the object of determining whether it has the character of an empirical or scientific theory, or whether it is, for example, tautological. Thirdly, there is the comparison with other theories, chiefly with the aim of determining whether the theory would constitute a scientific advance should it survive our various tests. And finally, there is the testing of the theory by way of empirical applications of the conclusions which can be derived from it.

The purpose of this last kind of test is to find out how far the new consequences of the theory—whatever may be new in what it asserts—stand up to the demands of practice, whether raised by purely scientific experiments, or by practical technological applications. Here too the procedure of testing turns out to be deductive. With the help of other statements, previously accepted, certain singular statements—which we may call 'predictions'—are deduced from the theory; especially predictions that are easily testable or applicable. From among these statements, those are selected which are not derivable from the current theory, and more especially those which the current theory contradicts. Next we seek a decision as regards these (and other) derived statements by comparing them with the results of practical applications and experiments. If this decision is positive, that is, if the singular conclusions turn out to be acceptable, or *verified,* then the theory has, for the time being, passed its test: we have found no reason to discard it. But if the decision is negative, or in other words, if the conclusions have been *falsified,* then their falsification also falsifies the theory from which they were logically deduced.

It should be noticed that a positive decision can only temporarily support the theory, for subsequent negative decisions may always overthrow it. 172 So long as a theory withstands detailed and severe tests and is not superseded by another theory in the course of scientific progress, we may say that it has 'proved its mettle' or that it is '*corroborated*'. . . .

Nothing resembling inductive logic appears in the procedure here outlined. I never assume that we can argue from the truth of singular statements to the truth of theories. I never assume that by force of 'verified' conclusions, theories can be established as 'true', or even as merely 'probable'.

. . . .

NOTES

1. H. Reichenbach, *Erkenntnis* 1, 1930, p. 186 (*cf.* also p. 64 f.).

2. Reichenbach, *ibid.*, p. 67.

3. *Cf.* J. M. Keynes, *A Treatise on Probability* (1921); O. Külpe, *Vorlesungen über Logic* (ed. by Selz, 1923); Reichenbach (who uses the term 'probability implications'), *Axiomatik der Wahrscheinlichkeitrechnung, Mathem. Zeitschr.* 34 (1932); and in many other places.

4. Reichenbach, *Erkenntnis* 1, 1930, p. 186.

Mill's Inductivism and Debate with Whewell

From

A System of Logic, Book 3 / John Stuart Mill

Chapter I. Preliminary Observations on Induction in General

§1. The portion of the present inquiry upon which we are now about to enter may be considered as the principal, both from its surpassing in intricacy all the other branches, and because it relates to a process which has been shown in the preceding Book to be that in which the investigation of nature essentially consists. We have found that all Inference, consequently all Proof, and all discovery of truths not self-evident, consists of inductions, and the interpretation of inductions; that all our knowledge, not intuitive, comes to us exclusively from that source. What Induction is, therefore, and what conditions render it legitimate, cannot but be deemed the main question of the science of logic—the question which includes all others. It is, however, one which professed writers on logic have almost entirely passed over. The generalities of the subject have not been altogether neglected by metaphysicians; but, for want of sufficient acquaintance with the processes by which science has actually succeeded in establishing general truths, their

---

Reprinted from *A System of Logic, Ratiocinative and Inductive,* 8th ed. (London: Longmans, 1872; new impression 1959), 185–200, 253–66, 299–305, 322–33.

analysis of the inductive operation, even when unexceptionable as to correctness, has not been specific enough to be made the foundation of practical rules, which might be for induction itself what the rules of the syllogism are for the interpretation of induction; while those by whom physical science has been carried to its present state of improvement—and who, to arrive at a complete theory of the process, needed only to generalise, and adapt to all varieties of problems, the methods which they themselves employed in their habitual pursuits—never until very lately made any serious attempt to philosophise on the subject, nor regarded the mode in which they arrived at their conclusions as deserving of study, independently of the conclusions themselves.

174

§2. For the purposes of the present inquiry, Induction may be defined, the operation of discovering and proving general propositions. It is true that (as already shown) the process of indirectly ascertaining individual facts is as truly inductive as that by which we establish general truths. But it is not a different kind of induction; it is a form of the very same process: since, on the one hand, generals are but collections of particulars, definite in kind but indefinite in number; and on the other hand, whenever the evidence which we derive from observation of known cases justifies us in drawing an inference respecting even one unknown case, we should on the same evidence be justified in drawing a similar inference with respect to a whole class of cases. The inference either does not hold at all, or it holds in all cases of a certain description; in all cases which, in certain definable respects, resemble those we have observed.

If these remarks are just; if the principles and rules of inference are the same whether we infer general propositions or individual facts; it follows that a complete logic of the sciences would be also a complete logic of practical business and common life. Since there is no case of legitimate inference from experience, in which the conclusion may not legitimately be a general proposition, an analysis of the process by which general truths are arrived at is virtually an analysis of all induction whatever. Whether we are inquiring into a scientific principle or into an individual fact, and whether we proceed by experiment or by ratiocination, every step in the train of inferences is essentially inductive, and the legitimacy of the induction depends in both cases on the same conditions.

True it is that in the case of the practical inquirer, who is endeavouring to ascertain facts not for the purposes of science but for those of business, such, for instance, as the advocate or the judge, the chief difficulty is one in which the principles of induction will afford him no assistance. It lies not in making his inductions, but in the selection of them; in choosing from among all general propositions ascertained to be true, those which furnish marks by which he may trace whether the given subject possesses or not the predicate in question. In arguing a doubtful question of fact before a jury, the general propositions or principles to which the advocate appeals are mostly, in themselves, sufficiently trite, and assented to as soon as stated: his skill lies in bringing his case under those propositions or principles; in calling to mind such of the known or received maxims of probability as admit of application to the case in hand, and selecting from among them those best adapted to his object. Success is here dependent on natural or acquired sagacity, aided by knowledge of the particular subject and of subjects allied with it. Invention, though it can be cultivated, cannot be reduced to rule; there is no science which will enable a man to bethink himself of that which will suit his purpose.

But when he *has* thought of something, science can tell him whether that which he has thought of will suit his purpose or not. The inquirer or arguer must be guided by his own knowledge and sagacity in the choice of the inductions out of which he will construct his argument. But the validity of the argument when constructed depends on principles and must be tried by tests which are the same for all descriptions of inquiries, whether the result be to give A an estate, or to enrich science with a new general truth. In the one case and in the other, the senses, or testimony, must decide on the individual facts; the rules of the syllogism will determine whether, those facts being supposed correct, the case really falls within the formulæ of the different inductions under which it has been successively brought; and finally, the legitimacy of the inductions themselves must be decided by other rules, and these it is now our purpose to investigate. If this third part of the operation be, in many of the questions of practical life, not the most, but the least arduous portion of it, we have seen that this is also the case in some great departments of the field of science; in all those which are principally deductive, and most of all in mathematics, where the inductions themselves are few in number, and so obvious and elementary that they seem to stand in no need of the evidence of experience, while to combine them so as to prove a given theorem or solve a problem may call for the utmost powers of invention and contrivance with which our species is gifted.

If the identity of the logical processes which prove particular facts and those which establish general scientific truths required any additional confirmation, it would be sufficient to consider that in many branches of science single facts have to be proved, as well as principles; facts as completely individual as any that are debated in a court of justice, but which are proved in the same manner as the other truths of the science, and without disturbing in any degree the homogeneity of its method. A remarkable example of this is afforded by astronomy. The individual facts on which that science grounds its most important deductions, such facts as the magnitudes of the bodies of the solar system, their distances from one another, the figure of the earth, and its rotation, are scarcely any of them accessible to our means of direct observation: they are proved indirectly by the aid of inductions founded on other facts which we can more easily reach. For example, the distance of the moon from the earth was determined by a very circuitous process. The share which direct observation had in the work consisted in ascertaining, at one and the same instant, the zenith distances of the moon,

as seen from two points very remote from one another on the earth's surface. The ascertainment of these angular distances ascertained their supplements; and since the angle at the earth's centre subtended by the distance between the two places of observation was deducible by spherical trigonometry from the latitude and longitude of those places, the angle at the moon subtended by the same line became the fourth angle of a quadrilateral of which the other three angles were known. The four angles being thus ascertained, and two sides of the quadrilateral being radii of the earth; the two remaining sides and the diagonal, or in other words, the moon's distance from the two places of observation, and from the centre of the earth, could be ascertained, at least in terms of the earth's radius, from elementary theorems of geometry. At each step in this demonstration a new induction is taken in, represented in the aggregate of its results by a general proposition.

Not only is the process by which an individual astronomical fact was thus ascertained exactly similar to those by which the same science establishes its general truths, but also (as we have shown to be the case in all legitimate reasoning) a general proposition might have been concluded instead of a single fact. In strictness, indeed, the result of the reasoning *is* a general proposition; a theorem respecting the distance, not of the moon in particular, but of any inaccessible object; showing in what relation that distance stands to certain other quantities. And although the moon is almost the only heavenly body the distance of which from the earth can really be thus ascertained, this is merely owing to the accidental circumstances of the other heavenly bodies, which render them incapable of affording such data as the application of the theorem requires; for the theorem itself is as true of them as it is of the moon.[1]

We shall fall into no error, then, if, in treating of Induction, we limit our attention to the establishment of general propositions. The principles and rules of Induction as directed to this end, are the principles and rules of all Induction; and the logic of Science is the universal Logic, applicable to all inquiries in which man can engage.

Chapter 2. Of Inductions Improperly So Called

§1. Induction, then, is that operation of the mind by which we infer that what we know to be true in a particular case or cases, will be true in all cases which resemble the former in certain assignable respects. In other words, Induction is the process by which we conclude that what is true of certain individuals of a class is true of the whole class, or that what is true at certain times will be true in similar circumstances at all times.

This definition excludes from the meaning of the term Induction, various logical operations, to which it is not unusual to apply that name.

Induction, as above defined, is a process of inference; it proceeds from the known to the unknown; and any operation involving no inference, any process in which what seems the conclusion is no wider than the premises from which it is drawn, does not fall within the meaning of the term. Yet in the common books of Logic we find this laid down as the most perfect, indeed the only quite perfect, form of induction. In those books, every process which sets out from a less general and terminates in a more general expression,—which admits of being stated in the form, "This and that A are B, therefore every A is B,"—is called an induction, whether anything be really concluded or not: and the induction is asserted not to be perfect, unless every single individual of the class A is included in the antecedent, or premise: that is, unless what we affirm of the class has already been ascertained to be true of every individual in it, so that the nominal conclusion is not really a conclusion, but a mere reassertion of the premises. If we were to say, All the planets shine by the sun's light, from observation of each separate planet, or all the Apostles were Jews, because this is true of Peter, Paul, John, and every other apostle,—these, and such as these, would, in the phraseology in question, be called perfect, and the only perfect, Inductions. This, however, is a totally different kind of induction from ours; it is not an inference from facts known to facts unknown, but a mere shorthand registration of facts known. The two simulated arguments which we have quoted are not generalisations; the propositions purporting to be conclusions from them are not really general propositions. A general proposition is one in which the predicate is affirmed or denied of an unlimited number of individuals; namely, all, whether few or many, existing or capable of existing, which possess the properties connoted by the subject of the proposition. "All men are mortal" does not mean all now living, but all men past, present, and to come. When the signification of the term is limited so as to render it a name not for any and every individual falling under a certain general description, but only for each of a number of individuals designated as such, and as it were counted off individually, the proposition, though it may be general in its language, is no general proposition, but merely that number of singular propositions, written in an abridged character. The operation may be very useful, as most forms of abridged notation are; but it is no part of the investigation of truth, though often bearing an important part in the preparation of the materials for that investigation.

As we may sum up a definite number of singular propositions in one proposition, which will be apparently, but not really, general, so we may sum up a definite number of general propositions in one proposition, which will be apparently, but not really, more general. If by a separate induction applied to every distinct species of animals, it has been established that each possesses a nervous system, and we affirm thereupon that all animals have a

178

nervous system; this looks like a generalisation, though as the conclusion merely affirms of all what has already been affirmed of each, it seems to tell us nothing but what we knew before. A distinction however must be made. If in concluding that all animals have a nervous system, we mean the same thing and no more as if we had said "all known animals," the proposition is not general, and the process by which it is arrived at is not induction. But if our meaning is that the observations made of the various species of animals have discovered to us a law of animal nature, and that we are in a condition to say that a nervous system will be found even in animals yet undiscovered, this indeed is an induction; but in this case the general proposition contains more than the sum of the special propositions from which it is inferred. The distinction is still more forcibly brought out when we consider, that if this real generalisation be legitimate at all, its legitimacy probably does not require that we should have examined without exception every known species. It is the number and nature of the instances, and not their being the whole of those which happen to be known, that makes them sufficient evidence to prove a general law: while the more limited assertion, which stops at all known animals, cannot be made unless we have rigorously verified it in every species. In like manner (to return to a former example) we might have inferred, not that all *the* planets, but that all *planets,* shine by reflected light: the former is no induction; the latter is an induction, and a bad one, being disproved by the case of double stars—self-luminous bodies which are properly planets, since they revolve round a centre.

§2. There are several processes used in mathematics which require to be distinguished from Induction, being not unfrequently called by that name, and being so far similar to Induction properly so called, that the propositions they lead to are really general propositions. For example, when we have proved with respect to the circle that a straight line cannot meet it in more than two points, and when the same thing has been successively proved of the ellipse, the parabola, and the hyperbola, it may be laid down as an universal property of the sections of the cone. The distinction drawn in the two previous examples can have no place here, there being no differences between all *known* sections of the cone and *all* sections, since a cone demonstrably cannot be intersected by a plane except in one of these four lines. It would be difficult, therefore, to refuse to the proposition arrived at the name of a generalisation, since there is no room for any generalisation beyond it. But there is no induction, because there is no inference: the conclusion is a mere summing up of what was asserted in the various propositions from which it is drawn. A case somewhat, though not altogether, similar, is the proof of a geometrical theorem by means of a diagram. Whether the diagram be on paper or only in the imagination, the demonstration . . . does not prove directly the general theorem; it proves only that the conclusion, which the theorem asserts generally, is true of the particular triangle or circle exhibited in the diagram; but since we perceive that in the same way in which we have proved it of that circle, it might also be proved of any other circle, we gather up into one general expression all the singular propositions susceptible of being thus proved, and embody them in an universal proposition. Having shown that the three angles of the triangle ABC are together equal to two right angles, we conclude that this is true of every other triangle, not because it is true of ABC, but for the same reason which proved it to be true of ABC. If this were to be called Induction, an appropriate name for it would be, induction by parity of reasoning. But the term cannot properly belong to it; the characteristic quality of Induction is wanting, since the truth obtained, though really general, is not believed on the evidence of particular instances. We do not conclude that all triangles have the property because some triangles have, but from the ulterior demonstrative evidence which was the ground of our conviction in the particular instances.

There are, nevertheless, in mathematics, some examples of so-called Induction, in which the conclusion does bear the appearance of a generalisation grounded on some of the particular cases included in it. A mathematician, when he has calculated a sufficient number of the terms of an algebraical or arithmetical series to have ascertained what is called the *law* of the series, does not hesitate to fill up any number of the succeeding terms without repeating the calculations. But I apprehend he only does so when it is apparent from *à priori* considerations (which might be exhibited in the form of demonstration) that the mode of formation of the subsequent terms, each from that which preceded it, must be similar to the formation of the terms which have been already calculated. And when the attempt has been hazarded without the sanction of such general considerations, there are instances on record in which it has led to false results.

It is said that Newton discovered the binomial theorem by induction; by raising a binomial successively to a certain number of powers, and comparing those powers with one another until he detected the relation in which the algebraic formula of each power stands to the exponent of that power, and to the two terms of the binomial. The fact is not improbable, but a mathematician like Newton, who seemed to arrive *per saltum* at principles and conclusions that ordinary mathematicians only reached by a succession of steps, certainly could not have performed the comparison in question without being led by it to the *à priori* ground of the law; since any one who understands sufficiently the nature of multiplication to venture upon multiplying several lines of symbols at one operation, cannot but perceive that in raising a binomial to a power, the co-efficients must depend on the laws of permutation and combination, and as soon as this is recognised, the theorem is demonstrated. Indeed, when once it was seen that the law prevailed in a few of the lower powers, its identity with the law of permutation would at once suggest the considerations which prove it to obtain universally. Even, therefore, such cases as these, are but examples of what I have called Induction by parity of reasoning, that is, not really Induction, because not involving inference of a general proposition from particular instances.

§3. There remains a third improper use of the term Induction, which it is of real importance to clear up, because the theory of Induction has been, in no ordinary degree, confused by it, and because the confusion is exemplified in the most recent and elaborate treatise on the inductive philosophy which exists in our language. The error in question is that of confounding a mere description, by general terms, of a set of observed phenomena, with an induction from them.

Suppose that a phenomenon consists of parts, and that these parts are only capable of being observed separately, and as it were piecemeal. When the observations have been made, there is a convenience (amounting for many purposes to a necessity) in obtaining a representation of the phenomenon as a whole, by combining, or, as we may say, piecing these detached fragments together. A navigator sailing in the midst of the ocean discovers land: he cannot at first, or by any one observation, determine whether it is a continent or an island; but he coasts along it, and after a few days finds himself to have sailed completely round it: he then pronounces it an island. Now there was no particular time or place of observation at which he could perceive that this land was entirely surrounded by water; he ascertained the fact by a succession of partial observations, and then selected a general expression which summed up in two or three words the whole of what he so observed. But is there anything of the nature of an induction in this process? Did he infer anything that had not been observed, from something else which had? Certainly not. He had observed the whole of what the proposition asserts. That the land in question is an island, is not an inference from the partial facts which the navigator saw in the course of his circumnavigation; it is the facts themselves; it is a summary of those facts; the description of a complex fact, to which those simpler ones are as the parts of a whole.

Now there is, I conceive, no difference in kind between this simple operation, and that by which Kepler ascertained the nature of the planetary orbits; and Kepler's operation, all at least that was characteristic in it, was not more an inductive act than that of our supposed navigator.

The object of Kepler was to determine the real path described by each of the planets, or let us say by the planet Mars (since it was of that body that he first established the two of his three laws which did not require a comparison of planets). To do this there was no other mode than that of direct observation; and all which observation could do was to ascertain a great number of the successive places of the planet, or rather, of its apparent places. That the planet occupied successively all these positions, or at all events, positions which produced the same impressions on the eye, and that it passed from one of these to another insensibly, and without any apparent breach of continuity; thus much the senses, with the aid of the proper instruments, could ascertain. What Kepler did more than this, was to find what sort of a curve these different points would make, supposing them to be all joined together. He expressed the whole series of the observed places of Mars by what Dr. Whewell calls the general conception of an ellipse. This operation was far from being as easy as that of the navigator who expressed the series of his observations on successive points of the coast by the general conception of an island. But it is the very same sort of operation; and if the one is not an induction but a description, this must also be true of the other.

The only real induction concerned in the case consisted in inferring that because the observed places of Mars were correctly represented by points in an imaginary ellipse, therefore Mars would continue to revolve in that same ellipse; and in concluding (before the gap had been filled up by further observations) that the positions of the planet during the time which intervened between two observations, must have coincided with the intermediate points of the curve. For these were facts which had not been directly observed. They were inferences from the observations; facts inferred, as distinguished from facts seen. But these inferences were so far from being a part of Kepler's philosophical operation, that they had been drawn long before he was born. Astronomers had long known that the planets periodically returned to the same places. When this had been ascertained, there was no induction left for Kepler to make, nor did he make any further induction. He merely applied his new conception to the facts inferred, as he did to the facts observed. Knowing already that the planets continued to move in the same paths; when he found that an ellipse correctly represented the past path he knew that it would represent the future path. In finding a compendious expression for the one set of facts, he found one for the other: but he found the expression only, not the inference; nor did he (which is the true test of a general truth) add anything to the power of prediction already possessed.

§4. The descriptive operation which enables a number of details to be summed up in a single proposition, Dr. Whewell, by an aptly chosen expression, has termed the Colligation of Facts. In most of his observations 182 concerning that mental process I fully agree, and would gladly transfer all that portion of his book into my own pages. I only think him mistaken in setting up this kind of operation, which, according to the old and received meaning of the term, is not induction at all, as the type of induction generally; and laying down, throughout his work, as principles of induction, the principles of mere colligation.

Dr. Whewell maintains that the general proposition which binds together the particular facts, and makes them, as it were, one fact, is not the mere sum of those facts, but something more, since there is introduced a conception of the mind, which did not exist in the facts themselves. "The particular facts," says he,[2] "are not merely brought together, but there is a new element added to the combination by the very act of thought by which they are combined. . . . When the Greeks, after long observing the motions of the planets, saw that these motions might be rightly considered as produced by the motion of one wheel revolving in the inside of another wheel, these wheels were creations of their minds, added to the facts which they perceived by sense. And even if the wheels were no longer supposed to be material, but were reduced to mere geometrical spheres or circles, they were not the less products of the mind alone,—something additional to the facts observed. The same is the case in all other discoveries. The facts are known, but they are insulated and unconnected, till the discoverer supplies from his own store a principle of connection. The pearls are there, but they will not hang together till some one provides the string."

Let me first remark that Dr. Whewell, in this passage, blends together, indiscriminately, examples of both the processes which I am endeavouring to distinguish from one another. When the Greeks abandoned the supposition that the planetary motions were produced by the revolutions of material wheels, and fell back upon the idea of "mere geometrical spheres or circles," there was more in this change of opinion than the mere substitution of an ideal curve for a physical one. There was the abandonment of a theory, and the replacement of it by a mere description. No one would think of calling the doctrine of material wheels a mere description. That doctrine was an attempt to point out the force by which the planets were acted upon, and compelled to move in their orbits. But when, by a great step in philosophy, the materiality of the wheels was discarded, and the geometrical forms alone retained, the attempt to account for the motions was given up, and what was left of the theory was a mere description of the orbits. The assertion that the planets were carried round by wheels revolving in the inside of other wheels, gave place to the proposition that they moved in the same lines which would be traced by bodies so carried: which was a mere mode of representing the sum of the observed facts; as Kepler's was another and a better mode of representing the same observations.

It is true that for these simply descriptive operations, as well as for the erroneous inductive one, a conception of the mind was required. The conception of an ellipse must have presented itself to Kepler's mind before he could identify the planetary orbits with it. According to Dr. Whewell, the conception was something added to the facts. He expresses himself as if Kepler had put something into the facts by his mode of conceiving them. But Kepler did no such thing. The ellipse was in the facts before Kepler recognised it; just as the island was an island before it had been sailed round. Kepler did not *put* what he had conceived into the facts, but *saw* it in them. A conception implies, and corresponds to, something conceived: and though the conception itself is not in the facts, but in our mind, yet if it is to convey any knowledge relating to them it must be a conception *of* something which really is in the facts, some property which they actually possess, and which they could manifest to our senses if our senses were able to take cognisance of it. If, for instance, the planet left behind it in space a visible track, and if the observer were in a fixed position at such a distance from the plane of the orbit as would enable him to see the whole of it at once, he would see it to be an ellipse; and if gifted with appropriate instruments and powers of locomotion, he would prove it to be such by measuring its different dimensions. Nay, further: if the track were visible, and he were so placed that he could see all parts of it in succession, but not all of them at once, he might be able, by piecing together his successive observations, to discover both that it was an ellipse and that the planet moved in it. The case would then exactly resemble that of the navigator who discovers the land to be an island by sailing round it. If the path was visible, no one I think would dispute that to identify it with an ellipse is to describe it: and I cannot see why any difference should be made by its not being directly an object of sense, when every point in it is as exactly ascertained as if it were so.

Subject to the indispensable condition which has just been stated, I do not conceive that the part which conceptions have in the operation of studying facts has ever been overlooked or undervalued. No one ever disputed that in order to reason about anything we must have a conception of it; or that when we include a multitude of things under a general expression, there is implied in the expression a conception of something common to those things. But it by no means follows that the conception is necessarily pre-existent, or constructed by the mind out of its own materials. If the facts are rightly classed under the conception, it is because there is in the facts themselves something of which the conception is itself a copy; and which if we cannot directly perceive, it is because of the limited power of our organs, and not because the thing itself is not there. The conception itself is often obtained by abstraction from the very facts which, in Dr. Whewell's language, it is afterwards called in to connect. This he himself admits, when he observes, (which he does on several occasions), how great a service would be rendered to the science of physiology by the philosopher "who should establish a precise, tenable, and consistent conception of life."[3] Such a conception can only be abstracted from the phenomena of life itself; from the very facts which it is put in requisition to connect. In other cases, no doubt, instead of collecting the conception from the very phenomena which we are attempting to colligate, we select it from among those which have been previously collected by abstraction from other facts. In the instance of Kepler's laws, the latter was the case. The facts being out of the reach of being observed in any such manner as would have enabled the senses to identify directly the path of the planet, the conception requisite for framing a general description of that path could not be collected by abstraction from the observations themselves; the mind had to supply hypothetically, from among the conceptions it had obtained from other portions of its experience, some one which would correctly represent the series of the observed facts. It had to frame a supposition respecting the general course of the phenomenon, and ask itself, If this be the general description, what will the details be? and then compare these with the details actually observed. If they agreed, the hypothesis would serve for a description of the phenomenon: if not, it was necessarily abandoned, and another tried. It is such a case as this which gives rise to the doctrine that the mind, in framing the descriptions, adds something of its own which it does not find in the facts.

Yet it is a fact surely that the planet does describe an ellipse; and a fact which we could see if we had adequate visual organs and a suitable position. Not having these advantages, but possessing the conception of an ellipse, or (to express the meaning in less technical language) knowing what an ellipse was, Kepler tried whether the observed places of the planet were consistent with such a path. He found they were so; and he, consequently, asserted as a fact that the planet moved in an ellipse. But this fact, which Kepler did not add to, but found in, the motions of the planet, namely, that it occupied in succession the various points in the circumference of a given ellipse, was the very fact the separate parts of which had been separately observed; it was the sum of the different observations.

Having stated this fundamental difference between my opinion and that of Dr. Whewell, I must add, that his account of the manner in which a conception is selected suitable to express the facts appears to me perfectly just. The experience of all thinkers will, I believe, testify that the process is tentative; that it consists of a succession of guesses; many being rejected, until one at last occurs fit to be chosen. We know from Kepler himself that before hitting upon the "conception" of an ellipse, he tried nineteen other imaginary paths, which, finding them inconsistent with the observations, he was obliged to reject. But, as Dr. Whewell truly says, the successful hypothesis, though a guess, ought generally to be called, not a lucky, but a skilful guess. The guesses which serve to give mental unity and wholeness to a chaos of scattered particulars are accidents which rarely occur to any minds but those abounding in knowledge and disciplined in intellectual combinations.

How far this tentative method, so indispensable as a means to the colligation of facts for purposes of description, admits of application to Induction itself, and what functions belong to it in that department, will be considered in the chapter of the present Book which relates to Hypotheses. On the present occasion we have chiefly to distinguish this process of Colligation from Induction properly so called; and that the distinction may be made clearer, it is well to advert to a curious and interesting remark, which is as strikingly true of the former operation, as it appears to me unequivocally false of the latter.

In different stages of the progress of knowledge, philosophers have employed, for the colligation of the same order of facts, different conceptions. The early rude observations of the heavenly bodies, in which minute precision was neither attained nor sought, presented nothing inconsistent with the representation of the path of a planet as an exact circle, having the earth for its centre. As observations increased in accuracy, facts were disclosed which were not reconcilable with this simple supposition: for the colligation of those additional facts, the supposition was varied; and varied again and again as facts became more numerous and precise. The earth was removed from the centre to some other point within the circle; the planet was supposed to revolve in a smaller circle called an epicycle, round an imaginary point which revolved in a circle round the earth: in proportion as observation elicited fresh facts contradictory to these representations, other epicycles and other excentrics were added, producing additional complications; until at last Kepler swept all these circles away, and substituted the conception of an exact ellipse. Even this is found not to represent with complete correctness the accurate observations of the present day, which disclose many slight deviations from an orbit exactly elliptical. Now Dr. Whewell has remarked that these successive general expressions, though apparently so conflicting, were all correct: they all answered the purpose of colligation; they all enabled the mind to represent to itself with facility, and by a simultaneous glance, the whole body of facts at the time ascertained: each in its turn served as a correct description of the phenomena, so far as the senses had up to that time taken cognisance of them. If a necessity afterwards arose for discarding one of these general descriptions of the planet's orbit, and framing a different imaginary line, by which to express the series of observed positions, it was because a number of new facts had now been added, which it was necessary to combine with the old facts into one general description. But this did not affect the correctness of the former expression, considered as a general statement of the only facts which it was intended to represent. And so true is this, that, as is well remarked by M. Comte, these ancient generalisations, even the rudest and most imperfect of them, that of uniform movement in a circle, are so far from being entirely false, that they are even now habitually employed by astronomers when only a rough approximation to correctness is required. . . .[4]

Dr. Whewell's remark, therefore, is philosophically correct. Successive expressions for the colligation of observed facts, or, in other words, successive descriptions of a phenomenon as a whole, which has been observed only in parts, may, though conflicting, be all correct as far as they go. But it would surely be absurd to assert this of conflicting inductions.

The scientific study of facts may be undertaken for three different purposes: the simple description of the facts; their explanation; or their prediction: meaning by prediction, the determination of the conditions under which similar facts may be expected again to occur. To the first of these three operations the name of Induction does not properly belong: to the other two it does. Now Dr. Whewell's observation is true of the first alone. Considered as a mere description, the circular theory of the heavenly motions represents perfectly well their general features: and by adding epicycles without limit, those motions, even as now known to us, might be expressed with any degree of accuracy that might be required. The elliptical theory, as a mere description, would have a great advantage in point of simplicity, and in the consequent facility of conceiving it and reasoning about it; but it would not really be more true than the other. Different descriptions, therefore, may be all true: but not, surely, different explanations. The doctrine that the heavenly bodies moved by a virtue inherent in their celestial nature; the doctrine that they were moved by impact, (which led to the hypothesis of vortices as the only impelling force capable of whirling bodies in circles,) and the Newtonian doctrine that they are moved by the composition of a centripetal with an original projectile force; all these are explanations collected by real induction from supposed parallel cases; and they were all successively received by philosophers, as scientific truths on the subject of the heavenly bodies. Can it be said of these, as was said of the different descriptions, that they are all true as far as they go? Is it not clear that only one can be true in any degree, and the other two must be altogether false? So much for explanations: let us now compare different predictions: the first, that eclipses will occur when one planet or satellite is so situated as to cast its shadow upon another; the second, that they will occur when some great calamity is impending over mankind. Do these two doctrines only differ in the degree of their truth as expressing real facts with unequal degrees of accuracy? Assuredly the one is true, and the other absolutely false.[5]

In every way, therefore, it is evident that to explain induction as the colligation of facts by means of appropriate conceptions, that is, conceptions which will really express them, is to confound mere descriptions of the observed facts with inference from those facts, and ascribe to the latter what is a characteristic property of the former.

There is, however, between Colligation and Induction a real correlation, which it is important to conceive correctly. Colligation is not always induction; but induction is always colligation. The assertion that the planets move in ellipses was but a mode of representing observed facts; it was but a colligation; while the assertion that they are drawn or tend towards the sun was the statement of a new fact, inferred by induction. But the induction, once made, accomplishes the purposes of colligation likewise. It brings the same facts, which Kepler had connected by his conception of an ellipse, under the additional conception of bodies acted upon by a central force, and serves therefore as a new bond of connection for those facts; a new principle for their classification.

Further, the descriptions which are improperly confounded with induction are nevertheless a necessary preparation for induction; no less necessary than correct observation of the facts themselves. Without the previous colligation of detached observations by means of one general conception, we could never have obtained any basis for an induction, except in the case of phenomena of very limited compass. We should not be able to affirm any predicates at all of a subject incapable of being observed otherwise than piecemeal: much less could we extend those predicates by induction to other similar subjects. Induction, therefore, always presupposes, not only that the necessary observations are made with the necessary accuracy, but also that the results of these observations are, so far as practicable, connected together by general descriptions, enabling the mind to represent to itself as wholes whatever phenomena are capable of being so represented.

§5. Dr. Whewell has replied at some length to the preceding observations, re-stating his opinions, but without (as far as I can perceive) adding anything material to his former arguments. Since, however, mine have not had the good fortune to make any impression upon him, I will subjoin a few remarks, tending to show more clearly in what our difference of opinion consists, as well as, in some measure, to account for it.

Nearly all the definitions of induction, by writers of authority, make it consist in drawing inferences from known cases to unknown; affirming of a class a predicate which has been found true of some cases belonging to the class; concluding, because some things have a certain property, that other things which resemble them have the same property—or because a thing has manifested a property at a certain time, that it has and will have that property at other times.

It can scarcely be contended that Kepler's operation was an Induction in this sense of the term. The statement that Mars moves in an elliptical orbit was no generalisation from individual cases to a class of cases. Neither was it an extension to all time of what had been found true at some particular time. The whole amount of generalisation which the case admitted of was already completed, or might have been so. Long before the elliptic theory was thought of, it had been ascertained that the planets returned periodically to the same apparent places; the series of these places was, or might have been, completely determined, and the apparent course of each planet marked out on the celestial globe in an uninterrupted line. Kepler did not extend an observed truth to other cases than those in which it had been observed: he did not widen the *subject* of the proposition which expressed the observed facts. The alteration he made was in the predicate. Instead of saying, the successive places of Mars are so and so, he summed them up in the statement, that the successive places of Mars are points in an ellipse. It is true this statement, as Dr. Whewell says, was not the sum of the observations *merely*; it was the sum of the observations *seen under a new point of view.*[6] But it was not the sum of *more* than the observations, as a real induction is. It took in no cases but those which had been actually observed, or which could have been inferred from the observations before the new point of view presented itself. There was not that transition from known cases to unknown which constitutes Induction in the original and acknowledged meaning of the term.

Old definitions, it is true, cannot prevail against new knowledge: and if the Keplerian operation, as a logical process, be really identical with what takes place in acknowledged induction, the definition of induction ought to be so widened as to take it in; since scientific language ought to adapt itself to the true relations which subsist between the things it is employed to designate. Here then it is that I am at issue with Dr. Whewell. He does think the operations identical. He allows of no logical process in any case of induction other than what there was in Kepler's case, namely, guessing until a guess is found which tallies with the facts; and accordingly, as we shall see hereafter, he rejects all canons of induction, because it is not by means of them that we guess. Dr. Whewell's theory of the logic of science would be very perfect if it did not pass over altogether the question of Proof. But in my apprehension there is such a thing as proof, and inductions differ altogether from descriptions in their relation to that element. Induction is proof; it is inferring something unobserved from something observed: it requires, therefore, an appropriate test of proof; and to provide that test is the special purpose of inductive logic. When, on the contrary, we merely collate known observations, and, in Dr. Whewell's phraseology, connect them by means of a new conception; if the conception does serve to connect the observations, we have all we want. As the proposition in which it is embodied pretends to no other truth than what it may share with many other modes of representing the same facts, to be consistent with the facts is all it requires: it neither needs nor admits of proof; though it may serve to prove other things, inasmuch as, by placing the facts in mental connection with other facts not previously seen to resemble them, it assimilates the case to another class of phenomena, concerning which real Inductions have already been made. Thus Kepler's so-called law brought the orbit of Mars into the class ellipse, and by doing so, proved all the properties of an ellipse to be true of the orbit: but in this proof Kepler's law supplied the minor premise, and not (as is the case with real Inductions) the major.

Dr. Whewell calls nothing Induction where there is not a new mental conception introduced, and everything induction where there is. But this is to confound two very different things, Invention and Proof. The introduction of a new conception belongs to Invention: and invention may be required in any operation, but is the essence of none. A new conception may be introduced for descriptive purposes, and so it may for inductive purposes. But it is so far from constituting induction, that induction does not necessarily stand in need of it. Most inductions require no conception but what was present in every one of the particular instances on which the induction is grounded. That all men are mortal is surely an inductive conclusion; yet no new conception is introduced by it. Whoever knows that any man has died, has all the conceptions involved in the inductive generalisation. But Dr. Whewell considers the process of invention, which consists in framing a new conception consistent with the facts, to be not merely a necessary part of all induction, but the whole of it.

The mental operation which extracts from a number of detached observations certain general characters in which the observed phenomena resemble one another, or resemble other known facts, is what Bacon, Locke, and most subsequent metaphysicians, have understood by the word Abstraction. A general expression obtained by abstraction, connecting known facts by means of common characters, but without concluding from them to unknown, may, I think, with strict logical correctness, be termed a Description; nor do I know in what other way things can ever be described. My position, however, does not depend on the employment of that particular word: I am quite content to use Dr. Whewell's term Colligation, or the more general phrases, "mode of representing, or of expressing, phenomena;" provided it be clearly seen that the process is not Induction, but something radically different.

What more may usefully be said on the subject of Colligation, or of the correlative expression invented by Dr. Whewell, the Explication of Conceptions, and generally on the subject of ideas and mental representations as connected with the study of facts, will find a more appropriate place in the Fourth Book, on the Operations Subsidiary to Induction: to which I must refer the reader for the removal of any difficulty which the present discussion may have left.

. . . .

Chapter 8. Of the Four Methods of Experimental Inquiry

§1. The simplest and most obvious modes of singling out from among the circumstances which precede or follow a phenomenon those with which it is really connected by an invariable law are two in number. One is, by comparing together different instances in which the phenomenon occurs. The other is, by comparing instances in which the phenomenon does occur, with instances in other respects similar in which it does not. These two methods may be respectively denominated the Method of Agreement and the Method of Difference.

In illustrating these methods, it will be necessary to bear in mind the twofold character of inquiries into the laws of phenomena, which may be either inquiries into the cause of a given effect, or into the effects or properties of a given cause. We shall consider the methods in their application to either order of investigation, and shall draw our examples equally from both.

We shall denote antecedents by the large letters of the alphabet, and the consequents corresponding to them by the small. Let A, then, be an agent or cause, and let the object of our inquiry be to ascertain what are the effects of this cause. If we can either find or produce the agent A in such varieties of circumstances that the different cases have no circumstance in common except A, then whatever effect we find to be produced in all our trials is indicated as the effect of A. Suppose, for example, that A is tried along with

B and C, and that the effect is *a b c;* and suppose that A is next tried with D and E, but without B and C, and that the effect is *a d e.* Then we may reason thus: *b* and *c* are not effects of A, for they were not produced by it in the second experiment; nor are *d* and *e,* for they were not produced in the first. Whatever is really the effect of A must have been produced in both instances; now this condition is fulfilled by no circumstance except *a.* The phenomenon *a* cannot have been the effect of B or C, since it was produced where they were not; nor of D or E, since it was produced where they were not. Therefore it is the effect of A.

For example, let the antecedent A be the contact of an alkaline substance and an oil. This combination being tried under several varieties of circumstances, resembling each other in nothing else, the results agree in the production of a greasy and detersive or saponaceous substance: it is therefore concluded that the combination of an oil and an alkali causes the production of a soap. It is thus we inquire, by the Method of Agreement, into the effect of a given cause.

In a similar manner we may inquire into the cause of a given effect. Let *a* be the effect. Here, as shown in the last chapter, we have only the resource of observation without experiment: we cannot take a phenomenon of which we know not the origin, and try to find its mode of production by producing it: if we succeeded in such a random trial it could only be by accident. But if we can observe *a* in two different combinations, *a b c* and *a d e;* and if we know, or can discover, that the antecedent circumstances in these cases respectively were A B C and A D E, we may conclude by a reasoning similar to that in the preceding example, that A is the antecedent connected with the consequent *a* by a law of causation. B and C, we may say, cannot be causes of *a,* since on its second occurrence they were not present; nor are D and E, for they were not present on its first occurrence. A, alone of the five circumstances, was found among the antecedents of *a* in both instances.

For example, let the effect *a* be crystallisation. We compare instances in which bodies are known to assume crystalline structure, but which have no other point of agreement; and we find them to have one, and, as far as we can observe, only one, antecedent in common: the deposition of a solid matter from a liquid state, either a state of fusion or of solution. We conclude, therefore, that the solidification of a substance from a liquid state is an invariable antecedent of its crystallisation.

In this example we may go farther, and say, it is not only the invariable antecedent, but the cause, or at least the proximate event which completes the cause. For in this case we are able, after detecting the antecedent A, to produce it artificially, and by finding that *a* follows it, verify the result of our induction. The importance of thus reversing the proof was strikingly manifested when by keeping a phial of water charged with siliceous particles undisturbed for years, a chemist (I believe Dr. Wollaston) succeeded in obtaining crystals of quartz; and in the equally interesting experiment in which Sir James Hall produced artificial marble by the cooling of its materials from fusion under immense pressure; two admirable examples of the light which may be thrown upon the most secret processes of Nature by well-contrived interrogation of her.

But if we cannot artificially produce the phenomenon A, the conclusion that it is the cause of *a* remains subject to very considerable doubt. Though an invariable, it may not be the unconditional antecedent of *a,* but may precede it as day precedes night or night day. This uncertainty arises from the impossibility of assuring ourselves that A is the *only* immediate antecedent common to both the instances. If we could be certain of having ascertained all the invariable antecedents, we might be sure that the unconditional invariable antecedent or cause must be found somewhere among them. Unfortunately it is hardly ever possible to ascertain all the antecedents, unless the phenomenon is one which we can produce artificially. Even then, the difficulty is merely lightened, not removed: men knew how to raise water in pumps long before they adverted to what was really the operating circumstance in the means they employed, namely, the pressure of the atmosphere on the open surface of the water. It is, however, much easier to analyse completely a set of arrangements made by ourselves, than the whole complex mass of the agencies which nature happens to be exerting at the moment of the production of a given phenomenon. We may overlook some of the material circumstances in an experiment with an electrical machine; but we shall, at the worst, be better acquainted with them than with those of a thunderstorm.

The mode of discovering and proving laws of nature, which we have now examined, proceeds on the following axiom. Whatever circumstances can be excluded, without prejudice to the phenomenon, or can be absent notwithstanding its presence, is not connected with it in the way of causation. The causal circumstance being thus eliminated, if only one remains, that one is the cause which we are in search of: if more than one, they either are, or contain among them, the cause; and so, *mutatis mutandis,* of the effect. As this method proceeds by comparing different instances to ascertain in what they agree, I have termed it the Method of Agreement; and we may adopt as its regulating principle the following canon:—

First Canon

If two or more instances of the phenomenon under investigation have only one circumstance in common, the circumstance in which alone all the instances agree is the cause (or effect) of the given phenomenon.

Quitting for the present the Method of Agreement, to which we shall almost immediately return, we proceed to a still more potent instrument of the investigation of nature, the Method of Difference.

§2. In the Method of Agreement, we endeavoured to obtain instances which agreed in the given circumstance but differed in every other: in the present method we require, on the contrary, two instances resembling one another in every other respect, but differing in the presence or absence of the phenomenon we wish to study. If our object be to discover the effects of an agent A, we must procure A in some set of ascertained circumstances, as A B C, and having noted the effects produced, compare them with the effect of the remaining circumstances B C, when A is absent. If the effect of A B C is *a b c*, and the effect of B C, *b c*, it is evident that the effect of A is *a*. So again, if we begin at the other end, and desire to investigate the cause of an effect *a*, we must select an instance, as *a b c*, in which the effect occurs, and in which the antecedents were A B C, and we must look out for another instance in which the remaining circumstances, *b c*, occur without *a*. If the antecedents, in that instance, are B C, we know that the cause of *a* must be A: either A alone, or A in conjunction with some of the other circumstances present.

It is scarcely necessary to give examples of a logical process to which we owe almost all the inductive conclusions we draw in early life. When a man is shot through the heart, it is by this method we know that it was the gunshot which killed him: for he was in the fulness of life immediately before, all circumstances being the same, except the wound.

The axioms implied in this method are evidently the following. Whatever antecedent cannot be excluded without preventing the phenomenon, is the cause, or a condition of that phenomenon: Whatever consequent can be excluded, with no other difference in the antecedents than the absence of a particular one, is the effect of that one. Instead of comparing different instances of a phenomenon, to discover in what they agree, this method compares an instance of its occurrence with an instance of its non-occurrence, to discover in what they differ. The canon which is the regulating principle of the Method of Difference may be expressed as follows:—

Second Canon

If an instance in which the phenomenon under investigation occurs, and an instance in which it does not occur, have every circumstance in common save one, that one occurring only in the former; the circumstance in which alone the two instances differ is the effect, or the cause, or an indispensable part of the cause, of the phenomenon.

§3. The two methods which we have now stated have many features of resemblance, but there are also many distinctions between them. Both are methods of *elimination.* This term (employed in the theory of equations to denote the process by which one after another of the elements of a question is excluded, and the solution made to depend on the relation between the remaining elements only) is well suited to express the operation, analogous to this, which has been understood since the time of Bacon to be the foundation of experimental inquiry, namely, the successive exclusion of the various circumstances which are found to accompany a phenomenon in a given instance, in order to ascertain what are those among them which can

194

be absent consistently with the existence of the phenomenon. The Method of Agreement stands on the ground that whatever can be eliminated is not connected with the phenomenon by any law. The Method of Difference has for its foundation, that whatever cannot be eliminated is connected with the phenomenon by a law.

Of these methods, that of Difference is more particularly a method of artificial experiment; while that of Agreement is more especially the resource employed where experimentation is impossible. A few reflections will prove the fact, and point out the reason of it.

It is inherent in the peculiar character of the Method of Difference that the nature of the combinations which it requires is much more strictly defined than in the Method of Agreement. The two instances which are to be compared with one another must be exactly similar in all circumstances except the one which we are attempting to investigate: they must be in the relation of A B C and B C, or of *a b c* and *b c.* It is true that this similarity of circumstances needs not extend to such as are already known to be immaterial to the result. And in the case of most phenomena we learn at once, from the commonest experience, that most of the co-existent phenomena of the universe may be either present or absent without affecting the given phenomenon; or, if present, are present indifferently when the phenomenon does not happen and when it does. Still, even limiting the identity which is required between the two instances, A B C and B C, to such circumstances as are not already known to be indifferent; it is very seldom that nature affords two instances, of which we can be assured that they stand in this precise relation to one another. In the spontaneous operations of nature there is generally such complication and such obscurity, they are mostly either on so overwhelmingly large or on so inaccessibly minute a scale, we are so ignorant of a great part of the facts which really take place, and even those of which we are not ignorant are so multitudinous, and therefore so seldom exactly alike in any two cases, that a spontaneous experiment, of the kind required by the Method of Difference, is commonly not to be found. When, on the contrary, we obtain a phenomenon by an artificial experiment, a pair of instances such as the method requires is obtained almost as a matter of course, provided the process does not last a long time. A certain state of surrounding circumstances existed before we commenced the experiment; this is B C. We then introduce A; say, for instance, by merely bringing an object from another part of the room, before there has been time for any change in the other elements. It is, in short, (as M. Comte observes,) the very nature of an experiment to introduce into the pre-existing state of circumstances a change perfectly definite. We choose a previous state of things with which we are well acquainted, so that no unforeseen alteration in that state is likely to pass unobserved; and into this we introduce, as rapidly as possible, the phenomenon which we wish to study; so that in general we are entitled to feel complete assurance that the pre-existing state, and the state which we have produced, differ in nothing except the presence or absence of that phenomenon. If a bird is taken from a cage, and instantly plunged into carbonic acid gas, the experimentalist may be fully assured (at all events after one or two repetitions) that no circumstance capable of causing suffocation had supervened in the interim, except the change from immersion in the atmosphere to immersion in carbonic acid gas. There is one doubt, indeed, which may remain in some cases of this description; the effect may have been produced not by the change, but by the means employed to produce the change. The possibility, however, of this last supposition generally admits of being conclusively tested by other experiments. It thus appears that in the study of the various kinds of phenomena which we can, by our voluntary agency, modify or control, we can in general satisfy the requisitions of the Method of Difference; but that by the spontaneous operations of nature those requisitions are seldom fulfilled.

The reverse of this is the case with the Method of Agreement. We do not here require instances of so special and determinate a kind. Any instances whatever, in which nature presents us with a phenomenon, may be examined for the purposes of this method; and if all such instances agree in anything, a conclusion of considerable value is already attained. We can seldom, indeed, be sure that the one point of agreement is the only one; but this ignorance does not, as in the Method of Difference, vitiate the conclusion; the certainty of the result, as far as it goes, is not affected. We have ascertained one invariable antecedent or consequent, however many other invariable antecedents or consequents may still remain unascertained. If A B C, A D E, A F G, are all equally followed by *a,* then *a* is an invariable consequent of A. If *a b c, a d e, a f g,* all number A among their antecedents, then A is connected as an antecedent, by some invariable law, with *a.* But to determine whether this invariable antecedent is a cause, or this invariable consequent an effect, we must be able, in addition, to produce the one by means of the other; or, at least, to obtain that which alone constitutes our assurance of having produced anything, namely, an instance in which the effect, *a,* has come into existence, with no other change in the pre-existing circumstances than the addition of A. And this, if we can do it, is an application of the Method of Difference, not of the Method of Agreement.

It thus appears to be by the Method of Difference alone that we can ever, in the way of direct experience, arrive with certainty at causes. The Method of Agreement leads only to laws of phenomena, (as some writers call them, but improperly, since laws of causation are also laws of phenomena,) that is, to uniformities, which either are not laws of causation, or in which the question of causation must for the present remain undecided. The Method of Agreement is chiefly to be resorted to as a means of suggesting applications of the Method of Difference, (as in the last example the comparison of A B C, A D E, A F G, suggested that A was the antecedent on which to try the experiment whether it could produce *a,*) or as an inferior resource in case the Method of Difference is impracticable; which, as we before showed, generally arises from the impossibility of artificially producing the phenomena. And hence it is that the Method of Agreement, though applicable in principle to either case, is more emphatically the method of investigation on those subjects where artificial experimentation is impossible; because on those it is generally our only resource of a directly inductive nature; while, in the phenomena which we can produce at pleasure, the Method of Difference generally affords a more efficacious process, which will ascertain causes as well as mere laws.

§4. There are, however, many cases in which, though our power of producing the phenomenon is complete, the Method of Difference either cannot be made available at all, or not without a previous employment of the Method of Agreement. This occurs when the agency by which we can produce the phenomenon is not that of one single antecedent, but a combination of antecedents, which we have no power of separating from each other and exhibiting apart. For instance, suppose the subject of inquiry to be the cause of the double refraction of light. We can produce this phenomenon at pleasure by employing any one of the many substances which are known to refract light in that peculiar manner. But if, taking one of those substances, as Iceland spar, for example, we wish to determine on which of the properties of Iceland spar this remarkable phenomenon depends, we can make no use for that purpose of the Method of Difference; for we cannot find another substance precisely resembling Iceland spar except in some one property. The only mode, therefore, of prosecuting this inquiry is that afforded by the Method of Agreement; by which, in fact, through a comparison of all the known substances which have the property of doubly refracting light, it was ascertained that they agree in the circumstance of being crystalline substances; and though the converse does not hold, though all crystalline substances have not the property of double refraction, it was concluded, with reason, that there is a real connection between these two properties; that either crystalline structure, or the cause which gives rise to that structure, is one of the conditions of double refraction.

Out of this employment of the Method of Agreement arises a peculiar modification of that method, which is sometimes of great avail in the investigation of nature. In cases similar to the above, in which it is not possible to obtain the precise pair of instances which our second canon requires—instances agreeing in every antecedent except A, or in every consequent except *a*—we may yet be able, by a double employment of the Method of Agreement, to discover in what the instances which contain A or *a* differ from those which do not.

If we compare various instances in which *a* occurs, and find that they all have in common the circumstance A, and (as far as can be observed) no other circumstance, the Method of Agreement, so far, bears testimony to a connection between A and *a.* In order to convert this evidence of connection into proof of causation by the direct Method of Difference, we ought to be able, in some one of these instances, as, for example, A B C, to leave out A, and observe whether by doing so *a* is prevented. Now supposing (what is often the case) that we are not able to try this decisive experiment, yet, provided we can by any means discover what would be its result if we could try it, the advantage will be the same. Suppose, then, that as we previously examined a variety of instances in which *a* occurred, and found them to agree in containing A, so we now observe a variety of instances in which *a* does not occur, and find them agree in not containing A; which establishes, by the Method of Agreement, the same connection between the absence of A and the absence of *a,* which was before established between their presence. As, then, it had been shown that whenever A is present *a* is present, so it being now shown that when A is taken away *a* is removed along with it, we have by the one proposition A B C, *a b c,* by the other B C, *b c,* the positive and negative instances which the Method of Difference requires.

This method may be called the Indirect Method of Difference, or the Joint Method of Agreement and Difference, and consists in a double employment of the Method of Agreement, each proof being independent of the other, and corroborating it. But it is not equivalent to a proof by the direct Method of Difference. For the requisitions of the Method of Difference are not satisfied unless we can be quite sure either that the instances affirmative of *a* agree in no antecedent whatever but A, or that the instances negative of *a* agree in nothing but the negation of A. Now if it were possible, which it never is, to have this assurance, we should not need the joint method; for either of the two sets of instances separately would then be sufficient to prove causation. This indirect method, therefore, can only be regarded as a great extension and improvement of the Method of Agreement, but not as participating in the more cogent nature of the Method of Difference. The following may be stated as its canon:—

Third Canon

198

If two or more instances in which the phenomenon occurs have only one circumstance in common, while two or more instances in which it does not occur have nothing in common save the absence of that circumstance, the circumstance in which alone the two sets of instances differ is the effect, or the cause, or an indispensable part of the cause, of the phenomenon.

We shall presently see that the Joint Method of Agreement and Difference constitutes, in another respect not yet adverted to, an improvement upon the common Method of Agreement, namely, in being unaffected by a characteristic imperfection of that method, the nature of which still remains to be pointed out. But as we cannot enter into this exposition without introducing a new element of complexity into this long and intricate discussion, I shall postpone it to a subsequent chapter, and shall at once proceed to a statement of two other methods, which will complete the enumeration of the means which mankind possess for exploring the laws of nature by specific observation and experience.

§5. The first of these has been aptly denominated the Method of Residues. Its principle is very simple. Subducting from any given phenomenon all the portions which, by virtue of preceding inductions, can be assigned to known causes, the remainder will be the effect of the antecedents which had been overlooked, or of which the effect was as yet an unknown quantity.

Suppose, as before, that we have the antecedents A B C, followed by the consequents *a b c,* and that by previous inductions (founded, we will suppose, on the Method of Difference) we have ascertained the causes of some of these effects, or the effects of some of these causes; and are thence apprised that the effect of A is *a,* and that the effect of B is *b.* Subtracting the sum of these effects from the total phenomenon, there remains *c,* which now, without any fresh experiments, we may know to be the effect of C. This Method of Residues is in truth a peculiar modification of the Method of Difference. If the instance A B C, *a b c,* could have been compared with a single instance A B, *a b,* we should have proved C to be the cause of *c,* by the common process of the Method of Difference. In the present case, however, instead of a single instance A B, we have had to study separately the causes A and B, and to infer from the effects which they produce separately what effect they must produce in the case A B C where they act together. Of the two instances, therefore, which the Method of Difference requires,—the one positive, the other negative,—the negative one, or that in which the given phenomenon is absent, is not the direct result of observation and experiment, but has been arrived at by deduction. As one of the forms of the Method of Difference, the Method of Residues partakes of its rigorous certainty, provided the previous inductions, those which gave the effects of A and B, were obtained by the same infallible method, and provided we are certain that C is the *only* antecedent to which the residual phenomenon *c* can be referred; the only agent of which we had not already calculated and subducted the effect. But as we can never be quite certain of this, the evidence derived from the Method of Residues is not complete unless we can obtain C artificially and try it separately, or unless its agency, when once suggested, can be accounted for, and proved deductively, from known laws.

Even with these reservations, the Method of Residues is one of the most important among our instruments of discovery. Of all the methods of investigating laws of nature, this is the most fertile in unexpected results: often informing us of sequences in which neither the cause nor the effect were sufficiently conspicuous to attract of themselves the attention of observers. The agent C may be an obscure circumstance, not likely to have been perceived unless sought for, nor likely to have been sought for until attention had been awakened by the insufficiency of the obvious causes to account for the whole of the effect. And *c* may be so disguised by its intermixture with *a* and *b*, that it would scarcely have presented itself spontaneously as a subject of separate study. Of these uses of the method we shall presently cite some remarkable examples. The canon of the Method of Residues is as follows:—

Fourth Canon

Subduct from any phenomenon such part as is known by previous inductions to be the effect of certain antecedents, and the residue of the phenomenon is the effect of the remaining antecedents.

§6. There remains a class of laws which it is impracticable to ascertain by any of the three methods which I have attempted to characterise, namely, the laws of those Permanent Causes, or indestructible natural agents, which it is impossible either to exclude or to isolate; which we can neither hinder from being present, nor contrive that they shall be present alone. It would appear at first sight that we could by no means separate the effects of these agents from the effects of those other phenomena with which they cannot be prevented from co-existing. In respect, indeed, to most of the permanent causes, no such difficulty exists; since, though we cannot eliminate them as co-existing facts, we can eliminate them as influencing agents, by simply trying our experiment in a local situation beyond the limits of their influence. The pendulum, for example, has its oscillations disturbed by the vicinity of a mountain: we remove the pendulum to a sufficient distance from the mountain, and the disturbance ceases: from these data we can determine by the Method of Difference the amount of effect due to the mountain; and beyond a certain distance everything goes on precisely as it would do if the mountain exercised no influence whatever, which, accordingly, we, with sufficient reason, conclude to be the fact.

The difficulty, therefore, in applying the methods already treated of to determine the effects of Permanent Causes, is confined to the cases in which it is impossible for us to get out of the local limits of their influence. The pendulum can be removed from the influence of the mountain, but it cannot be removed from the influence of the earth: we cannot take away the earth from the pendulum, nor the pendulum from the earth, to ascertain whether it would continue to vibrate if the action which the earth exerts upon it were withdrawn. On what evidence, then, do we ascribe its vibrations to the earth's influence? Not on any sanctioned by the Method of Difference; for one of the two instances, the negative instance, is wanting. Nor by the Method of Agreement; for though all pendulums agree in this, that during their oscillations the earth is always present, why may we not as well ascribe the phenomenon to the sun, which is equally a co-existent fact in all the experiments? It is evident that to establish even so simple a fact of causation as this, there was required some method over and above those which we have yet examined.

As another example, let us take the phenomenon Heat. Independently of all hypothesis as to the real nature of the agency so called, this fact is certain, that we are unable to exhaust any body of the whole of its heat. It is equally certain that no one ever perceived heat not emanating from a body. Being unable, then, to separate Body and Heat, we cannot effect such a variation of circumstances as the foregoing three methods require; we cannot ascertain, by those methods, what portion of the phenomena exhibited by any body is due to the heat contained in it. If we could observe a body with its heat, and the same body entirely divested of heat, the Method of Difference would show the effect due to the heat, apart from that due to the body. If we could observe heat under circumstances agreeing in nothing but heat, and therefore not characterised also by the presence of a body, we could ascertain the effects of heat, from an instance of heat with a body and an instance of heat without a body, by the Method of Agreement; or we could determine by the Method of Difference what effect was due to the body, when the remainder which was due to the heat would be given by the Method of Residues. But we can do none of these things; and without them the application of any of the three methods to the solution of this problem would be illusory. It would be idle, for instance, to attempt to ascertain the effect of heat by subtracting from the phenomena exhibited by a body all that is due to its other properties; for as we have never been able to observe any bodies without a portion of heat in them, effects due to that heat might form a part of the very results which we were affecting to subtract in order that the effect of heat might be shown by the residue.

If, therefore, there were no other methods of experimental investigation than these three, we should be unable to determine the effects due to heat as a cause. But we have still a resource. Though we cannot exclude an antecedent altogether, we may be able to produce, or nature may produce for us, some modification in it. By a modification is here meant a change in it, not amounting to its total removal. If some modification in the antecedent A is always followed by a change in the consequent *a,* the other consequents *b* and *c* remaining the same; or *vice versa,* if every change in *a* is found to have been preceded by some modification in A, none being observable in any of the other antecedents; we may safely conclude that *a* is, wholly or in part, an effect traceable to A, or at least in some way connected with it through causation. For example, in the case of heat, though we cannot expel it altogether from any body, we can modify it in quantity, we can increase or diminish it; and doing so, we find by the various methods of experimentation or observation already treated of, that such increase or diminution of heat is followed by expansion or contraction of the body. In this manner we arrive at the conclusion, otherwise unattainable by us, that one of the effects of heat is to enlarge the dimensions of bodies; or what is the same thing in other words, to widen the distances between their particles.

A change in a thing, not amounting to its total removal, that is, a change which leaves it still the same thing it was, must be a change either in its quantity, or in some of its variable relations to other things, of which variable relations the principal is its position in space. In the previous example, the modification which was produced in the antecedent was an alteration in its quantity. Let us now suppose the question to be, what influence the moon exerts on the surface of the earth. We cannot try an experiment in the absence of the moon, so as to observe what terrestrial phenomena her annihilation would put an end to; but when we find that all the variations in the *position* of the moon are followed by corresponding variations in the time and place of high water, the place being always either the part of the earth which is nearest to, or that which is most remote from, the moon, we have ample evidence that the moon is, wholly or partially, the cause which determines the tides. It very commonly happens, as it does in this instance, that the variations of an effect are correspondent, or analogous, to those of its cause; as the moon moves farther towards the east, the high-water point does the same: but this is not an indispensable condition, as may be seen in the same example; for along with that high-water point there is at the same instant another high-water point diametrically opposite to it, and which, therefore, of necessity, moves towards the west, as the moon, followed by the nearer of the tide-waves, advances towards the east: and yet both these motions are equally effects of the moon's motion.

That the oscillations of the pendulum are caused by the earth is proved 202 by similar evidence. Those oscillations take place between equidistant points on the two sides of a line, which, being perpendicular to the earth, varies with every variation in the earth's position, either in space or relatively to the object. Speaking accurately, we only know by the method now characterised that all terrestrial bodies tend to the earth, and not to some unknown fixed point lying in the same direction. In every twenty-four hours, by the earth's rotation, the line drawn from the body at right angles to the earth coincides successively with all the radii of a circle, and in the course of six months the place of that circle varies by nearly two hundred millions of miles; yet in all these changes of the earth's position, the line in which bodies tend to fall continues to be directed towards it: which proves that terrestrial gravity is directed to the earth, and not, as was once fancied by some, to a fixed point of space.

The method by which these results were obtained may be termed the Method of Concomitant Variations: it is regulated by the following canon:—

Fifth Canon

Whatever phenomenon varies in any manner whenever another phenomenon varies in some particular manner, is either a cause or an effect of that phenomenon, or is connected with it through some fact of causation.

The last clause is subjoined because it by no means follows, when two phenomena accompany each other in their variations, that the one is cause and the other effect. The same thing may, and indeed must happen, supposing them to be two different effects of a common cause: and by this method alone it would never be possible to ascertain which of the suppositions is the true one. The only way to solve the doubt would be that which we have so often adverted to, viz. by endeavouring to ascertain whether we can produce the one set of variations by means of the other. In the case of heat, for example, by increasing the temperature of a body we increase its bulk, but by increasing its bulk we do not increase its temperature; on the contrary, (as in the rarefaction of air under the receiver of an air-pump,) we generally diminish it: therefore heat is not an effect, but a cause, of increase of bulk. If we cannot ourselves produce the variations, we must endeavour, though it is an attempt which is seldom successful, to find them produced by nature in some case in which the pre-existing circumstances are perfectly known to us.

It is scarcely necessary to say, that in order to ascertain the uniform concomitants of variations in the effect with variations in the cause, the same precautions must be used as in any other case of the determination of an invariable sequence. We must endeavour to retain all the other antecedents unchanged, while that particular one is subjected to the requisite series of variations; or, in other words, that we may be warranted in inferring causation from concomitance of variations, the concomitance itself must be proved by the Method of Difference.

It might at first appear that the Method of Concomitant Variations assumes a new axiom, or law of causation in general, namely, that every modification of the cause is followed by a change in the effect. And it does usually happen that when a phenomenon A causes a phenomenon *a,* any variation in the quantity or in the various relations of A is uniformly followed by a variation in the quantity or relations of *a.* To take a familiar instance, that of gravitation. The sun causes a certain tendency to motion in the earth; here we have cause and effect; but that tendency is *towards* the sun, and therefore varies in direction as the sun varies in the relation of position; and moreover the tendency varies in intensity, in a certain numerical correspondence to the sun's distance from the earth, that is, according to another relation of the sun. Thus we see that there is not only an invariable connection between the sun and the earth's gravitation, but that two of the relations of the sun, its position with respect to the earth and its distance from the earth, are invariably connected as antecedents with the quantity and direction of the earth's gravitation. The cause of the earth's gravitating at all is simply the sun; but the cause of its gravitating with a given intensity and in a given direction is the existence of the sun in a given direction and at a given distance. It is not strange that a modified cause, which is in truth a different cause, should produce a different effect.

Although it is for the most part true that a modification of the cause is followed by a modification of the effect, the Method of Concomitant Variations does not, however, pre-suppose this as an axiom. It only requires the converse proposition, that anything on whose modifications, modifications of an effect are invariably consequent, must be the cause (or connected with the cause) of that effect; a proposition, the truth of which is evident; for if the thing itself had no influence on the effect, neither could the modifications of the thing have any influence. If the stars have no power over the fortunes of mankind, it is implied in the very terms that the conjunctions or oppositions of different stars can have no such power.

Although the most striking applications of the Method of Concomitant Variations take place in the cases in which the Method of Difference, strictly so called, is impossible, its use is not confined to those cases; it may often usefully follow after the Method of Difference, to give additional precision to a solution which that has found. When by the Method of Difference it has first been ascertained that a certain object produces a certain effect, the Method of Concomitant Variations may be usefully called in to determine according to what law the quantity or the different relations of the effect follow those of the cause.

§7. The case in which this method admits of the most extensive employment is that in which the variations of the cause are variations of quantity. Of such variations we may in general affirm with safety that they will be attended not only with variations, but with similar variations of the effect: the proposition, that more of the cause is followed by more of the effect, being a corollary from the principle of the Composition of Causes, which, as we have seen, is the general rule of causation; cases of the opposite description, in which causes change their properties on being conjoined with one another, being, on the contrary, special and exceptional. Suppose, then, that when A changes in quantity, *a* also changes in quantity, and in such a manner that we can trace the numerical relation which the changes of the one bear to such changes of the other as take place within our limits of observation. We may then, with certain precautions, safely conclude that the same numerical relation will hold beyond those limits. If, for instance, we find that when A is double, *a* is double; that when A is treble or quadruple, *a* is treble or quadruple; we may conclude that if A were a half or a third, *a* would be a half or a third; and finally, that if A were annihilated, *a* would be annihilated; and that *a* is wholly the effect of A, or wholly the effect of the same cause with A. And so with any other numerical relation according to which A and *a* would vanish simultaneously; as, for instance, if *a* were proportional to the square of A. If, on the other hand, *a* is not wholly the effect of A, but yet varies when A varies, it is probably a mathematical function not of A alone, but of A and something else; its changes, for example, may be such as would occur if part of it remained constant, or varied on some other principle, and the remainder varied in some numerical relation to the variations of A. In that case, when A diminishes, *a* will be seen to approach not towards zero, but towards some other limit; and when the series of variations is such as to indicate what that limit is, if constant, or the law of its variation if variable, the limit will exactly measure how much of *a* is the effect of some other and independent cause, and the remainder will be the effect of A (or of the cause of A).

These conclusions, however, must not be drawn without certain precautions. In the first place, the possibility of drawing them at all manifestly supposes that we are acquainted not only with the variations, but with the absolute quantities both of A and *a*. If we do not know the total quantities, we cannot, of course, determine the real numerical relation according to which those quantities vary. It is therefore an error to conclude, as some have concluded, that because increase of heat expands bodies, that is, increases the distance between their particles, therefore the distance is wholly the effect of heat, and that if we could entirely exhaust the body of its heat, the particles would be in complete contact. This is no more than a guess, and of the most hazardous sort, not a legitimate induction; for since we neither know how much heat there is in any body, nor what is the real distance between any two of its particles, we cannot judge whether the contraction of the distance does or does not follow the diminution of the quantity of heat according to such a numerical relation that the two quantities would vanish simultaneously.

In contrast with this, let us consider a case in which the absolute quantities are known—the case contemplated in the first law of motion, viz. that all bodies in motion continue to move in a straight line with uniform velocity until acted upon by some new force. This assertion is in open opposition to first appearances; all terrestrial objects, when in motion, gradually abate their velocity and at last stop; which accordingly the ancients, with their *inductio per enumerationem simplicem* [induction by simple enumeration], imagined to be the law. Every moving body, however, encounters various obstacles, as friction, the resistance of the atmosphere, &c., which we know by daily experience to be causes capable of destroying motion. It was suggested that the whole of the retardation might be owing to these causes. How was this inquired into? If the obstacles could have been entirely removed, the case would have been amenable to the Method of Difference. They could not be removed, they could only be diminished, and the case therefore admitted only of the Method of Concomitant Variations. This accordingly being employed, it was found that every diminution of the obstacles diminished the retardation of the motion; and inasmuch as in this case (unlike the case of heat) the total quantities both of the antecedent and of the consequent were known, it was practicable to estimate, with an approach to accuracy, both the amount of the retardation and the amount of the retarding causes or resistances, and to judge how near they both were to being exhausted; and it appeared that the effect dwindled as rapidly, and at each step was as far on the road towards annihilation, as the cause was. The simple oscillation of a weight suspended from a fixed point, and moved a little out of the perpendicular, which in ordinary circumstances lasts but a few minutes, was prolonged in Borda's experiments to more than thirty hours, by diminishing as much as possible the friction at the point of suspension, and by making the body oscillate in a space exhausted as nearly as possible of its air. There could therefore be no hesitation in assigning the whole of the retardation of motion to the influence of the obstacles; and since, after subducting this retardation from the total phenomenon, the remainder was an uniform velocity, the result was the proposition known as the first Law of Motion.

There is also another characteristic uncertainty affecting the inference that the law of variation, which the quantities observe within our limits of observation, will hold beyond those limits. There is, of course, in the first instance, the possibility that beyond the limits, and in circumstances therefore of which we have no direct experience, some counteracting cause might develop itself; either a new agent, or a new property of the agents concerned, which lies dormant in the circumstances we are able to observe. This is an element of uncertainty which enters largely into all our predictions of effects; but it is not peculiarly applicable to the Method of Concomitant Variations. The uncertainty, however, of which I am about to speak is characteristic of that method, especially in the cases in which the extreme limits of our observation are very narrow in comparison with the possible variations in the quantities of the phenomena. Any one who has the slightest acquaintance with mathematics is aware that very different laws of variation may produce numerical results which differ but slightly from one another within narrow limits; and it is often only when the absolute amounts of variation are considerable that the difference between the results given by one law and by another becomes appreciable. When, therefore, such variations in the quantity of the antecedents as we have the means of observing are small in comparison with the total quantities, there is much danger lest we should mistake the numerical law, and be led to miscalculate the variations which would take place beyond the limits; a miscalculation which would vitiate any conclusion respecting the dependence of the effect upon the cause, that could be founded on those variations. Examples are not wanting of such mistakes. "The formulæ," says Sir John Herschel,[7] "which have been empirically deduced for the elasticity of steam, (till very recently,) and those for the resistance of fluids, and other similar subjects," when relied on beyond the limits of the observations from which they were deduced, "have almost invariably failed to support the theoretical structures which have been erected on them."

In this uncertainty, the conclusion we may draw from the concomitant variations of *a* and A, to the existence of an invariable and exclusive connection between them, or to the permanency of the same numerical relation between their variations when the quantities are much greater or smaller than those which we have had the means of observing, cannot be considered to rest on a complete induction. All that in such a case can be regarded as proved on the subject of causation is, that there is some connection between the two phenomena; that A, or something which can influence A, must be *one* of the causes which collectively determine *a*. We may, however, feel assured that the relation which we have observed to exist between the variations of A and *a*, will hold true in all cases which fall between the same extreme limits; that is, wherever the utmost increase or diminution in which the result has been found by observation to coincide with the law, is not exceeded.

The four methods which it has now been attempted to describe are the only possible modes of experimental inquiry—of direct induction *à posteriori,* as distinguished from deduction: at least, I know not, nor am able to imagine, any others. And even of these, the Method of Residues, as we have seen, is not independent of deduction; though, as it also requires specific experience, it may, without impropriety, be included among methods of direct observation and experiment.

These, then, with such assistance as can be obtained from Deduction, compose the available resources of the human mind for ascertaining the laws of the succession of phenomena. Before proceeding to point out certain circumstances by which the employment of these methods is subjected to an immense increase of complication and of difficulty, it is expedient to illustrate the use of the methods by suitable examples drawn from actual physical investigations. These, accordingly, will form the subject of the succeeding chapter.

. . . .

Chapter 11. Of the Deductive Method

§1. The mode of investigation which, from the proved inapplicability of direct methods of observation and experiment, remains to us as the main source of the knowledge we possess or can acquire respecting the conditions and laws of recurrence of the more complex phenomena, is called, in its most general expression, the Deductive Method, and consists of three operations—the first, one of direct induction; the second, of ratiocination; the third, of verification.

I call the first step in the process an inductive operation, because there must be a direct induction as the basis of the whole, though in many particular investigations the place of the induction may be supplied by a prior deduction; but the premises of this prior deduction must have been derived from induction.

The problem of the Deductive Method is to find the law of an effect from the laws of the different tendencies of which it is the joint result. The first requisite, therefore, is to know the laws of those tendencies—the law of each of the concurrent causes; and this supposes a previous process of observation or experiment upon each cause separately, or else a previous deduction, which also must depend for its ultimate premises on observation or experiment. Thus, if the subject be social or historical phenomena, the

premises of the Deductive Method must be the laws of the causes which determine that class of phenomena; and those causes are human actions, together with the general outward circumstances under the influence of which mankind are placed, and which constitute man's position on the earth. The Deductive Method applied to social phenomena must begin, therefore, by investigating, or must suppose to have been already investigated, the laws of human action, and those properties of outward things by which the actions of human beings in society are determined. Some of these general truths will naturally be obtained by observation and experiment, others by deduction; the more complex laws of human action, for example, may be deduced from the simpler ones, but the simple or elementary laws will always and necessarily have been obtained by a directly inductive process.

To ascertain, then, the laws of each separate cause which takes a share in producing the effect is the first desideratum of the Deductive Method. To know what the causes are which must be subjected to this process of study may or may not be difficult. In the case last mentioned, this first condition is of easy fulfilment. That social phenomena depend on the acts and mental impressions of human beings never could have been a matter of any doubt, however imperfectly it may have been known either by what laws those impressions and actions are governed, or to what social consequences their laws naturally lead. Neither, again, after physical science had attained a certain development, could there be any real doubt where to look for the laws on which the phenomena of life depend, since they must be the mechanical and chemical laws of the solid and fluid substances composing the organised body and the medium in which it subsists, together with the peculiar vital laws of the different tissues constituting the organic structure. In other cases really far more simple than these, it was much less obvious in what quarter the causes were to be looked for, as in the case of the celestial phenomena. Until, by combining the laws of certain causes, it was found that those laws explained all the facts which experience had proved concerning the heavenly motions, and led to predictions which it always verified, mankind never knew that those *were* the causes. But whether we are able to put the question before or not until after we have become capable of answering it, in either case it must be answered; the laws of the different causes must be ascertained before we can proceed to deduce from them the conditions of the effect.

The mode of ascertaining those laws neither is nor can be any other than the fourfold method of experimental inquiry, already discussed. A few remarks on the application of that method to cases of the Composition of Causes are all that is requisite.

It is obvious that we cannot expect to find the law of a tendency by an induction from cases in which the tendency is counteracted. The laws of motion could never have been brought to light from the observation of bodies kept at rest by the equilibrium of opposing forces. Even where the tendency is not, in the ordinary sense of the word, counteracted, but only modified, by having its effects compounded with the effects arising from some other tendency or tendencies, we are still in an unfavourable position for tracing, by means of such cases, the law of the tendency itself. It would have been scarcely possible to discover the law that every body in motion tends to continue moving in a straight line, by an induction from instances in which the motion is deflected into a curve, by being compounded with the effect of an accelerating force. Notwithstanding the resources afforded in this description of cases by the Method of Concomitant Variations, the principles of a judicious experimentation prescribe that the law of each of the tendencies should be studied, if possible, in cases in which that tendency operates alone, or in combination with no agencies but those of which the effect can, from previous knowledge, be calculated and allowed for.

Accordingly, in the cases, unfortunately very numerous and important, in which the causes do not suffer themselves to be separated and observed apart, there is much difficulty in laying down with due certainty the inductive foundation necessary to support the deductive method. This difficulty is most of all conspicuous in the case of physiological phenomena: it being seldom possible to separate the different agencies which collectively compose an organised body, without destroying the very phenomena which it is our object to investigate:

> "Following life, in creatures we dissect,
> We lose it in the moment we detect."

And for this reason I am inclined to the opinion that physiology (greatly and rapidly progressive as it now is) is embarrassed by greater natural difficulties, and is probably susceptible of a less degree of ultimate perfection than even the social science, inasmuch as it is possible to study the laws and operations of one human mind apart from other minds much less imperfectly than we can study the laws of one organ or tissue of the human body apart from the other organs or tissues.

It has been judiciously remarked that pathological facts, or, to speak in common language, diseases in their different forms and degrees, afford in the case of physiological investigation the most valuable equivalent to experimentation properly so called, inasmuch as they often exhibit to us a definite disturbance in some one organ or organic function, the remaining organs and functions being, in the first instance at least, unaffected. It is true that from the perpetual actions and reactions which are going on among all parts of the organic economy there can be no prolonged disturbance in any one function without ultimately involving many of the others; and when once it has done so, the experiment for the most part loses its scientific value. All depends on observing the early stages of the derangement, which, unfortunately, are of necessity the least marked. If, however, the organs and functions not disturbed in the first instance, become affected in a fixed order of succession, some light is thereby thrown upon the action which one organ exercises over another, and we occasionally obtain a series of effects which we can refer with some confidence to the original local derangement; but for this it is necessary that we should know that the original derangement *was* local. If it was what is termed constitutional, that is, if we do not know in what part of the animal economy it took its rise, or the precise nature of the disturbance which took place in that part, we are unable to determine which of the various derangements was cause and which effect; which of them were produced by one another, and which by the direct, though perhaps tardy, action of the original cause.

Besides natural pathological facts, we can produce pathological facts artificially; we can try experiments, even in the popular sense of the term, by subjecting the living being to some external agent, such as the mercury of our former example, or the section of a nerve to ascertain the functions of different parts of the nervous system. As this experimentation is not intended to obtain a direct solution of any practical question, but to discover general laws, from which afterwards the conditions of any particular effect may be obtained by deduction, the best cases to select are those of which the circumstances can be best ascertained: and such are generally not those in which there is any practical object in view. The experiments are best tried, not in a state of disease, which is essentially a changeable state, but in the condition of health, comparatively a fixed state. In the one, unusual agencies are at work, the results of which we have no means of predicting; in the other, the course of the accustomed physiological phenomena would, it may generally be presumed, remain undisturbed, were it not for the disturbing cause which we introduce.

Such, with the occasional aid of the Method of Concomitant Variations, (the latter not less encumbered than the more elementary methods by the peculiar difficulties of the subject,) are our inductive resources for ascertaining the laws of the causes considered separately, when we have it not in our power to make trial of them in a state of actual separation. The insufficiency of these resources is so glaring, that no one can be surprised at the backward state of the science of physiology in which indeed our knowledge of causes is so imperfect, that we can neither explain, nor could without specific experience have predicted, many of the facts which are certified to us by the most ordinary observation. Fortunately, we are much better informed as to the empirical laws of the phenomena, that is, the uniformities respecting which we cannot yet decide whether they are cases of causation or mere results of it. Not only has the order in which the facts of organisation and life successively manifest themselves, from the first germ of existence to death, been found to be uniform, and very accurately ascertainable; but, by a great application of the Method of Concomitant Variations to the entire facts of comparative anatomy and physiology, the characteristic organic structure corresponding to each class of functions has been determined with considerable precision. Whether these organic conditions are the whole of the conditions, and in many cases whether they are conditions at all, or mere collateral effects of some common cause, we are quite ignorant; nor are we ever likely to know, unless we could construct an organised body, and try whether it would live.

Under such disadvantages do we, in cases of this description, attempt the initial or inductive step in the application of the Deductive Method to complex phenomena. But such, fortunately, is not the common case. In general, the laws of the causes on which the effect depends may be obtained by an induction from comparatively simple instances, or, at the worst, by deduction from the laws of simpler causes, so obtained. By simple instances are meant, of course, those in which the action of each cause was not intermixed or interfered with, or not to any great extent, by other causes whose laws were unknown; and only when the induction which furnished the premises to the Deductive Method rested on such instances has the application of such a method to the ascertainment of the laws of a complex effect been attended with brilliant results.

§2. When the laws of the causes have been ascertained, and the first stage of the great logical operation now under discussion satisfactorily accomplished, the second part follows; that of determining from the laws of the causes what effect any given combination of those causes will produce. This is a process of calculation, in the wider sense of the term, and very often involves processes of calculation in the narrowest sense. It is a ratiocination; and when our knowledge of the causes is so perfect as to extend to the exact numerical laws which they observe in producing their effects, the ratiocination may reckon among its premises the theorems of the science of number, in the whole immense extent of that science. Not only are the most advanced truths of mathematics often required to enable us to compute an effect the numerical law of which we already know, but, even by the aid of those most advanced truths, we can go but a little way. In so simple a case as the common problem of three bodies gravitating towards one another, with a force directly as their mass and inversely as the square of the distance, all the resources of the calculus have not hitherto sufficed to obtain any general solution but an approximate one. In a case a little more complex, but still one of the simplest which arise in practice, that of the motion of a projectile, the causes which affect the velocity and range (for example) of a cannon-ball may be all known and estimated; the force of the gunpowder, the angle of elevation, the density of the air, the strength and direction of the wind; but it is one of the most difficult of mathematical problems to combine all these, so as to determine the effect resulting from their collective action.

Besides the theorems of number, those of geometry also come in as premises, where the effects take place in space, and involve motion and extension, as in mechanics, optics, acoustics, astronomy. But when the complication increases, and the effects are under the influence of so many and such shifting causes as to give no room either for fixed numbers or for straight lines and regular curves, (as in the case of physiological, to say nothing of mental and social phenomena,) the laws of number and extension are applicable, if at all, only on that large scale on which precision of details becomes unimportant. Although these laws play a conspicuous part in the most striking examples of the investigation of nature by the Deductive Method, as, for example, in the Newtonian theory of the celestial motions, they are by no means an indispensable part of every such process. All that is essential in it is reasoning from a general law to a particular case, that is, determining by means of the particular circumstances of that case what result is required in that instance to fulfil the law. Thus in the Torricellian experiment, if the fact that air has weight had been previously known, it would have been easy, without any numerical data, to deduce from the general law of equilibrium that the mercury would stand in the tube at such a height that the column of mercury would exactly balance a column of the atmosphere of equal diameter; because, otherwise, equilibrium would not exist.

By such ratiocinations from the separate laws of the causes we may, to a certain extent, succeed in answering either of the following questions: Given a certain combination of causes, what effect will follow? and, What combination of causes, if it existed, would produce a given effect? In the one case, we determine the effect to be expected in any complex circumstances of which the different elements are known: in the other case we learn, according to what law—under what antecedent conditions—a given complex effect will occur.

§3. But (it may here be asked) are not the same arguments by which the methods of direct observation and experiment were set aside as illusory when applied to the laws of complex phenomena, applicable with equal force against the Method of Deduction? When in every single instance a multitude, often an unknown multitude, of agencies, are clashing and combining, what security have we that in our computation *à priori* we have taken all these into our reckoning? How many must we not generally be ignorant of? Among those which we know, how probable that some have been overlooked; and, even were all included, how vain the pretence of summing up the effects of many causes, unless we know accurately the numerical law of each,—a condition in most cases not to be fulfilled; and even when it is fulfilled, to make the calculation transcends, in any but very simple cases, the utmost power of mathematical science with all its most modern improvements.

These objections have real weight, and would be altogether unanswerable, if there were no test by which, when we employ the Deductive Method, we might judge whether an error of any of the above descriptions had been committed or not. Such a test, however, there is; and its application forms, under the name of Verification, the third essential component part of the Deductive Method, without which all the results it can give have little other value than that of conjecture. To warrant reliance on the general conclusions arrived at by deduction, these conclusions must be found, on careful comparison, to accord with the results of direct observation wherever it can be had. If, when we have experience to compare with them, this experience confirms them, we may safely trust to them in other cases of which our specific experience is yet to come. But if our deductions have led to the conclusion that from a particular combination of causes a given effect would result, then in all known cases where that combination can be shown to have existed, and where the effect has not followed, we must be able to show (or at least to make a probable surmise) what frustrated it: if we cannot, the theory is imperfect, and not yet to be relied upon. Nor is the verification complete, unless some of the cases in which the theory is borne out by the observed result, are of at least equal complexity with any other cases in which its application could be called for.

If direct observation and collation of instances have furnished us with any empirical laws of the effect, (whether true in all observed cases, or only true for the most part,) the most effectual verification of which the theory could be susceptible would be, that it led deductively to those empirical laws; that the uniformities, whether complete or incomplete, which were observed to exist among the phenomena were accounted for by the laws of the causes—were such as could not but exist if those be really the causes by which the phenomena are produced. Thus it was very reasonably deemed an essential requisite of any true theory of the causes of the celestial motions, that it should lead by deduction to Kepler's laws; which, accordingly, the Newtonian theory did.

In order, therefore, to facilitate the verification of theories obtained by deduction, it is important that as many as possible of the empirical laws of the phenomena should be ascertained by a comparison of instances, conformably to the Method of Agreement, as well as (it must be added) that the phenomena themselves should be described, in the most comprehensive as well as accurate manner possible, by collecting from the observation of parts the simplest possible correct expressions for the corresponding wholes: as when the series of the observed places of a planet was first expressed by a circle, then by a system of epicycles, and subsequently by an ellipse.

It is worth remarking, that complex instances which would have been of no use for the discovery of the simple laws into which we ultimately analyse their phenomena, nevertheless, when they have served to verify the analysis, become additional evidence of the laws themselves. Although we could not have got at the law from complex cases, still when the law, got at otherwise, is found to be in accordance with the result of a complex case, that case becomes a new experiment on the law, and helps to confirm what it did not assist to discover. It is a new trial of the principle in a different set of circumstances; and occasionally serves to eliminate some circumstance not previously excluded, and the exclusion of which might require an experiment impossible to be executed. This was strikingly conspicuous in the example formerly quoted, in which the difference between the observed and the calculated velocity of sound was ascertained to result from the heat extricated by the condensation which takes place in each sonorous vibration. This was a trial, in new circumstances, of the law of the development of heat by compression; and it added materially to the proof of the universality of that law. Accordingly any law of nature is deemed to have gained in point of certainty by being found to explain some complex case which had not previously been thought of in connection with it; and this indeed is a consideration to which it is the habit of scientific inquirers to attach rather too much value than too little.

To the Deductive Method, thus characterised in its three constituent parts, Induction, Ratiocination, and Verification, the human mind is indebted for its most conspicuous triumphs in the investigation of nature. To it we owe all the theories by which vast and complicated phenomena are embraced under a few simple laws, which, considered as the laws of those great phenomena, could never have been detected by their direct study. We may form some conception of what the method has done for us from the case of the celestial motions, one of the simplest among the greater instances of the Composition of Causes, since (except in a few cases not of primary importance) each of the heavenly bodies may be considered, without material inaccuracy, to be never at one time influenced by the attraction of more than two bodies, the sun and one other planet or satellite; making with the reaction of the body itself, and the force generated by the body's own motion and acting in the direction of the tangent, only four different agents on the concurrence of which the motions of that body depend; a much smaller number, no doubt, than that by which any other of the great phenomena of nature is determined or modified. Yet how could we ever have ascertained the combination of forces on which the motions of the earth and planets are dependent by merely comparing the orbits or velocities of different planets, or the different velocities or positions of the same planet? Notwithstanding the regularity which manifests itself in those motions, in a degree so rare among the effects of concurrence of causes; and although the periodical recurrence of exactly the same effect affords positive proof that all the combinations of causes which occur at all, recur periodically; we should not have known what the causes were, if the existence of agencies precisely similar on our own earth had not, fortunately, brought the causes themselves within the reach of experimentation under simple circumstances. As we shall have occasion to analyse, farther on, this great example of the Method of Deduction, we shall not occupy any time with it here, but shall proceed to that secondary application of the Deductive Method the result of which is not to prove laws of phenomena, but to explain them.

. . . .

Chapter 14. Of the Limits to the Explanation of Laws of Nature, and of Hypotheses

. . . .

§4. An hypothesis is any supposition which we make (either without actual evidence, or on evidence avowedly insufficient) in order to endeavour to deduce from it conclusions in accordance with facts which are known to be real; under the idea that if the conclusions to which the hypothesis leads are known truths, the hypothesis itself either must be, or at least is likely to be, true. If the hypothesis relates to the cause or mode of production of a phenomenon, it will serve, if admitted, to explain such facts as are found capable of being deduced from it. And this explanation is the purpose of many, if not most, hypotheses. Since explaining, in the scientific sense, means resolving an uniformity which is not a law of causation into the laws of causation from which it results, or a complex law of causation into simpler and more general ones from which it is capable of being deductively inferred; if there do not exist any known laws which fulfil this requirement, we may feign or imagine some which would fulfil it; and this is making an hypothesis.

An hypothesis being a mere supposition, there are no other limits to hypotheses than those of the human imagination; we may, if we please, imagine, by way of accounting for an effect, some cause of a kind utterly unknown, and acting according to a law altogether fictitious. But as hypotheses of this sort would not have any of the plausibility belonging to those which ally themselves by analogy with known laws of nature, and besides would not supply the want which arbitrary hypotheses are generally invented to satisfy, by enabling the imagination to represent to itself an obscure phenomenon in a familiar light, there is probably no hypothesis in the history of science in which both the agent itself and the law of its operation were fictitious. Either the phenomenon assigned as the cause is real, but the law according to which it acts merely supposed, or the cause is fictitious, but is supposed to produce its effects according to laws similar to those of some known class of phenomena. An instance of the first kind is afforded by the different suppositions made respecting the law of the planetary central force anterior to the discovery of the true law, that the force varies as the inverse square of the distance; which also suggested itself to Newton, in the first instance, as an hypothesis, and was verified by proving that it led deductively to Kepler's laws. Hypotheses of the second kind are such as the vortices of Descartes, which were fictitious, but were supposed to obey the known laws of rotatory motion; or the two rival hypotheses respecting the nature of light, the one ascribing the phenomena to a fluid emitted from all luminous bodies, the other (now generally received) attributing them to vibratory motions among the particles of an ether pervading all space. Of the existence of either fluid there is no evidence, save the explanation they are calculated to afford of some of the phenomena; but they are supposed to produce their effects according to known laws; the ordinary laws of continued locomotion in the one case, and in the other, those of the propagation of undulatory movements among the particles of an elastic fluid.

According to the foregoing remarks, hypotheses are invented to enable the Deductive Method to be earlier applied to phenomena. But[8] in order to discover the cause of any phenomenon by the Deductive Method, the process must consist of three parts—induction, ratiocination, and verification. Induction, (the place of which, however, may be supplied by a prior deduction,) to ascertain the laws of the causes; ratiocination, to compute from those laws how the causes will operate in the particular combination known to exist in the case in hand; verification, by comparing this calculated effect with the actual phenomenon. No one of these three parts of the process can be dispensed with. In the deduction which proves the identity of gravity with the central force of the solar system, all the three are found. First, it is proved from the moon's motions that the earth attracts her with a force varying as the inverse square of the distance. This (though partly dependent on prior deductions) corresponds to the first or purely inductive step, the ascertainment of the law of the cause. Secondly, from this law, and from the knowledge previously obtained of the moon's mean distance from the earth, and of the actual amount of her deflection from the tangent, it is ascertained with what rapidity the earth's attraction would cause the moon to fall, if she were no farther off and no more acted upon by extraneous forces than terrestrial bodies are; that is the second step, the ratiocination. Finally, this calculated velocity being compared with the observed velocity with which all heavy bodies fall, by mere gravity, towards the surface of the earth (sixteen feet in the first second, forty-eight in the second, and so forth, in the ratio of the odd numbers, 1, 3, 5, &c.), the two quantities are found to agree. The order in which the steps are here presented was not that of their discovery; but it is their correct logical order, as portions of the proof that the same attraction of the earth which causes the moon's motion causes also the fall of heavy bodies to the earth, a proof which is thus complete in all its parts.

Now, the Hypothetical Method suppresses the first of the three steps, the induction to ascertain the law, and contents itself with the other two operations, ratiocination and verification, the law which is reasoned from being assumed instead of proved.

This process may evidently be legitimate on one supposition, namely, if the nature of the case be such that the final step, the verification, shall amount to and fulfil the conditions of a complete induction. We want to be assured that the law we have hypothetically assumed is a true one; and its leading deductively to true results will afford this assurance, provided the case be such that a false law cannot lead to a true result—provided no law except the very one which we have assumed can lead deductively to the same conclusions which that leads to. And this proviso is often realised. For example, in the very complete specimen of deduction which we just cited, the original major premise of the ratiocination, the law of the attractive force, was ascertained in this mode, by this legitimate employment of the Hypothetical Method. Newton began by an assumption that the force which at each instant deflects a planet from its rectilineal course, and makes it describe a curve round the sun, is a force tending directly towards the sun. He then proved that if this be so the planet will describe, as we know by Kepler's first law that it does describe, equal areas in equal times; and, lastly, he proved that if the force acted in any other direction whatever, the planet would not describe equal areas in equal times. It being thus shown that no other hypothesis would accord with the facts, the assumption was proved; the hypothesis became an inductive truth. Not only did Newton ascertain by this hypothetical process the direction of the deflecting force, he proceeded in exactly the same manner to ascertain the law of variation of the quantity of that force. He assumed that the force varied inversely as the square of the distance, showed that from this assumption the remaining two of Kepler's laws might be deduced, and, finally, that any other law of variation would give results inconsistent with those laws, and inconsistent, therefore, with the real motions of the planets, of which Kepler's laws were known to be a correct expression.

I have said that in this case the verification fulfils the conditions of an induction; but an induction of what sort? On examination we find that it conforms to the canon of the Method of Difference. It affords the two instances, A B C, *a b c,* and B C, *b c.* A represents central force; A B C, the planets *plus* a central force; B C, the planets apart from a central force. The planets with a central force give *a,* areas proportional to the times; the planets without a central force give *b c* (a set of motions) without *a,* or with something else instead of *a.* This is the Method of Difference in all its strictness. It is true, the two instances which the method requires are obtained in this case, not by experiment, but by a prior deduction. But that is of no consequence. It is immaterial what is the nature of the evidence from which we derive the assurance that A B C will produce *a b c,* and B C only *b c;* it is enough that we have that assurance. In the present case, a process of reasoning furnished Newton with the very instances which, if the nature of the case had admitted of it, he would have sought by experiment.

It is thus perfectly possible, and indeed is a very common occurrence, that what was an hypothesis at the beginning of the inquiry, becomes a proved law of nature before its close. But in order that this should happen, we must be able, either by deduction or experiment, to obtain *both* the instances which the Method of Difference requires. That we are able from the hypothesis to deduce the known facts, gives only the affirmative instance, A B C, *a b c.* It is equally necessary that we should be able to obtain, as Newton did, the negative instance B C, *b c,* by showing that no antecedent, except the one assumed in the hypothesis, would in conjunction with B C produce *a*.

Now it appears to me that this assurance cannot be obtained when the cause assumed in the hypothesis is an unknown cause, imagined solely to account for *a*. When we are only seeking to determine the precise law of a cause already ascertained, or to distinguish the particular agent which is in fact the cause, among several agents of the same kind, one or other of which it is already known to be, we may then obtain the negative instance. An inquiry which of the bodies of the solar system causes by its attraction some particular irregularity in the orbit or periodic time of some satellite or comet, would be a case of the second description. Newton's was a case of the first. If it had not been previously known that the planets were hindered from moving in straight lines by some force tending towards the interior of their orbit, though the exact direction was doubtful; or if it had not been known that the force increased in some proportion or other as the distance diminished, and diminished as it increased, Newton's argument would not have proved his conclusion. These facts, however, being already certain, the range of admissible suppositions was limited to the various possible directions of a line, and the various possible numerical relations between the variations of the distance, and the variations of the attractive force: now among these it was easily shown that different suppositions could not lead to identical consequences.

Accordingly, Newton could not have performed his second great scientific operation, that of identifying terrestrial gravity with the central force of the solar system, by the same hypothetical method. When the law of the moon's attraction had been proved from the data of the moon itself, then on finding the same law to accord with the phenomena of terrestrial gravity, he was warranted in adopting it as the law of those phenomena likewise; but it would not have been allowable for him, without any lunar data, to assume that the moon was attracted towards the earth with a force as the inverse square of the distance, merely because that ratio would enable him to account for terrestrial gravity: for it would have been impossible for him to prove that the observed law of the fall of heavy bodies to the earth could not result from any force, save one extending to the moon, and proportional to the inverse square.

It appears, then, to be a condition of the most genuinely scientific hypothesis, that it be not destined always to remain an hypothesis, but be of such a nature as to be either proved or disproved by comparison with observed facts. This condition is fulfilled when the effect is already known to depend on the very cause supposed, and the hypothesis relates only to the precise mode of dependence; the law of the variation of the effect according to the variations in the quantity or in the relations of the cause.

With these may be classed the hypotheses which do not make any supposition with regard to causation, but only with regard to the law of correspondence between facts which accompany each other in their variations, though there may be no relation of cause and effect between them. Such were the different false hypotheses which Kepler made respecting the law of the refraction of light. It was known that the direction of the line of refraction varied with every variation in the direction of the line of incidence, but it was not known how; that is, what changes of the one corresponded to the different changes of the other. In this case any law, different from the true one, must have led to false results. And, lastly, we must add to these all hypothetical modes of merely representing, or *describing*, phenomena; such as the hypothesis of the ancient astronomers that the heavenly bodies moved in circles; the various hypotheses of excentrics, deferents, and epicycles, which were added to that original hypothesis; the nineteen false hypotheses which Kepler made and abandoned respecting the form of the planetary orbits; and even the doctrine in which he finally rested, that those orbits are ellipses, which was but an hypothesis like the rest until verified by facts.

In all these cases, verification is proof; if the supposition accords with the phenomena, there needs no other evidence of it. But in order that this may be the case, I conceive it to be necessary, when the hypothesis relates to causation, that the supposed cause should not only be a real phenomenon, something actually existing in nature, but should be already known to exercise, or at least to be capable of exercising, an influence of some sort over the effect. In any other case, it is no sufficient evidence of the truth of the hypothesis that we are able to deduce the real phenomena from it.

Is it, then, never allowable, in a scientific hypothesis, to assume a cause; but only to ascribe an assumed law to a known cause? I do not assert this. I only say, that in the latter case alone can the hypothesis be received as true merely because it explains the phenomena. In the former case it may be very useful by suggesting a line of investigation which may possibly terminate in obtaining real proof. But, for this purpose, as is justly remarked by M. Comte, it is indispensable that the cause suggested by the hypothesis should be in its own nature susceptible of being proved by other evidence. This seems to be the philosophical import of Newton's maxim, (so often cited with approbation by subsequent writers,) that the cause assigned for any phenomenon must not only be such as, if admitted, would explain the phenomenon, but must also be a *vera causa.* What he meant by a *vera causa* Newton did not indeed very explicitly define; and Dr. Whewell, who dissents from the propriety of any such restriction upon the latitude of framing hypotheses, has had little difficulty in showing[9] that his conception of it was neither precise nor consistent with itself: accordingly his optical theory was a signal instance of the violation of his own rule. It is certainly not necessary that the cause assigned should be a cause already known; otherwise we should sacrifice our best opportunities of becoming acquainted with new causes. But what is true in the maxim is, that the cause, though not known previously, should be capable of being known thereafter; that its existence should be capable of being detected, and its connection with the effect ascribed to it should be susceptible of being proved, by independent evidence. The hypothesis, by suggesting observations and experiments, puts us on the road to that independent evidence if it be really attainable; and till it be attained, the hypothesis ought only to count for a more or less plausible conjecture.

§5. This function, however, of hypotheses, is one which must be reckoned absolutely indispensable in science. When Newton said, "Hypotheses non fingo," he did not mean that he deprived himself of the facilities of investigation afforded by assuming in the first instance what he hoped ultimately to be able to prove. Without such assumptions, science could never have attained its present state: they are necessary steps in the progress to something more certain; and nearly everything which is now theory was once hypothesis. Even in purely experimental science, some inducement is necessary for trying one experiment rather than another; and though it is abstractly possible that all the experiments which have been tried might have been produced by the mere desire to ascertain what would happen in certain circumstances, without any previous conjecture as to the result; yet, in point of fact, those unobvious, delicate, and often cumbrous and tedious processes of experiment, which have thrown most light upon the general constitution of nature, would hardly ever have been undertaken by the persons or at the time they were, unless it had seemed to depend on them whether some general doctrine or theory which had been suggested, but not yet proved, should be admitted or not. If this be true even of merely experimental inquiry, the conversion of experimental into deductive truths could still less have been effected without large temporary assistance from hypotheses. The process of tracing regularity in any complicated, and at first sight confused set of appearances, is necessarily tentative: we begin by making any supposition, even a false one, to see what consequences will follow from it; and by observing how these differ from the real phenomena, we learn what corrections to make in our assumption. The simplest supposition which accords with the more obvious facts is the best to begin with, because its consequences are the most easily traced. This rude hypothesis is then rudely corrected, and the operation repeated; and the comparison of the consequences deducible from the corrected hypothesis with the observed facts suggests still further correction, until the deductive results are at last made to tally with the phenomena. "Some fact is as yet little understood, or some law is unknown; we frame on the subject an hypothesis as accordant as possible with the whole of the data already possessed; and the science, being thus enabled to move forward freely, always ends by leading to new consequences capable of observation, which either confirm or refute, unequivocally, the first supposition." Neither induction nor deduction would enable us to understand even the simplest phenomena, "if we did not often commence by anticipating on the results; by making a provisional supposition, at first essentially conjectural, as to some of the very notions which constitute the final object of the inquiry."[10]

Let any one watch the manner in which he himself unravels a complicated mass of evidence; let him observe how, for instance, he elicits the true history of any occurrence from the involved statements of one or of many witnesses: he will find that he does not take all the items of evidence into his mind at once, and attempt to weave them together: he extemporises, from a few of the particulars, a first rude theory of the mode in which the facts took place, and then looks at the other statements one by one, to try whether they can be reconciled with that provisional theory, or what alterations or additions it requires to make it square with them. In this way, which has been justly compared to the Methods of Approximation of mathematicians, we arrive, by means of hypotheses, at conclusions not hypothetical.[11]

§6. It is perfectly consistent with the spirit of the method, to assume in this provisional manner not only an hypothesis respecting the law of what we already know to be the cause, but an hypothesis respecting the cause itself. It is allowable, useful, and often even necessary, to begin by asking ourselves what cause *may* have produced the effect, in order that we may know in what direction to look out for evidence to determine whether it actually *did*. The vortices of Descartes would have been a perfectly legitimate hypothesis, if it had been possible, by any mode of exploration which we could entertain the hope of ever possessing, to bring the reality of the vortices, as a fact in nature, conclusively to the test of observation. The vice of the hypothesis was that it could not lead to any course of investigation capable of converting it from an hypothesis into a proved fact. It might chance to be *dis*proved, either by some want of correspondence with the phenomena it purported to explain, or (as actually happened) by some extraneous fact. "The free passage of comets through the spaces in which these vortices should have been, convinced men that these vortices did not exist."[12] But the hypothesis would have been false, though no such direct evidence of its falsity had been procurable. Direct evidence of its truth there could not be.

The prevailing hypothesis of a luminiferous ether, in other respects not without analogy to that of Descartes, is not in its own nature entirely cut off from the possibility of direct evidence in its favour. It is well known that the difference between the calculated and the observed times of the periodical return of Encke's comet, has led to a conjecture that a medium capable of opposing resistance to motion is diffused through space. If this surmise should be confirmed, in the course of ages, by the gradual accumulation of a similar variance in the case of the other bodies of the solar system, the luminiferous ether would have made a considerable advance towards the character of a *vera causa,* since the existence would have been ascertained of a great cosmical agent, possessing some of the attributes which the hypothesis assumes; though there would still remain many difficulties, and the identification of the ether with the resisting medium would even, I imagine, give rise to new ones. At present, however, this supposition cannot be looked upon as more than a conjecture; the existence of the ether still rests on the possibility of deducing from its assumed laws a considerable number of actual phenomena; and this evidence I cannot regard as conclusive, because we cannot have, in the case of such an hypothesis, the assurance that if the hypothesis be false it must lead to results at variance with the true facts.

Accordingly, most thinkers of any degree of sobriety allow, that an hypothesis of this kind is not to be received as probably true because it accounts for all the known phenomena, since this is a condition sometimes fulfilled tolerably well by two conflicting hypotheses; while there are probably many others which are equally possible, but which, for want of anything analogous in our experience, our minds are unfitted to conceive. But it seems to be thought that an hypothesis of the sort in question is entitled to a more favourable reception, if, besides accounting for all the facts previously known, it has led to the anticipation and prediction of others which experience afterwards verified; as the undulatory theory of light led to the prediction, subsequently realised by experiment, that two luminous rays might meet each other in such a manner as to produce darkness. Such predictions and their fulfilment are, indeed, well calculated to impress the uninformed, whose faith in science rests solely on similar coincidences between its prophecies and what comes to pass. But it is strange that any considerable stress should be laid upon such a coincidence by persons of scientific attainments. If the laws of the propagation of light accord with those of the vibrations of an elastic fluid in as many respects as is necessary to make the hypothesis afford a correct expression of all or most of the phenomena known at the time, it is nothing strange that they should accord with each other in one respect more. Though twenty such coincidences should occur, they would not prove the reality of the undulatory ether; it would not follow that the phenomena of light were results of the laws of elastic fluids, but at most that they are governed by laws partially identical with these; which, we may observe, is already certain, from the fact that the hypothesis in question could be for a moment tenable.[13] Cases may be cited, even in our imperfect acquaintance with nature, where agencies that we have good reason to consider as radically distinct produce their effects, or some of their effects, according to laws which are identical. The law, for example, of the inverse square of the distance, is the measure of the intensity not only of gravitation, but (it is believed) of illumination, and of heat diffused from a centre. Yet no one looks upon this identity as proving similarity in the mechanism by which the three kinds of phenomena are produced. According to Dr. Whewell, the coincidence of results predicted

from an hypothesis with facts afterwards observed amounts to a conclusive proof of the truth of the theory. "If I copy a long series of letters, of which the last half-dozen are concealed, and if I guess these aright, as is found to be the case when they are afterwards uncovered, this must be because I have made out the import of the inscription. To say, that because I have copied all that I could see, it is nothing strange that I should guess those which I cannot see, would be absurd, without supposing such a ground for guessing."[14] If any one, from examining the greater part of a long inscription, can interpret the characters so that the inscription gives a rational meaning in a known language, there is a strong presumption that his interpretation is correct; but I do not think the presumption much increased by his being able to guess the few remaining letters without seeing them: for we should naturally expect (when the nature of the case excludes chance) that even an erroneous interpretation which accorded with all the visible parts of the inscription would accord also with the small remainder; as would be the case, for example, if the inscription had been designedly so contrived as to admit of a double sense. I assume that the uncovered characters afford an amount of coincidence too great to be merely casual: otherwise the illustration is not a fair one. No one supposes the agreement of the phenomena of light with the theory of undulations to be merely fortuitous. It must arise from the actual identity of some of the laws of undulations with some of those of light; and if there be that identity, it is reasonable to suppose that its consequences would not end with the phenomena which first suggested the identification, nor be even confined to such phenomena as were known at the time. But it does not follow, because some of the laws agree with those of undulations, that there are any actual undulations; no more than it followed because some (though not so many) of the same laws agreed with those of the projection of particles, that there was actual emission of particles. Even the undulatory hypothesis does not account for all the phenomena of light. The natural colours of objects, the compound nature of the solar ray, the absorption of light, and its chemical and vital action, the hypothesis leaves as mysterious as it found them; and some of these facts are, at least apparently, more reconcilable with the emission theory than with that of Young and Fresnel. Who knows but that some third hypothesis, including all these phenomena, may in time leave the undulatory theory as far behind as that has left the theory of Newton and his successors?

To the statement that the condition of accounting for all the known phenomena is often fulfilled equally well by two conflicting hypotheses, Dr. Whewell makes answer that he knows "of no such case in the history of science, where the phenomena are at all numerous and complicated."[15] Such an affirmation, by a writer of Dr. Whewell's minute acquaintance with the history of science, would carry great authority, if he had not, a few pages before, taken pains to refute it,[16] by maintaining that even the exploded scientific hypotheses might always, or almost always, have been so modified as to make them correct representations of the phenomena. The hypothesis of vortices, he tells us, was, by successive modifications, brought to coincide in its results with the Newtonian theory and with the facts. The vortices did not indeed explain all the phenomena which the Newtonian theory was ultimately found to account for, such as the precession of the equinoxes; but this phenomenon was not, at the time, in the contemplation of either party, as one of the facts to be accounted for. All the facts which they did contemplate we may believe on Dr. Whewell's authority to have accorded as accurately with the Cartesian hypothesis, in its finally improved state, as with Newton's.

But it is not, I conceive, a valid reason for accepting any given hypothesis that we are unable to imagine any other which will account for the facts. There is no necessity for supposing that the true explanation must be one which, with only our present experience, we could imagine. Among the natural agents with which we are aquainted, the vibrations of an elastic fluid may be the only one whose laws bear a close resemblance to those of light; but we cannot tell that there does not exist an unknown cause, other than an elastic ether diffused through space, yet producing effects identical in some respects with those which would result from the undulations of such an ether. To assume that no such cause can exist appears to me an extreme case of assumption without evidence. And at the risk of being charged with want of modesty, I cannot help expressing astonishment that a philosopher of Dr. Whewell's abilities and attainments should have written an elaborate treatise on the philosophy of induction, in which he recognises absolutely no mode of induction except that of trying hypothesis after hypothesis until one is found which fits the phenomena; which one, when found, is to be assumed as true, with no other reservation than that if on re-examination it should appear to assume more than is needful for explaining the phenomena, the superfluous part of the assumption should be cut off. And this without the slightest distinction between the cases in which it may be known beforehand that two different hypotheses cannot lead to the same result, and those in which, for aught we can ever know, the range of suppositions, all equally consistent with the phenomena, may be infinite.[17]

Nevertheless, I do not agree with M. Comte in condemning those who employ themselves in working out into detail the application of these hypotheses to the explanation of ascertained facts, provided they bear in mind that the utmost they can prove is, not that the hypothesis *is,* but that it *may* be true. The ether hypothesis has a very strong claim to be so followed out, a claim greatly strengthened since it has been shown to afford a mechanism which would explain the mode of production not of light only, but also of heat. Indeed the speculation has a smaller element of hypothesis in its application to heat than in the case for which it was originally framed. We have proof by our senses of the existence of molecular movement among the particles of all heated bodies, while we have no similar experience in the case of light. When, therefore, heat is communicated from the sun to the earth, across apparently empty space, the chain of causation has molecular motion both at the beginning and end. The hypothesis only makes the motion continuous by extending it to the middle. Now motion in a body is known to be capable of being imparted to another body contiguous to it; and the intervention of a hypothetical elastic fluid occupying the space between the sun and the earth supplies the contiguity which is the only condition wanting, and which can be supplied by no supposition but that of an intervening medium. The supposition, notwithstanding, is at best a probable conjecture, not a proved truth; for there is no proof that contiguity is absolutely required for the communication of motion from one body to another. Contiguity does not always exist, to our senses at least, in the cases in which motion produces motion. The forces which go under the name of attraction, especially the greatest of all, gravitation, are examples of motion producing motion without apparent contiguity. When a planet moves, its distant satellites accompany its motion. The sun carries the whole solar system along with it in the progress which it is ascertained to be executing through space. And even if we were to accept as conclusive the geometrical reasonings (strikingly similar to those by which the Cartesians defended their vortices) by which it has been attempted to show that the motions of the ether may account for gravitation itself, even then it would only have been proved that the supposed mode of production may be, but not that no other mode can be, the true one.

§7. It is necessary, before quitting the subject of hypotheses, to guard against the appearance of reflecting upon the scientific value of several branches of physical inquiry, which, though only in their infancy, I hold to be strictly inductive. There is a great difference between inventing agencies to account for classes of phenomena, and endeavouring, in conformity with known laws, to conjecture what former collocations of known agents may have given birth to individual facts still in existence. The latter is the legitimate operation of inferring from an observed effect the existence, in time past, of a cause similar to that by which we know it to be produced in all cases in which we have actual experience of its origin. This, for example, is the scope of the inquiries of geology; and they are no more illogical or visionary than judicial inquiries, which also aim at discovering a past event by inference from those of its effects which still subsist. As we can ascertain whether a man was murdered or died a natural death from the indications exhibited by the corpse, the presence or absence of signs of struggling on the ground or on the adjacent objects, the marks of blood, the footsteps of the supposed murderers, and so on, proceeding throughout on uniformities ascertained by a perfect induction without any mixture of hypothesis, so if we find, on and beneath the surface of our planet, masses exactly similar to deposits from water, or to results of the cooling of matter melted by fire, we may justly conclude that such has been their origin; and if the effects, though similar in kind, are on a far larger scale than any which are now produced, we may rationally and without hypothesis conclude, either that the causes existed formerly with greater intensity, or that they have operated during an enormous length of time. Further than this no geologist of authority has, since the rise of the present enlightened school of geological speculation, attempted to go.

In many geological inquiries it doubtless happens that though the laws to which the phenomena are ascribed are known laws, and the agents known agents, those agents are not known to have been present in the particular case. In the speculation respecting the igneous origin of trap or granite, the fact does not admit of direct proof, that those substances have been actually subjected to intense heat. But the same thing might be said of all judicial inquiries which proceed on circumstantial evidence. We can conclude that a man was murdered, though it is not proved by the testimony of eyewitnesses that some person who had the intention of murdering him was present on the spot. It is enough, for most purposes, if no other known cause could have generated the effects shown to have been produced.

The celebrated speculation of Laplace concerning the origin of the earth and planets participates essentially in the inductive character of modern geological theory. The speculation is, that the atmosphere of the sun originally extended to the present limits of the solar system; from which, by the process of cooling, it has contracted to its present dimensions; and since, by the general principles of mechanics, the rotation of the sun and of its accompanying atmosphere must increase in rapidity as its volume diminishes, the increased centrifugal force generated by the more rapid rotation, overbalancing the action of gravitation, has caused the sun to abandon successive rings of vaporous matter, which are supposed to have condensed by cooling, and to have become the planets. There is in this theory no unknown substance introduced on supposition, nor any unknown property or law ascribed to a known substance. The known laws of matter authorise us to suppose that a body which is constantly giving out so large an amount of heat as the sun is must be progressively cooling, and that, by the process of cooling, it must contract; if, therefore, we endeavour, from the present state of that luminary, to infer its state in a time long past, we must necessarily suppose that its atmosphere extended much farther than at present, and we are entitled to suppose that it extended as far as we can trace effects such as it might naturally leave behind it on retiring; and such the planets are. These suppositions being made, it follows from known laws that successive zones of the solar atmosphere might be abandoned; that these would continue to revolve round the sun with the same velocity as when they formed part of its substance; and that they would cool down, long before the sun itself, to any given temperature, and consequently to that at which the greater part of the vaporous matter of which they consisted would become liquid or solid. The known law of gravitation would then cause them to agglomerate in masses, which would assume the shape our planets actually exhibit; would acquire, each about its own axis, a rotatory movement; and would in that state revolve, as the planets actually do, about the sun, in the same direction with the sun's rotation, but with less velocity, because in the same periodic time which the sun's rotation occupied when his atmosphere extended to that point. There is thus, in Laplace's theory, nothing, strictly speaking, hypothetical; it is an example of legitimate reasoning from a present effect to a possible past cause, according to the known laws of that cause. The theory therefore is, as I have said, of a similar character to the theories of geologists, but considerably inferior to them in point of evidence. Even if it were proved (which it is not) that the conditions necessary for determining the breaking off of successive rings would certainly occur; there would still be a much greater chance of error in assuming that the existing laws of nature are the same which existed at the origin of the solar system, than in merely presuming (with geologists) that those laws have lasted through a few revolutions and transformations of a single one among the bodies of which that system is composed.

NOTES

1. Dr. Whewell thinks it improper to apply the term Induction to any operation not terminating in the establishment of a general truth. Induction, he says, (*Philosophy of Discovery* [London, 1860], p. 245,) "is not the same thing as experience and observation.

Induction is experience or observation *consciously* looked at in a *general* form. This consciousness and generality are necessary parts of that knowledge which is science." And he objects (p. 241) to the mode in which the word Induction is employed in this work, as an undue extension of that term "not only to the cases in which the general induction is consciously applied to a particular instance, but to the cases in which the particular instance is dealt with by means of experience in that rude sense in which experience can be asserted of brutes, and in which of course we can in no way imagine that the law is possessed or understood as a general proposition." This use of the term he deems a "confusion of knowledge with practical tendencies."

I disclaim, as strongly as Dr. Whewell can do, the application of such terms as induction, inference, or reasoning to operations performed by mere instinct, that is, from an animal impulse, without the exertion of any intelligence. But I perceive no ground for confining the use of those terms to cases in which the inference is drawn in the forms and with the precautions required by scientific propriety. To the idea of science, an express recognition and distinct apprehension of general laws, as such, is essential; but nine-tenths of the conclusions drawn from experience in the course of practical life are drawn without any such recognition: they are direct inferences from known cases to a case supposed to be similar. I have endeavoured to show that this is not only as legitimate an operation, but substantially the same operation as that of ascending from known cases to a general proposition; except that the latter process has one great security for correctness which the former does not possess. In science the inference must necessarily pass through the intermediate stage of a general proposition, because Science wants its conclusions for record, and not for instantaneous use. But the inferences drawn for the guidance of practical affairs by persons who would often be quite incapable of expressing in unexceptional terms the corresponding generalisations may, and frequently do, exhibit intellectual powers quite equal to any which have ever been displayed in science: and if these inferences are not inductive, what are they? The limitation imposed on the term by Dr. Whewell seems perfectly arbitrary; neither justified by any fundamental distinction between what he includes and what he desires to exclude, nor sanctioned by usage, at least from the time of Reid and Stewart, the principal legislators (as far as the English language is concerned) of modern metaphysical terminology.

2. *Novum Organum Renovatum* [London, 1858], pp. 72, 73.

3. *Novum Organum Renovatum,* p. 32.

4. *Cours de Philosophie Positive* [6 vols., Paris, 1830–42], vol. ii. p. 202.

5. Dr. Whewell, in his reply, contests the distinction here drawn, and maintains, that not only different descriptions, but different explanations of a phenomenon, may all be true. Of the three theories respecting the motions of the heavenly bodies, he says (*Philosophy of Discovery,* p. 231): "Undoubtedly all these explanations may be true and consistent with each other, and would be so if each had been followed out so as to show in what manner it could be made consistent with the facts. And this was in reality in a great measure done. The doctrine that the heavenly bodies were moved by vortices was successfully modified, so that it came to coincide in its results with the doctrine of an inverse-quadratic centripetal force. . . . When this point was reached, the vortex was merely a machinery, well or ill devised, for producing such a centripetal force, and therefore did not contradict the doctrine of a centripetal force. Newton himself does not appear to have been averse to explaining gravity by impulse. So little is it true that if one theory be true the other must be false. The attempt to explain gravity by the impulse of streams of particles flowing through the universe in all directions, which I have mentioned in the *Philosophy,* is so far from being inconsistent with the Newtonian theory, that

it is founded entirely upon it. And even with regard to the doctrine that the heavenly bodies move by an inherent virtue, if this doctrine had been maintained in any such way that it was brought to agree with the facts, the inherent virtue must have had its laws determined; and then it would have been found that the virtue had a reference to the central body; and so the 'inherent virtue' must have coincided in its effect with the Newtonian force; and then the two explanations would agree, except so far as the word 'inherent' was concerned. And if such a part of an earlier theory as this word *inherent* indicates is found to be untenable, it is of course rejected in the transition to later and more exact theories, in Inductions of this kind, as well as in what Mr. Mill calls Descriptions. There is, therefore, still no validity discoverable in the distinction which Mr. Mill attempts to draw between descriptions like Kepler's law of elliptical orbits, and other examples of induction."

If the doctrine of vortices had meant, not that vortices existed, but only that the planets moved *in the same manner* as if they had been whirled by vortices; if the hypothesis had been merely a mode of representing the facts, not an attempt to account for them; if, in short, it had been only a Description, it would, no doubt, have been reconcilable with the Newtonian theory. The vortices, however, were not a mere aid to conceiving the motions of the planets, but a supposed physical agent, actively impelling them; a material fact which might be true or not true, but could not be both true and not true. According to Descartes' theory it was true, according to Newton's it was not true. Dr. Whewell probably means that since the phrases, centripetal and projectile force, do not declare the nature but only the direction of the forces, the Newtonian theory does not absolutely contradict any hypothesis which may be framed respecting the mode of their production. The Newtonian theory, regarded as a mere *description* of the planetary motions, does not; but the Newtonian theory as an *explanation* of them does. For in what does the explanation consist? In ascribing those motions to a general law which obtains between all particles of matter, and in identifying this with the law by which bodies fall to the ground. If the planets are kept in their orbits by a force which draws the particles composing them towards every other particle of matter in the solar system, they are not kept in those orbits by the impulsive force of certain streams of matter which whirl them round. The one explanation absolutely excludes the other. Either the planets are not moved by vortices, or they do not move by a law common to all matter. It is impossible that both opinions can be true. As well might it be said that there is no contradiction between the assertions, that a man died because somebody killed him, and that he died a natural death.

So, again, the theory that the planets move by a virtue inherent in their celestial nature, is incompatible with either of the two others: either that of their being moved by vortices, or that which regards them as moving by a property which they have in common with the earth and all terrestrial bodies. Dr. Whewell says that the theory of an inherent virtue agrees with Newton's when the word inherent is left out, which of course it would be (he says) if "found to be untenable." But leave that out, and where is the theory? The word inherent *is* the theory. When that is omitted, there remains nothing except that the heavenly bodies move "by a virtue," *i.e.* by a power of some sort, or by virtue of their celestial nature, which directly contradicts the doctrine that terrestrial bodies fall by the same law.

If Dr. Whewell is not yet satisfied, any other subject will serve equally well to test his doctrine. He will hardly say that there is no contradiction between the emission theory and the undulatory theory of light; or that there can be both one and two electricities; or that the hypothesis of the production of the higher organic forms by development from the lower, and the supposition of separate and successive acts of creation, are quite reconcilable; or that the theory that volcanoes are fed from a central fire, and the doctrines which ascribe them to chemical action at a comparatively small depth below the earth's surface, are consistent with one another, and all true as far as they go.

If different explanations of the same fact cannot both be true, still less, surely, can different predictions. Dr. Whewell quarrels (on what ground it is not necessary here to consider) with the example I had chosen on this point, and thinks an objection to an illustration a sufficient answer to a theory. Examples not liable to his objection are easily found, if the proposition that conflicting predictions cannot both be true can be made clearer by any examples. Suppose the phenomenon to be a newly discovered comet, and that one astronomer predicts its return once in every 300 years—another once in every 400: can they both be right? When Columbus predicted that by sailing constantly westward he should in time return to the point from which he set out, while others asserted that he could never do so except by turning back, were both he and his opponents true prophets? Were the predictions which foretold the wonders of railways and steamships, and those which averred that the Atlantic could never be crossed by steam navigation, nor a railway train propelled ten miles an hour, both (in Dr. Whewell's words) "true and consistent with one another?"

Dr. Whewell sees no distinction between holding contradictory opinions of a question of fact, and merely employing different analogies to facilitate the conception of the same fact. The case of different Inductions belongs to the former class, that of different Descriptions to the latter.

6. *Phil. of Discov.*, p. 256.

7. [*Preliminary*] *Discourse on the Study of Natural Philosophy* [London, 1830], p. 179.

8. *Vide supra,* book iii. ch. xi.

9. *Philosophy of Discovery,* pp. 185 et seq.

10. Comte, *Philosophie Positive,* ii. 434–437.

11. As an example of legitimate hypothesis according to the test here laid down, has been justly cited that of Broussais, who, proceeding on the very rational principle that every disease must originate in some definite part or other of the organism, boldly assumed that certain fevers, which not being known to be local were called constitutional, had their origin in the mucous membrane of the alimentary canal. The supposition was indeed, as is now generally admitted, erroneous; but he was justified in making it, since by deducing the consequences of the supposition, and comparing them with the facts of those maladies, he might be certain of disproving his hypothesis if it was ill-founded, and might expect that the comparison would materially aid him in framing another more conformable to the phenomena.

The doctrine now universally received that the earth is a natural magnet, was originally an hypothesis of the celebrated Gilbert.

Another hypothesis, to the legitimacy of which no objection can lie, and which is well calculated to light the path of scientific inquiry, is that suggested by several recent writers, that the brain is a voltaic pile, and that each of its pulsations is a discharge of electricity through the system. It has been remarked that the sensation felt by the hand from the beating of a brain bears a strong resemblance to a voltaic shock. And the hypothesis, if followed to its consequences, might afford a plausible explanation of many physiological facts, while there is nothing to discourage the hope that we may in time sufficiently understand the conditions of voltaic phenomena to render the truth of the hypothesis amenable to observation and experiment.

The attempt to localise, in different regions of the brain, the physical organs of our different mental faculties and propensities, was, on the part of its original author, a legitimate example of a scientific hypothesis; and we ought not, therefore, to blame him for the extremely slight grounds on which he often proceeded in an operation which could only be tentative, though we may regret that materials barely sufficient for a first rude hypothesis should have been hastily worked up into the vain semblance of a science. If there be really a connection between the scale of mental endowments and the various degrees of complication in the cerebral system, the nature of that connection was in no other way so likely to be brought to light as by framing, in the first instance, an hypothesis similar to that of Gall. But the verification of any such hypothesis is attended, from the peculiar nature of the phenomena, with difficulties which phrenologists have not shown themselves even competent to appreciate, much less to overcome.

Mr. Darwin's remarkable speculation on the Origin of Species is another unimpeachable example of a legitimate hypothesis. What he terms "natural selection" is not only a

232

vera causa, but one proved to be capable of producing effects of the same kind with those which the hypothesis ascribes to it: the question of possibility is entirely one of degree. It is unreasonable to accuse Mr. Darwin (as has been done) of violating the rules of Induction. The rules of Induction are concerned with the conditions of Proof. Mr. Darwin has never pretended that his doctrine was proved. He was not bound by the rules of Induction, but by those of Hypothesis. And these last have seldom been more completely fulfilled. He has opened a path of inquiry full of promise, the results of which none can foresee. And is it not a wonderful feat of scientific knowledge and ingenuity to have rendered so bold a suggestion, which the first impulse of every one was to reject at once, admissible and discussable, even as a conjecture?

12. Whewell's *Phil. of Discovery,* pp. 275, 276.

13. What has most contributed to accredit the hypothesis of a physical medium for the conveyance of light, is the certain fact that light *travels,* (which cannot be proved of gravitation;) that its communication is not instantaneous, but requires time; and that it is intercepted (which gravitation is not) by intervening objects. These are analogies between its phenomena and those of the mechanical motion of a solid or fluid substance. But we are not entitled to assume that mechanical motion is the only power in nature capable of exhibiting those attributes.

14. *Phil. of Disc.,* p. 274.

15. P. 271.

16. P. 251 and the whole of Appendix G.

17. In Dr. Whewell's latest version of this theory (*Philosophy of Discovery,* p. 331) he makes a concession respecting the medium of the transmission of light, which, taken in conjunction with the rest of his doctrine on the subject, is not, I confess, very intelligible to me, but which goes far towards removing, if it does not actually remove, the whole of the difference between us. He is contending, against Sir William Hamilton, that all matter has weight. Sir William, in proof of the contrary, cited the luminiferous ether and the calorific and electric fluids, "which," he said, "we can neither denude of their character of substance nor clothe with the attribute of weight." "To which," continues Dr. Whewell, "my reply is, that precisely because I cannot clothe these agents with the attribute of Weight, I *do* denude them of the character of Substance. They are not substances, but agencies. These Imponderable Agents are not properly called Imponderable Fluids. This I conceive that I have proved." Nothing can be more philosophical. But if the luminiferous ether is not matter, and fluid matter too, what is the meaning of its undulations? Can an agency undulate? Can there be alternate motion forward and backward of the particles of an agency? And does not the whole mathematical theory of the undulations imply them to be material? Is it not a series of deductions from the known

properties of elastic fluids? *This* opinion of Dr. Whewell reduces the undulations to a figure of speech, and the undulatory theory to the proposition, which all must admit, that the transmission of light takes place according to laws which present a very striking and remarkable agreement with those of undulations. If Dr. Whewell is prepared to stand by this doctrine, I have no difference with him on the subject.

14

The Wave Theory of Light and the Mill–Whewell Debate

Waves and Scientific Method / Peter Achinstein

1. Introduction

In 1802 a youthful Thomas Young, British physician and scientist, had the audacity to resuscitate the wave theory of light (Young 1802). For this he was excoriated by Henry Brougham (1803) in the *Edinburgh Review.* Brougham, a defender of the Newtonian particle theory, asserted that Young's paper was "destitute of every species of merit" because it was not based on inductions from observations but involved simply the formulation of hypotheses to explain various optical phenomena. And, Brougham continued:

> A discovery in mathematics, or a successful induction of facts, when once completed, cannot be too soon given to the world. But . . . an hypothesis is a work of fancy, useless in science, and fit only for the amusement of a vacant hour. (1803, p. 451)

This dramatic confrontation between Young and Brougham, it has been claimed, is but one example of a general methodological gulf between 19th century wave theorists and 18th and 19th century particle theorists. The wave theorists, it has been urged by Larry Laudan (1981) and Geoffrey Cantor (1975), employed a method of hypothesis in defending their theory.

Reprinted from *PSA 1992: Proceedings of the Biennial Meeting of the Philosophy of Science Association* 2 (1992): 193–204.

This method was firmly rejected by particle theorists, who insisted, with Brougham, that the only way to proceed in physics is to make inductions from observations and experiments.

In a recent work (Achinstein 1991), I argue, contra Laudan and Cantor, that 19th century wave theorists, both in their practice and in their philosophical reflections on that practice, employed a method that is different from the method of hypothesis in important respects; moreover, there are strong similarities between the method the wave theorists practiced and preached and that of 19th century particle theorists such as Brougham and David Brewster. In the present paper I will focus just on the wave theorists. My aims are these: to review my claims about how in fact wave theorists typically argued for their theory; to see whether, or to what extent, this form of reasoning corresponds to the method of hypothesis or to inductivism in sophisticated versions of these doctrines offered by William Whewell and John Stuart Mill; and finally to deal with a problem of anomalies which I did not develop in *Particles and Waves* and might be said to pose a difficulty for my account.

2. The Method of Hypothesis and Inductivism

According to a simple version of the method of hypothesis, if the observed phenomena are explained by, or derived from, an hypothesis, then one may infer the truth or probability of that hypothesis. Laudan maintains that by the 1830s an important shift occurred in the use of this method. An hypothesis was inferable not simply if it explained known phenomena that prompted it in the first place, but only if it also explained and/or predicted phenomena of a kind different from those it was invented to explain. This version received its most sophisticated formulation in the works of William Whewell, a defender of the wave theory. In what follows I will employ Whewell's version of the method of hypothesis as a foil for my discussion of the wave theorist's argument.

Whewell (1967, pp. 60–74) offered four conditions which, if satisfied, will make an hypothesis inferable with virtual certainty. First, it should explain all the phenomena which initially prompted it. Second, it should predict new phenomena. Third, it should explain and/or predict phenomena of a "kind different from those which were contemplated in the formation of . . . [the] hypothesis" (p. 65). If this third condition is satisfied Whewell says that there is a "consilience of inductions." Whewell's fourth condition derives from the idea that hypotheses are part of a theoretical system the components of which are not framed all at once, but are developed over time. The condition is that as the theoretical system evolves it becomes simpler and more coherent.

Since both Laudan and Cantor claim that the wave theorists followed the method of hypothesis while the particle theorists rejected this method in favor of inductivism, it will be useful to contrast Whewell's version of the former with Mill's account of the latter. This contrast should be of special interest for two reasons. Both Whewell and Mill discuss the wave theory, which Whewell supports and Mill rejects; and each criticizes the other's methodology.

One of the best places to note the contrast in Mill is in his discussion of the "deductive method" (which he distinguishes from the "hypothetical method" or method of hypothesis) (Mill 1959, pp. 299–305). Mill asserts that the deductive method is to be used in situations where causes subject to various laws operate, in other words, in solving typical problems in physics as well as other sciences. It consists of three steps. First, there is a direct induction from observed phenomena to the various causes and laws governing them. Mill defines induction as "the process by which we conclude that what is true of certain individuals of a class is true of the whole class, or that what is true at certain times will be true in similar circumstances at all times" (p. 188). This concept of inductive generalization is used together with his four famous canons of causal inquiry to infer the causes operating and the laws that govern them. The second part of the deductive method Mill calls "ratiocination." It is a process of calculation, deduction, or explanation: from the causes and laws we calculate what effects will follow. Third, and finally, there is "verification": "the conclusions [derived by ratiocination] must be found, on careful comparison, to accord with the result of direct observation wherever it can be had" (p. 303).

Now, in rejecting the method of hypothesis, Mill writes:

> The Hypothetical Method suppresses the first of the three steps, the induction to ascertain the law, and contents itself with the other two operations, ratiocination and verification, the law which is reasoned from being assumed instead of proved. (p. 323)

Mill's major objection to the method of hypothesis is that various conflicting hypotheses are possible from which the phenomena can be derived and verified. In his discussion of the wave theory of light, Mill rejects the hypothesis of the luminiferous ether on these grounds. He writes:

> This supposition cannot be looked upon as more than a conjecture; the existence of the ether still rests on the possibility of deducing from its assumed laws a considerable number of actual phenomena. . . . [M]ost thinkers of any degree of sobriety allow, that an hypothesis of this kind is not to be received as probably true because it accounts for all the known

> phenomena, since this is a condition sometimes fulfilled tolerably well by two conflicting hypotheses; while there are probably many others which are equally possible, but which, for want of anything analogous in our experience, our minds are unfitted to conceive. (p. 328)

With Whewell's ideas about prediction and consilience in mind, Mill continues:

> But it seems to be thought that an hypothesis of the sort in question is entitled to a more favourable reception if, besides accounting for all the facts previously known it has led to the anticipation and prediction of others which experience afterwards verified. . . . Such predictions and their fulfillment are, indeed, well calculated to impress the uninformed, whose faith in science rests solely on similar coincidences between its prophecies and what comes to pass. . . . Though twenty such coincidences should occur they would not prove the reality of the undulatory ether. . . . (pp. 328–9)

Although in these passages Mill does not discuss Whewell's ideas about coherence and the evolution of theories, it is clear that Mill would not regard Whewell's four conditions as sufficient to infer an hypothesis with virtual certainty or even high probability. The reason is that Whewell's conditions omit the first crucial step of the deductive method, the induction to the causes and laws.

If Laudan and Cantor are correct in saying that 19th century wave theorists followed the method of hypothesis and rejected inductivism, then, as these opposing methodologies are formulated by Whewell and Mill, this would mean the following: 19th century wave theorists argued for the virtual certainty or high probability of their theory by first assuming, without argument, various hypotheses of the wave theory; then showing how these will not only explain the known optical phenomena but will explain and/or predict ones of a kind different from those prompting the wave hypotheses in the first place; and finally arguing that as the theory has evolved it has become simpler and more coherent. Is this an adequate picture? Or, in addition, did wave theorists employ a crucial inductive step to their hypotheses at the outset? Or do neither of these methodologies adequately reflect the wave theorists' argument?

3. The Wave Theorists' Argument

Nineteenth century wave theorists frequently employed the following strategy in defense of their theory.

1. Start with the assumption that light consists either in a wave motion transmitted through a rare, elastic medium pervading the universe, or in a stream of particles emanating from luminous bodies. Thomas Young (1845) in his 1807 Lectures, Fresnel (1816) in his prize essay on diffraction, John Herschel (1845) in an 1827 review article of 246 pages, and Humphrey Lloyd (1834) in a 119 page review article,[1] all begin with this assumption in presentations of the wave theory.
2. Show how each theory explains various optical phenomena, including the rectilinear propagation of light, reflection, refraction, diffraction, Newton's rings, polarization, etc.

3. Argue that in explaining one or more of these phenomena the particle theory introduces improbable auxiliary hypotheses but the wave theory does not. For example, light is diffracted by small obstacles and forms bands both inside and outside the shadow. To explain diffraction particle theorists postulate both attractive and repulsive forces emanating from the obstacle and acting at a distance on the particles of light so as to turn some of them away from the shadow and others into it. Wave theorists such as Young and Fresnel argue that the existence of such forces is very improbable. By contrast, diffraction is explainable from the wave theory (on the basis of Huygens' principle that each point in a wave front can be considered a source of waves), without the introduction of any new improbable assumptions. Similar arguments are given for several other optical phenomena, including interference and the constant velocity of light.
4. Conclude from steps 1 through 3 that the wave theory is true, or very probably true.

This represents, albeit sketchily, the overall structure of the argument. More details are needed before seeing whether, or to what extent, it conforms to Whewell's conditions or Mill's. But even before supplying such details we can see that the strategy is not simply to present a positive argument for the wave theory via an induction to its hypotheses and/or by showing that it can explain various optical phenomena. Whether it does these things or not, the argument depends crucially on showing that the rival particle theory has serious problems.

To be sure, neither Whewell's methodology nor Mill's precludes comparative judgments. For example, Whewell explicitly claims that the wave theory is more consilient and coherent than the particle theory. And Mill (who believed that neither theory satisfied his crucial inductive step) could in principle allow the possibility that new phenomena could be discovered permitting an induction to one theory but not the other. I simply want to stress at the outset that the argument strategy of the wave theorists, as I have outlined it so far, is essentially comparative. The aim is to show at least that the wave theory is better, or more probable, than the rival particle theory.

Is the wave theorist's argument intended to be stronger than that? I believe that it is. Thomas Young, both in his 1802 and 1803 Bakerian lectures (reprinted in Crew 1900), makes it clear that he is attempting to show that hitherto performed experiments, and analogies with sound, and passages in Newton, provide strong support for the wave theory, not merely that the wave theory is better supported than its rival. A similar attitude is taken by Fresnel, whose aim is not simply to show that the wave theory is better in certain respects than the particle theory, but that it is acceptable because it can explain various phenomena, including diffraction, without introducing improbable assumptions; by contrast, the particle theory is not acceptable, since it cannot. Even review articles are not simply comparative. Although he does compare the merits of the wave and particle theories in his 1834 report, Humphrey Lloyd makes it clear that this comparison leads him to assert the truth of the wave theory. In that theory, he claims:

> there is thus established that connexion and harmony in its parts which is the never failing attribute of truth. . . . It may be confidently said that it possesses characters which no false theory ever possessed before. (1877, p. 79)[2]

Let us now look more closely at the three steps of the argument leading to the conclusion. Wave theorists who make the assumption that light consists either of waves or particles do not do so simply in order to see what follows. They offer reasons, which are generally of two sorts. First, there is an argument from authority: "Leading physicists support one or the other assumption." Second, there is an argument from some observed property of light. For example, Lloyd notes that light travels in space from one point to another with a finite velocity, and that in nature one observes motion from one point to another occurring by the motion of a body or by vibrations of a medium.

Whatever one might think of the validity of these arguments, I suggest that they were being offered in support of the assumption that light consists either of waves or of particles. This is not a mere supposition. Argument from authority was no stranger to optical theorists of this period. Young in his 1802 paper explicitly appeals to passages in Newton in defense of three of his four basic assumptions. And Brougham, a particle theorist, defends his theory in part also by appeal to the authority and success of Newton. Moreover, the second argument, if not the first, can reasonably be interpreted as an induction in Mill's sense, i.e., as claiming that all observed cases of finite motion are due to particles or waves, so in all probability this one is too.[3]

I suggest, then, that wave theorists offered grounds for supposing it to be very probable that light consists either of waves or particles. I will write their claim as

(1) p(W or P/O&b) ≈ 1,

where W is the wave theory, P is the particle theory, O includes certain observed facts about light including its finite motion, and b is background information including facts about modes of travel in other cases. (≈ means "is close to.")

This is the first step in the earlier argument. I will postpone discussion of the second step for a moment, and turn to the third. Here the wave theorists assert that in order to explain various optical phenomena the rival particle theorists introduce improbable auxiliary hypotheses. By contrast, the wave theorists can explain these phenomena without introducing auxiliary hypotheses, or at least any that are improbable. Why are the particle theorists' auxiliary hypotheses improbable? And even if they are, how does this cast doubt on the central assumptions of the particle theory?

Let us return to diffraction, which particle theorists explained by the auxiliary hypothesis that attractive and repulsive forces emanate from the diffracting obstacle and act at a distance on the light particles, bending some into the shadow and others away from it. By experiment Fresnel showed that the observed diffraction patterns do not vary with the mass or shape of the diffracting body. But known attractive and repulsive forces exerted by bodies do vary with the mass and shape of the body. So Fresnel concludes that the existence of such forces of diffraction is highly improbable. Again it seems plausible to construe this argument as an inductive one, making an inference from properties of known forces to what should be (but is not) a property of the newly postulated ones. Fresnel's experiments together with observations of other known forces provide inductive reasons for concluding that the particle theorists' auxiliary assumption about attractive and repulsive forces is highly improbable.

Even if this is so, how would it show that other assumptions of the particle theory are improbable? It would if the probability of the auxiliary force assumption given the other assumptions of the particle theory is much, much greater than the probability of this auxiliary assumption not given the rest of the particle theory, i.e., if

(2) p(A/P&O&b) >> p(A/O&b),

where A is the auxiliary assumption, O includes information about diffraction patterns and Fresnel's experimental result that these do not vary with the mass or shape of the diffractor, b includes information about other known forces, and >> means "is much, much greater than." If this condition is satisfied, it is provable that the other assumptions of the particle theory have a probability close to zero,[4] i.e.,

(3) $p(P/O\&b) \approx 0$.

Although wave theorists did not explicitly argue for (2) above, they clearly had grounds for doing so. If by the particle theory P light consists of particles subject to Newton's laws, and if by observational results O light is diffracted from its rectilinear path, then by Newton's first law a force or set of forces must be acting on the light particles. Since the light is being diffracted in the vicinity of the obstacle, it is highly probable that this obstacle is exerting a force or forces on the light particles. That is, with the assumptions of the particle theory, auxiliary hypothesis A is very probable. However, without these assumptions the situation is very different. Without them the fact that other known forces vary with the mass and shape of the body exerting the force, but diffraction patterns do not, makes it unlikely that such forces exist in the case of diffraction. Or at least their existence is much, much more likely on the assumption that light consists of particles obeying Newton's laws than without such an assumption, i.e., (2) above. An important part of the argument here is inductive, based as it is on information about other mechanical forces.

From (1) and (3) we infer:

(4) $p(W/O\&b) >> 1$,

that is, the probability of the wave theory is close to 1, given the background information and certain optical phenomena, including diffraction.

Now we can return to the second step of the original argument, the one in which the wave theorist shows that his theory can explain a range of optical phenomena, not just the finite velocity of light and diffraction. What inferential value does this have? The wave theorist wants to show that his theory is probable not just given some limited selection of optical phenomena but given all known optical phenomena. This he can do if he can explain these phenomena by deriving them from his theory. Where $O_1, \ldots, O_n$ represent known optical phenomena other than diffraction and the constant velocity of light—including rectilinear propagation, reflection, refraction, and interference—if the wave theorist can derive these from his theory, then the probability of that theory will be at least sustained if not increased. This is a simple fact about probabilities.

Accordingly, the explanatory step in which the wave theorist derives various optical phenomena $O_1, \ldots, O_n$ from his theory permits an inference from (4) above to:

$$(5)\ p(W/O_1, \ldots, O_n \& O \& b) \approx 1,$$

i.e., the high probability of the wave theory given a wide range of observed optical phenomena. This is the conclusion of the wave theorist's argument.

If the explanation of known optical phenomena sustains the high probability of the wave theory without increasing it, does this mean that such phenomena fail to constitute evidence for the wave theory? Not at all. According to a theory of evidence I have developed (Achinstein 1983, chs. 10–11), optical phenomena can count as evidence for the wave theory even if they do not increase its probability. I reject the usual increase-in-probability account of evidence in favor of conditions that require the high probability of the theory T given the putative evidence O_i, and the high probability of an explanatory connection between T and O_i, given T and O_i. Both conditions are satisfied in the case of the wave theory.

In formulating the steps of the argument in the probabilistic manner above, I have clearly gone beyond what wave theorists say. For one thing, they do not appeal to probability in the way I have done. More importantly perhaps, while they argue that auxiliary hypotheses of the particle theorists are very improbable, they do not say that these assumptions are much more probable given the rest of the particle theory than without it. The following points are, I think, reasonably clear. (i) Wave theorists suppose that it is very likely that the wave or the particle theory is true, an assumption for which they have arguments. (ii) They argue against the particle theory by criticizing auxiliary assumptions of that theory, which introduce forces (or whatever) that violate inductively supported principles. (iii) Wave theorists argue that their theory can explain various optical phenomena without introducing any such questionable assumptions. (iv) Their reasoning, although eliminative, is different from typical eliminative reasoning; their first step is not to canvass all possible theories of light, but only two, for which they give arguments; their reasoning is not of the typical eliminative form "these are the only possible explanations of optical phenomena, all of which but one lead to difficulties." Reconstructing the wave theorists' argument in the probabilistic way I have done captures these four points. Whether it introduces too many fanciful ideas is a question I leave for my critics.

Is the argument Whewellian or Millian? It does satisfy the first three of

Whewell's conditions. It invokes the fact that various optical phenomena are derived from the wave theory. These include ones that prompted the theory in the first place (rectilinear propagation, reflection, and refraction), hitherto unobserved phenomena that were predicted (e.g., the Poisson spot in diffraction), and phenomena of a kind different from those that prompted it (e.g., diffraction, interference, polarization). The argument does not, however, satisfy Whewell's fourth condition. It does not appeal to the historical tendency of the theory over time to become simpler and more coherent. But the latter is not what divides Whewell from Mill. Nor is it Whewell's first three conditions, each of which Mill allows for in the ratiocinative part of his deductive method. Mill's claim is only that Whewell's conditions are not sufficient to establish the truth or high probability of an hypothesis. They omit the crucial first step, the inductive one to the hypothesized causes and laws.

As I have reconstructed the wave theorists' argument, an appeal to the explanatory power of the theory is a part, but not the whole, of the reasoning. There is also reasoning of a type that Mill would call inductive. It enters at two points. It is used to argue that light is most probably composed either of waves or of particles (e.g., the "finite motion" argument of Lloyd). And it is used to show that light is probably not composed of particles, since auxiliary hypotheses introduced to explain various optical phenomena are very improbable. This improbability is established by inductive generalization (e.g., in the case of diffraction, by inductively generalizing from what observations and experiments show about diffraction effects, and from what they show about forces). My claims are that wave theorists did in fact employ such inductive reasoning; that with it the argument that I have constructed is valid; and that without it the argument is invalid, or at least an appeal to Whewell's explanatory conditions is not sufficient to establish the high probability of the theory (though this last claim requires much more than I say here; see Achinstein 1991, Essay 4).

4. Explanatory Anomalies

One objection critics of my account may raise is that it does not do justice to explanatory anomalies in the wave theory. That theory was not able to explain all known optical phenomena. Herschel (1845), e.g., notes dispersion as one such phenomenon—the fact that different colors are refracted at different angles. Now the wave theorist wants to show that his theory is probable given all known optical phenomena, not just some favorable subset. But if dispersion is not derivable from the theory, and if there is no inductive argument from dispersion to that theory, then on the account I offer, the wave theorist cannot reach his desired conclusion. He can say only that his theory is probable given other optical phenomena. And he can take a wait-and-see attitude with respect to the unexplained ones. This is essentially what Herschel himself does in the case of dispersion.[5]

Let me now say how wave theorists could in principle deal with such anomalies in a manner that invokes the probabilistic reconstruction I offer. The suggestion I will make is, I think, implicit in their writings, if not explicit. And, interestingly, it is a response that combines certain Whewellian and Millian ideas. In what follows, I restrict the anomalies to phenomena which have not yet been derived from the wave theory by itself or from that theory together with auxiliary assumptions whose probability is very much greater given the wave theory than without it.

As Cantor notes in his very informative book *Optics after Newton:*

> Probably the central, and certainly the most repeated, claim [by the 1830's] was that in comparison with its rival the wave theory was more successful in explaining optical phenomena. (Cantor 1983, p. 192)

Cantor goes on to cite a table constructed in 1833 by Baden Powell, a wave theorist, listing 23 optical phenomena and evaluating the explanations proposed by wave and particle theories as "perfect," "imperfect," or "none." In the no-explanation category there are 12 entries for the particle theory and only 2 for the wave theory; while here are 18 "perfects" for the wave theory and only 5 for the particle theory.

Appealing, then, to the explanatory success of the wave theory, a very simple argument is this:

(6) Optical phenomena $O_1, \ldots, O_n$ can be coherently explained by the wave theory.
O is another optical phenomenon.
So probably
O can be coherently explained by the wave theory.

By a "coherent" explanation I follow what I take to be Whewell's idea: either the phenomenon is explained from the theory without introducing any additional assumptions, or if they are introduced they cohere both with the theory and with other known phenomena. In particular, no auxiliary assumption is introduced whose probability given the theory is very high but whose probability on the phenomena alone is low. Or, more generally, no such assumption is employed whose probability on the theory is very much greater than its probability without it.

Commenting on argument (6), the particle theorist might offer a similar argument to the conclusion that the particle theory can also explain O. But this does not vitiate the previous argument. For one thing, by the 1830's, even though Powell's table was not constructed by a neutral observer, it was generally agreed that the number of optical phenomena known to be coherently explainable by the wave theory was considerably greater than the number explainable by the particle theory. So the wave theorist's argument for his conclusion would be stronger than the particle theorist's for his. But even more importantly, the conclusion of the argument is only that O can be coherently explained by the wave theory, not that it cannot be coherently explained by the particle theory. This is not eliminative reasoning.

Argument (6) above might be construed in Millian terms as inductive: concluding "that what is true of certain individuals of a class is true of the whole class," and hence of any other particular individual in that class (Mill 1959, p. 188; see note 3). Mill's definition is quite general and seems to permit an inference from the explanatory success of a theory to its continued explanatory success. Indeed, in his discussion of the wave theory he notes that "if the laws of propagation of light accord with those of the vibrations of an elastic fluid in as many respects as is necessary to make the hypothesis afford a correct expression of all or most of the phenomena known at the time, it is nothing strange that they should accord with each other in one respect more" (Mill 1959, p. 329). Mill seems to endorse this reasoning. What he objects to is concluding from it that the explanation is true or probable.

Argument (6) might also be construed as exhibiting certain Whewellian features. Whewell stresses the idea that a theory is an historical entity which changes over time and can "tend to simplicity and harmony." One of the important aspects of this tendency is that "the elements which we require for explaining a new class of facts are already contained in our system." He explicitly cites the wave theory, by contrast to the particle theory, as exhibiting this tendency. Accordingly, it seems reasonable to suppose that it will be able to coherently explain some hitherto unexplained optical phenomenon. The important difference between Whewell and Mill in this connection is not over whether the previous explanatory argument (6) is valid, but over whether from the continued explanatory success of the wave theory one can infer its truth. For Whewell one can, for Mill one cannot.

Let me assume, then, that some such argument as (6) was at least implicit in the wave theorists' thinking; and that it would have been endorsed by both Mill and Whewell. How, if at all, can it be used to supplement the probabilistic reconstruction of the wave theorists' argument that I offer earlier in the paper? More specifically, how does it relate to the question of determining the probability of the wave theory given all the known optical phenomena, not just some subset?

The conclusion of the explanatory success argument (6) is that the wave theory coherently explains optical phenomenon O. This conclusion is made probable by the fact that the wave theory coherently explains optical phenomena $O_1, \ldots, O_n$.

Accordingly, we have:

(7) p(W coherently explains optical phenomenon O/W coherently explains optical phenomena $O_1, \ldots, O_n) > k$

where k is some threshold of high probability, and W is the wave theory. If we construe such explanations as deductive, then

(8) "W coherently explains O" entails that $p(W/O\&O_1, \ldots, O_n) \geq p(W/O_1, \ldots, O_n)$

So from (7) and (8) we get the second-order probability statement

(9) $p(p(W/O\&O_1, \ldots, O_n) \geq p(W/O_1, \ldots, O_n)$/W coherently explains $O_1, \ldots, O_n) > k$

But the conclusion of the wave theorists' argument is

(10) $p(W/O_1, \ldots, O_n) \approx 1$,

where $O_1, \ldots, O_n$ includes all those phenomena for which the wave theorist supplies a coherent explanation (I suppress reference to background information here). If we add (10) to the conditional side of (9), then from (9) we get

(11) $p(p(W/O\&O_1, \ldots, O_n) \approx 1$/W coherently explains $O_1, \ldots, O_n$ and $p(W/O_1, \ldots, O_n) \approx 1) > k$.

This says that, given that the wave theory coherently explains optical phenomena $O_1, \ldots, O_n$, and that the probability of the wave theory is close to 1 on these phenomena, the probability is high that the wave theory's probability is close to 1 given O—the hitherto unexplained optical phenomenon—together with the other explained phenomena. If we put all the known but hitherto unexplained optical phenomena into O, then we can conclude that the probability is high that the wave theory's probability is close to 1 given all the known optical phenomena.

How is this to be understood? Suppose we construe the probabilities here as representing reasonable degrees of belief. Then the first-order probability can be understood as representing how much belief it is reasonable to have in W; while the second-order probability is interpreted as representing how reasonable it is to have that much belief. Accordingly, conclusion (11) says this:

> Given that the wave theory coherently explains optical phenomena $O_1, \ldots, O_n$, and that it is reasonable to have a degree of belief in the wave theory, on these explained phenomena, that is close to 1, there is a high degree of reasonableness (greater than k) in having a degree of belief in the wave theory W, on both the explained and the unexplained optical phenomena, that is close to 1.

This, of course, does not permit the wave theorist to conclude that $p(W/O\&O_1, \ldots, O_n) \approx 1$, i.e., that the probability of the wave theory on all known optical phenomena—explained and unexplained—is close to 1. But it does permit him to say something stronger than simply that his theory is probable given a partial set of known optical phenomena. It goes beyond a wait-and-see attitude with respect to the unexplained phenomena.

NOTES

1. Reprinted in Lloyd (1877). In what follows page references will be to this.

2. Herschel (1845) does not take as strong a position as Lloyd, although there are passages in which he says that the wave theory is confirmed by experiments (e.g., pp. 473, 486). In his 1830 work he is even more positive. For example:

> It may happen (and it has happened in the case of the undulatory doctrine of light) that such a weight of analogy and probability may become accumulated on the side of an hypothesis that we are compelled to admit one of two things; either that it is an actual statement of what really passes in nature, or that the reality, whatever it be, must run so close a parallel with it, as to admit of expression common to both, at least as far as the phenomena actually known are concerned. (Herschel 1830, pp. 196–7)

3. Although Mill defines induction as involving an inference from observed members of a class to the whole class, he clearly includes inferences to other unobserved members of the class. He writes: "It is true that (as already shown) the process of indirectly ascertaining individual facts is as truly inductive as that by which we establish general truths. But it is not a different kind of induction; it is a form of the same process . . ." (Mill 1959, p. 186).

4. For a proof see Achinstein 1991, pp. 85–6. It might be noted that the introduction of an auxiliary assumption with very low probability does not by itself suffice to show that the other assumptions of the theory are highly improbable.

5. Herschel writes: "We hold it better to state it [the difficulty in explaining dispersion] in its broadest terms, and call on the reader to suspend his condemnation of the doctrine for what it apparently will not explain, till he has become acquainted with the immense variety and complication of the phenomena which it will. The fact is, that neither the corpuscular nor the undulatory, nor any other system which has yet been devised, will furnish that complete and satisfactory explanation of all the phenomena of light which is desirable" (Herschel 1845, p. 450).

REFERENCES

Achinstein, P. (1983), *The Nature of Explanation.* New York: Oxford University Press.

—— (1991), *Particles and Waves.* New York: Oxford University Press.

Brougham, H. (1803), *Edinburgh Review* 1: 451ff.

Cantor, G. (1975), "The Reception of the Wave Theory of Light in Britain," *Historical Studies in the Physical Sciences* 6: 109–132.

—— (1983), *Optics after Newton.* Manchester: Manchester University Press.

Crew, H. (ed.) (1900), *The Wave Theory of Light.* New York: American Book Co.

Fresnel, A. (1816), "Memoir on the Diffraction of Light," reprinted in Crew (1900), 79–144.

Herschel, J. (1845), "Light," *Encyclopedia Metropolitana.*

—— (1830), *A Preliminary Discourse on the Study of Natural Philosophy.* Chicago: University of Chicago Press, 1987.

Laudan, L. (1981), "The Medium and its Message," in *Conceptions of the Ether,* G. Cantor and M. Hodge (eds.). Cambridge: Cambridge University Press.

Lloyd, H. (1834), "Report on the Progress and Present State of Physical Optics," *Reports of the British Association for the Advancement of Science,* 297ff.

—— (1877), *Miscellaneous Papers.* London: Longmans, Green, and Co.

Mill, J. (1959), *A System of Logic.* London: Longmans, Green, and Co.

Whewell, W. (1967), *The Philosophy of the Inductive Sciences,* vol. 2. New York: Johnson Reprint Corporation.

Young, T. (1802), "On the Theory of Light and Colours," reprinted in *Crew* (1900), 45–61.

—— (1845), *A Course of Lectures on Natural Philosophy and the Mechanical Arts.* London: Taylor and Walton.

PART IV

Realism vs. Antirealism & Molecular Reality

Although the views so far presented about scientific rules are very different, they have two important things in common. First, they all accept the idea that scientists who use such rules are justified in inferring the truth or falsity of scientific propositions, or at least inferring their *probable* truth or falsity.[1] Second, they all accept the idea that rules hold universally and objectively: they hold for all scientific reasoning and do not vary from one science or scientist to another. These assumptions will be critically discussed in the last two parts of this volume.

The first assumption, discussed in part 4, is the basis for a longstanding debate among scientists and philosophers of science between so-called scientific realists, who make this assumption, and scientific antirealists, who deny it or want to modify it seriously. The second assumption, dealt with in part 5, is one that came under fire in the middle of the twentieth century with the work of Paul Feyerabend and Thomas Kuhn.

Scientific Realism

According to one standard formulation of scientific realism, scientific theories describe the world using sentences that are either true or false. They are true if they use terms referring to things in the world and if the propositions expressed by the sentences in the theory correctly describe the properties and relations of these things; otherwise, they are false. Some things in the world are observable (e.g., the sun, trees, and rocks); others (e.g., quarks, very distant galaxies, and gravitational forces between celestial bodies) are not observable, owing to size, distance from us, or other facts. Theories can refer to both sorts of entities and make true (or false) claims about them. Scientific realists typically assume that theories in what they call "mature" sciences such as physics are usually true, or approximately true, not false; and hence that the terms in the theory for observables (e.g., "the sun") as well as terms for unobservables (e.g., "gravitational force") refer to things that really exist.

Scientific realists also usually assume that facts about the world can come to be known by scientists, and that such knowledge is possible by using one or the other of the methodologies discussed previously. Thus Descartes believes (1) that a physical world exists and is described by scientific theories that purport to be true; and (2) that if they are true (or if they are false), this fact can come to be known with complete certainty by use of intuition, deduction, and strategies introduced in his twenty-one rules. Newton believed (1) and a crucial modification of (2), according to which such facts can be known, though not with Cartesian certainty, by using causal and inductive reasoning of the sort expressed in his four rules.

Finally, scientific realists usually espouse what is called "metaphysical realism." They hold that scientific theories describe a world that exists independently of the thoughts and theories of scientists. So if space, time, motion, gravitational forces, and electrons exist and have certain properties, they do so whether or not there are any physicists or any theories about such things and their properties. Scientific theories purport to describe such an independently existing world.

In 1980 a prominent contemporary philosopher of science, Bas van Fraassen, published a book entitled *The Scientific Image*, which became influential. In it he proposed a new formulation of scientific realism (a doctrine he proceeded to criticize in the book), and of "constructive empiricism" (a form of antirealism that he defended). (An excerpt from van Fraassen's book is included below.) His new formulation of scientific realism is this: "*Science aims to give us, in its theories, a literally true story of what the world is like; and acceptance of a scientific theory involves the belief that it is true.* This is the correct statement of scientific realism" (chap. 2, sec. 1.1). On this formulation, scientific realism is a doctrine not about the world but about the aim of science. It does not say even that there is a world but only that scientists believe there is and aim to describe it literally and correctly. Further, when scientists accept a theory, they are not simply saying that it is a useful device for organizing their experiences, or for making predictions, or for practical applications. They are saying that they believe that the theory is a literally true description of the way the world is. As van Fraassen recognizes, his formulation of scientific realism is "quite minimal" (*ibid.*). Unlike the first version described above, it does not assert or presuppose the existence of a (theory-independent) world. Nor, of course, does it deny that such a world exists. It claims only that scientists believe that the world exists and seek to describe it.

Other formulations of scientific realism exist. Some of these will be considered in the material excerpted below from contributions by Wesley Salmon and by the editor.

Scientific Antirealism

As with realism, several versions of antirealism exist. On the extreme side is the view that there is no independently existing physical world. What we call such a world is entirely dependent on our minds and theories. One position, called "idealism," is the view of the eighteenth-century British philosopher George Berkeley

(1685–1753) that the only things that exist are minds and the ideas they contain. A contemporary version, defended by some sociologists of science, is called "social constructivism," according to which it is scientists in a community who decide what physical things exist in the universe, and in so deciding, they "create" the universe, or parts thereof, by adopting the theories they do.

In the present volume we will consider less extreme formulations. One of the most important is due to Pierre Duhem (1861–1916), a French physicist who worked in the fields of thermodynamics and physical chemistry. Duhem also wrote important works in the philosophy of science and the history of science. Perhaps his most famous book, published in 1906, is *The Aim and Structure of Physical Theory,* in which he spells out his antirealist position in a manner accessible to a nontechnical audience. (Parts of the first two chapters of that book are included in this volume.)

Duhem distinguishes observable physical phenomena, which he also calls "sensible appearances," from unobservable "material reality," which underlies the appearances. Unlike some more radical antirealists, Duhem is not out to deny that "under the sensible appearances, which are revealed in our perceptions, there is a reality distinct from these appearances" (chap. 1, sec. 2). His claim is simply that if science is to be *empirical,* which he thinks it is, it cannot determine the nature of this unobservable reality. He writes:

> Now these two questions—Does there exist a material reality distinct from sensible appearances? and What is the nature of this reality?—do not have their source in experimental method, which is acquainted only with sensible appearances and can discover nothing beyond them. The resolution of these questions transcends the methods used by physics; it is the object of metaphysics. (*Ibid.*)

For example, Duhem regarded atoms (as well as subatomic particles, such as electrons, which had been postulated by the time he was writing) as unobservable matter invoked to explain observable phenomena. But empirical science, he believed, has no access to such entities, only to the observable effects they are supposed to explain. Moreover, if we examine the history of atomic physics, he pointed out, we will see numerous conflicting theories of what atoms are and how they behave. Such theories include the idea that atoms are the fundamental, indivisible units of matter that interact only by contact; other theories claim that atoms are divisible, contain various parts, and do not interact by contact. Duhem regarded all such theories as metaphysical speculations whose truth cannot be determined empirically. Physics, he thought, has too many such theories. If science is to be empirical, its aim cannot be to uncover truths about hidden reality.

What, then, should the aim of science be? A physical theory is developed by starting with a set of laws established experimentally that heretofore have not been related. For example, in geometrical optics there is the law of rectilinear motion (that light travels in straight lines), the law of reflection (that light is reflected from a surface so that the angle of incidence is equal to the angle of reflection), the law of refraction (that light is refracted by passing through a refracting medium such as a prism so that the sines of the angles of incidence and refraction are proportional), and several others. These laws are established by experiment and observation. What a physicist should attempt to do is "represent" these laws with as much economy as possible by constructing a simple set of mathematically expressed propositions from which the experimental laws can all be deduced. These theoretical propositions are not to be thought of as describing an underlying unobservable reality; they are only convenient, economical devices for representing the experimental laws.

For example, the wave theory of light (discussed in part 3 of the present volume) is not to be construed as describing unobservable wave motions in an unobservable, underlying ether, but only as a simple, economical set of mathematically expressed propositions from which the various experimental laws of geometrical optics can be derived. If the theoretical set is simple and well organized, and if it does entail the experimental laws, the theory is a good one, and we can say that it is "true" (but only in this special sense). If the theoretical set is complex and not well organized, or if it is incompatible with the experimental laws, it is not a good theory, and we can say that it is "false."

Van Fraassen's version of antirealism, "constructive empiricism," is similar to Duhem's in certain ways. Here is his formulation: "Science aims to give us theories which are empirically adequate; and acceptance of a theory involves as belief only that it is empirically adequate" (*The Scientific Image*, chap. 2, sec. 1.3). By "empirically adequate" van Fraassen means that what the theory says about the observable things and events in the world is true. He makes it clear that this includes not just those things and events that have been or will be observed, but all observables, including ones that will never be observed. Van Fraassen, like Duhem, is not denying that unobservables exist. Nor is he claiming that scientists should use terms only if they denote observable entities. Nor is he saying that propositions about unobservables are neither true nor false. Rather, he is saying that to accept a scientific theory postulating unobservables does not commit one to believing that the postulated unobservables exist, or that what the theory says about them is true, but only to the belief that what the theory says about observables is true.

Methodological Rules

What implications does this debate between realists and antirealists have for the status of methodological rules of the sort championed by Descartes, Newton, and hypothetico-deductivist theorists such as Whewell? No changes in the formulation of such rules are required for realists such as these. Newton's four rules of method, for example, permit an inference to the truth of propositions whether or not those propositions postulate unobservable entities. From observed motions of the planets and their satellites, Newton, using his causal rules 1 and 2, infers that there is a unique inverse-square force causing these accelerations; he makes this inference, which he then generalizes by induction to all bodies, despite the fact that such a force is not observable, even if its effects are. For a realist such as Newton, one uses the same rules to infer the existence of unobservables as one uses to infer the existence of observables that have not yet been observed.

The situation is different for the antirealist. The antirealist can use methodological rules of the sorts we have considered only to infer the truth of propositions about observables, not unobservables. In the latter case, the rule, if it is to be used at all, must be significantly changed to permit the conclusion that the inference is to the "empirical adequacy" of the theory, not its truth. For example, from the observed motions of the planets and their satellites Newton should have inferred not that his universal law of gravity is true, but only that it "saves the phenomena"—that all observable phenomena regarded as effects of gravitational attraction are in accord with the law. Another way this might be put is that all bodies exhibiting observable accelerations behave *as if* they were acted on by an unobservable universal force satisfying Newton's law of gravity. The "as if" statement could be true whether or not such a force exists.

Similarly, an antirealist will say, from the fact that in the mid-nineteenth century the wave theory of light satisfied Whewell's criteria of predictivity, consilience, and coherence, Whewell should not have inferred that the theory is true—that is, that there really is an unobservable ether in which light produces unobservable wave motions. At most, Whewell should have inferred that the wave theory of light "saves the phenomena"—that its implications regarding observable phenomena such as reflection, refraction, interference, and so on are true. This could be the case, even if claims about the existence of an ether and light waves in it are false.

Accordingly, a resolution of the realism–antirealism debate will have important implications about how methodological rules are to be formulated, understood, and used.

Jean Perrin and Molecules: An Empirical Argument for Scientific Realism

In 1984 the distinguished philosopher of science Wesley C. Salmon (1927–2001) published a book, *Scientific Explanation and the Causal Structure of the World,* in which he proposed to settle the debate between realists and antirealists *experimentally.* Indeed, Salmon claimed, the issue had already been settled experimentally in favor of realism during the early years of the twentieth century by the French physical chemist Jean Perrin (1870–1942) in his work on Brownian motion.

In 1827 the English botanist Robert Brown discovered that small particles, visible through a microscope, when suspended in a liquid do not sink but exhibit rapid, haphazard motion. This became known as Brownian motion. Various theories were invoked to explain the cause of this motion during the nineteenth century. In 1905–6 Albert Einstein and Maryan von Smoluchowski developed a theoretical approach that could be used to relate the mean square displacement of the suspended Brownian particles to the mean molecular kinetic energy. But the issue was not settled until the experiments and arguments of Jean Perrin beginning in 1908. From his experiments Perrin drew the conclusion that the observable motion of the Brownian particles is being caused by haphazard collisions of these particles with unobservable molecules comprising the fluid.

From the time of the ancient Greek atomists, atoms—and, later, molecules—were postulated by various scientists to explain observed properties and combinations of matter. But this was a hypothesis, and their existence was not considered to be demonstrated conclusively by empirical means. Perrin claimed such a conclusive demonstration. In 1908 he conducted a series of experiments, later described in his book *Atoms* (1913), in which he carefully prepared microscopic particles of the same mass and density and introduced them into a cylinder of known height, containing a dilute liquid of known density and temperature. Using the assumption that molecules in a dilute solution will behave like molecules in a gas with respect to their vertical distribution, he derived an equation relating the number of microscopic Brownian particles per unit volume at the upper and lower levels of the cylinder to the mass and density of the particles, the density of the liquid, the height of the cylinder, the temperature of the liquid, and, most important, a number N, the number of molecules in a substance whose weight in grams equals its molecular weight. (N is called Avogadro's number, after the Italian physicist Amadeo Avogadro who in 1811 proposed the hypothesis that equal volumes of gases at the same pressure and temperature contain the same number of molecules.) In Perrin's experiments, all of the quantities except for N were directly measurable by observation. Perrin measured these quantities in different experiments, employing different types of microscopic particles with different masses and densities. Using the equation he derived, he determined that Avogadro's number N was the same in all cases, approximately 6×10^{23}. From this result Perrin concluded that molecules exist:

> Even if no other information were available as to the molecular magnitudes, *such constant results would justify the very suggestive hypotheses that have guided us* [including that molecules exist], and we should certainly accept as extremely probable the values obtained with such concordance for the masses of the molecules and atoms. . . . The objective reality of the molecules therefore becomes hard to deny. (*Atoms*, chap. 3, sec. 65)

Thus Perrin provided an experimental argument for the claim that molecules exist and cause the observed Brownian motion. Since molecules are unobservable (or a least were at the time of Perrin's experiments), Wesley Salmon concludes that Perrin provided an empirical argument for the claim that unobservable entities exist and cause observable effects, and hence an empirical argument for scientific realism. Salmon focuses on the fact that in his experiments on Brownian motion Perrin provided an empirical determination of Avogadro's number that matched those made from other, hitherto unrelated, phenomena (including X-ray diffraction, alpha particle decay, and blackbody radiation). He suggests that Perrin was using a "common cause principle" according to which, if similar effects have been produced (in this case, experiments involving different phenomena all yielding the same value for Avogadro's number), and if none of these effects causes any of the others, then it can be concluded that the effects all result from a common cause; if no observable common cause is found, one may infer that an unobservable one exists (in the present case, molecules).

Several questions are raised by Salmon's claims. Granting that a common cause argument of the sort proposed by Salmon was actually used by Perrin, is it the only one, or indeed the most important one? Whatever empirical arguments were used by Perrin, do they establish scientific realism in any reasonable sense of that term? If so, how exactly is that sense to be understood? Even if Perrin established that unobservable molecules exist, can scientific realism be understood simply as the view that unobservable entities exist and that what some theories say about them is true? Are there other important assumptions underlying scientific realism? If, as Salmon claims, scientific realism is established by Perrin's arguments, what does this indicate about rules of method? Suppose, indeed, that Perrin's argument shows that one can argue from experimental observations to the truth of propositions about unobservables. Still, the methodological rules for doing so might be different from those used for arguing from observable results to the truth of claims about other observables. We need to think about whether the rules are the same or different.

To deal with these issues raised by Salmon's claims, the selections that follow include excerpts from Perrin's book *Atoms* as well as from Salmon's discussion of Perrin and scientific realism. This part of the volume also contains material from a paper by the editor that distinguishes various types of scientific realism and argues that in a historically and conceptually important sense of that term Perrin's experimental argument provides an empirical basis for scientific realism.

NOTE

1. Popper is the only one who denies the validity of inductive inferences to (probable) truth, although he accepts deductive inferences to the falsity of hypotheses.

15

Duhem's Antirealism

From

The Aim and Structure of Physical Theory /
Pierre Duhem

Chapter 1. Physical Theory and Metaphysical Explanation

1. Physical Theory Considered as Explanation

The first question we should face is: What is the aim of a physical theory? To this question diverse answers have been made, but all of them may be reduced to two main principles:

"A physical theory," certain logicians have replied, "has for its object the *explanation* of a group of laws experimentally established."

"A physical theory," other thinkers have said, "is an abstract system whose aim is to *summarize* and *classify logically* a group of experimental laws without claiming to explain these laws."

We are going to examine these two answers one after the other, and weigh the reasons for accepting or rejecting each of them. We begin with the first, which regards a physical theory as an explanation.

Bracketed interpolations are the translator's.

But, first, what is an explanation?

To explain (explicate, *explicare*) is to strip reality of the appearances covering it like a veil, in order to see the bare reality itself.

The observation of physical phenomena does not put us into relation with the reality hidden under the sensible appearances, but enables us to apprehend the sensible appearances themselves in a particular and concrete form. Besides, experimental laws do not have material reality for their object, but deal with these sensible appearances, taken, it is true, in an abstract and general form. Removing or tearing away the veil from these sensible appearances, theory proceeds into and underneath them, and seeks what is really in bodies.

For example, string or wind instruments have produced sounds to which we have listened closely and which we have heard become stronger or weaker, higher or lower, in a thousand nuances productive in us of auditory sensations and musical emotions; such are the acoustic facts.

These particular and concrete sensations have been elaborated by our intelligence, following the laws by which it functions, and have provided us with such general and abstract notions as intensity, pitch, octave, perfect major or minor chord, timbre, etc. The experimental laws of acoustics aim at the enunciation of fixed relations among these and other equally abstract and general notions. A law, for example, teaches us what relation exists between the dimensions of two strings of the same metal which yield two sounds of the same pitch or two sounds an octave apart.

But these abstract notions—sound intensity, pitch, timbre, etc.—depict to our reason no more than the general characteristics of our sound perceptions; these notions get us to know sound as it is in relation to us, not as it is by itself in sounding bodies. This reality whose external veil alone appears in our sensations is made known to us by theories of acoustics. The latter are to teach us that where our perceptions grasp only that appearance we call sound, there is in reality a very small and very rapid periodic motion; that intensity and pitch are only external aspects of the amplitude and frequency of this motion; and that timbre is the apparent manifestation of the real structure of this motion, the complex sensation which results from the diverse vibratory motions into which we can analyze it. Acoustic theories are therefore explanations.

The explanation which acoustic theories give of experimental laws governing sound claims to give us certainty; it can in a great many cases make us see with our own eyes the motions to which it attributes these phenomena, and feel them with our fingers.

Most often we find that physical theory cannot attain that degree of perfection; it cannot offer itself as a *certain* explanation of sensible appearances, for it cannot render accessible to the senses the reality it proclaims as residing underneath those appearances. It is then content with proving that all our perceptions are produced *as if* the reality were what it asserts; such a theory is a hypothetical explanation.

Let us, for example, take the set of phenomena observed with the sense of sight. The rational analysis of these phenomena leads us to conceive certain abstract and general notions expressing the properties we come across in every perception of light: a simple or complex color, brightness, etc. Experimental laws of optics make us acquainted with fixed relations among these abstract and general notions as well as among other analogous notions. One law, for instance, connects the intensity of yellow light reflected by a thin plate with the thickness of the plate and the angle of incidence of the rays which illuminate it.

Of these experimental laws the vibratory theory of light gives a hypothetical explanation. It supposes that all the bodies we see, feel, or weigh are immersed in an imponderable, unobservable medium called the ether. To this ether certain mechanical properties are attributed; the theory states that all simple light is a transverse vibration, very small and very rapid, of this ether, and that the frequency and amplitude of this vibration characterize the color of this light and its brightness; and, without enabling us to perceive the ether, without putting us in a position to observe directly the back-and-forth motion of light vibration, the theory tries to prove that its postulates entail consequences agreeing at every point with the laws furnished by experimental optics.

2. According to the Foregoing Opinion, Theoretical Physics Is Subordinate to Metaphysics

When a physical theory is taken as an explanation, its goal is not reached until every sensible appearance has been removed in order to grasp the physical reality. For example, Newton's research on the dispersion of light has taught us to decompose the sensation we experience of light emanating from the sun; his experiments have shown us that this light is complex and resolvable into a certain number of simpler light phenomena, each associated with a determinate and invariable color. But these simple or monochromatic light data are abstract and general representations of certain sensations; they are sensible appearances, and we have only dissociated a complicated appearance into other simpler appearances. But we have not reached the real thing, we have not given an explanation of the color effects, we have not constructed an optical theory.

Thus, it follows that in order to judge whether a set of propositions constitutes a physical theory or not, we must inquire whether the notions connecting these propositions express, in an abstract and general form, the elements which really go to make up material things, or merely represent the universal properties perceived.

For such an inquiry to make sense or to be at all possible, we must first of all regard as certain the following affirmation: Under the sensible appearances, which are revealed in our perceptions, there is a reality distinct from these appearances.

This point granted, and without it the search for a physical explanation could not be conceived, it is impossible to recognize having reached such an explanation until we have answered this next question: What is the nature of the elements which constitute material reality?

Now these two questions—Does there exist a material reality distinct from sensible appearances? and What is the nature of this reality?—do not have their source in experimental method, which is acquainted only with sensible appearances and can discover nothing beyond them. The resolution of these questions transcends the methods used by physics; it is the object of metaphysics.

Therefore, *if the aim of physical theories is to explain experimental laws, theoretical physics is not an autonomous science; it is subordinate to metaphysics.*

3. According to the Foregoing Opinion, the Value of a Physical Theory Depends on the Metaphysical System One Adopts

The propositions which make up purely mathematical sciences are, to the highest degree, universally accepted truths. The precision of language and the rigor of the methods of demonstration leave no room for any permanent divergences among the views of different mathematicians; over the centuries doctrines are developed by continuous progress without new conquests causing the loss of any previously acquired domains.

There is no thinker who does not wish for the science he cultivates a growth as calm and as regular as that of mathematics. But if there is a science for which this wish seems particularly legitimate, it is indeed theoretical physics, for of all the well-established branches of knowledge it surely is the one which least departs from algebra and geometry.

Now, to make physical theories depend on metaphysics is surely not the way to let them enjoy the privilege of universal consent. In fact, no philosopher, no matter how confident he may be in the value of the methods used in dealing with metaphysical problems, can dispute the following empirical truth: Consider in review all the domains of man's intellectual activity; none of the systems of thought arising in different eras or the contemporary systems born of different schools will appear more profoundly distinct, more sharply separated, more violently opposed to one another, than those in the field of metaphysics.

If theoretical physics is subordinated to metaphysics, the divisions separating the diverse metaphysical systems will extend into the domain of physics. A physical theory reputed to be satisfactory by the sectarians of one metaphysical school will be rejected by the partisans of another school.

Consider, for example, the theory of the action exerted by a magnet on iron, and suppose for a moment that we are Aristotelians.

What does the metaphysics of Aristotle teach us concerning the real nature of bodies? Every substance—in particular, every material substance—results from the union of two elements: one permanent (matter) and one variable (form). Through its permanence, the piece of matter before me

remains always and in all circumstances the same piece of iron. Through the variations which its form undergoes, through the *alterations* that it experiences, the properties of this same piece of iron may change according to circumstances; it may be solid or liquid, hot or cold, and assume such and such a shape.

Placed in the presence of a magnet, this piece of iron undergoes a special alteration in its form, becoming more intense with the proximity of the magnet. This alteration corresponds to the appearance of two poles and gives the piece of iron a principle of movement such that one pole tends to draw near the pole opposite to it on the magnet and the other to be repelled by the one designated as the like pole on the magnet.

Such for the Aristotelian philosopher is the reality hidden under the magnetic phenomena; when we have analyzed all these phenomena by reducing them to the properties of the magnetic quality of the two poles, we have given a complete explanation and formulated a theory altogether satisfactory. It was such a theory that Niccolo Cabeo constructed in 1629 in his remarkable work on magnetic philosophy.[1]

If an Aristotelian declares he is satisfied with the theory of magnetism as Father Cabeo conceives it, the same will not be true of a Newtonian philosopher faithful to the cosmology of Father Boscovich.

According to the natural philosophy which Boscovich has drawn from the principles of Newton and his disciples,[2] to explain the laws of the action which the magnet exerts on the iron by a magnetic alteration of the substantial form of the iron is to explain nothing at all; we are really concealing our ignorance of reality under words that sound deep but are hollow.

Material substance is not composed of matter and form; it can be resolved into an immense number of points, deprived of extension and shape but having mass; between any two of these points is exerted a mutual attraction or repulsion proportional to the product of the masses and to a certain function of the distance separating them. Among these points there are some which form the bodies themselves. A mutual action takes place among the latter points, and as soon as the distances separating them exceed a certain limit, this action becomes the universal gravitation studied by Newton. Other points, deprived of this action of gravity, compose weightless fluids such as electric fluids and calorific fluid. Suitable assumptions about the masses of all these material points, about their distribution, and about the form of the functions of the distance on which their mutual actions depend are to account for all physical phenomena.

For example, in order to explain magnetic effects, we imagine that each molecule of iron carries equal masses of south magnetic fluid and north magnetic fluid; that the distribution of the fluids about this molecule is governed by the laws of mechanics; that two magnetic masses exert on one another an action proportional to the product of those masses and to the inverse square of the distance between them; finally, that this action is a repulsion or an attraction according to whether the masses are of the same or of different kinds. Thus was developed the theory of magnetism which, inaugurated by Franklin, Oepinus, Tobias Mayer, and Coulomb, came to full flower in the classical memoirs of Poisson.

Does this theory give an explanation of magnetic phenomena capable of satisfying an atomist? Surely not. Among some portions of magnetic fluid distant from one another, the theory admits the existence of actions of attraction or repulsion; for an atomist such actions at a distance amount to appearances which cannot be taken for realities.

According to the atomistic teachings, matter is composed of very small, hard, and rigid bodies of diverse shapes, scattered profusely in the void. Separated from each other, two such corpuscles cannot in any way influence each other; it is only when they come in contact with one another that their impenetrable natures clash and that their motions are modified according to fixed laws. The magnitudes, shapes, and masses of the atoms, and the rules governing their impact alone provide the sole satisfactory explanation which physical laws can admit.

In order to explain in an intelligible manner the various motions which a piece of iron undergoes in the presence of a magnet, we have to imagine that floods of magnetic corpuscles escape from the magnet in compressed, though invisible and intangible, streams, or else are precipitated toward the magnet. In their rapid course these corpuscles collide in various ways with the molecules of the iron, and from these collisions arise the forces which a superficial philosophy attributed to magnetic attraction and repulsion. Such is the principle of a theory of the magnet's properties already outlined by Lucretius, developed by Gassendi in the seventeenth century, and often taken up again since that time.

Shall we not find more minds, difficult to satisfy, who condemn this theory for not explaining anything at all and for taking appearances for reality? Here is where the Cartesians appear.

According to Descartes, matter is essentially identical with the extended in length, breadth, and depth, as the language of geometry goes; we have to consider only its various shapes and motions. Matter for the Cartesians is, if you please, a kind of vast fluid, incompressible and absolutely homogeneous. Hard, unbreakable atoms and the empty spaces separating them are merely so many appearances, so many illusions. Certain parts of the universal fluid may be animated by constant whirling or vortical motions; to the coarse eyes of the atomist these whirlpools or vortices will look like individual corpuscles. The intermediary fluid transmits from one vortex to the other forces which Newtonians, through insufficient analysis, will take for actions at a distance. Such are the principles of the physics first sketched by Descartes, which Malebranche investigated further, and to which W. Thomson, aided by the hydrodynamic researches of Cauchy and Helmholtz, has given the elaboration and precision characteristic of present-day mathematical doctrines.

This Cartesian physics cannot dispense with a theory of magnetism; Descartes had already tried to construct such a theory. The corkscrews of "subtle matter" with which Descartes, not without some naïveté, in his theory replaced the magnetic corpuscles of Gassendi were succeeded, among the Cartesians of the nineteenth century, by the vortices conceived more scientifically by Maxwell.

Thus we see each philosophical school glorifying a theory which reduces magnetic phenomena to the elements with which it composes the essence of matter, but the other schools rejecting this theory, in which their principles do not let them recognize a satisfactory explanation of magnetism.

4. The Quarrel over Occult Causes

There is one form of criticism which very often occurs when one cosmological school attacks another school: the first accuses the second of appealing to "occult causes."

The great cosmological schools, the Aristotelian, the Newtonian, the atomistic, and the Cartesian, may be arranged in an order such that each admits the existence in matter of a smaller number of essential properties than the preceding schools are willing to admit.

The Aristotelian school composes the substance of bodies out of only two elements, matter and form; but this form may be affected by qualities whose number is not limited. Each physical property can thus be attributed to a special quality: a *sensible* quality, directly accessible to our perception, like weight, solidity, fluidity, heat, or brightness; or else an *occult* quality whose effects alone will appear in an indirect manner, as with magnetism or electricity.

The Newtonians reject this endless multiplication of qualities in order to simplify, to a high degree, the notion of material substance: in the elements of matter they leave only masses, mutual actions, and shapes, when they do not go as far as Boscovich and several of his successors, who reduce the elements to unextended points.

The atomistic school goes further: its material elements preserve mass, shape, and hardness. But the forces through which the elements act on one another, according to the Newtonian school, disappear from the domain of realities; they are regarded merely as appearances and fictions.

Finally, the Cartesians push to the limit this tendency to strip material substances of various properties: they reject the hardness of atoms and even the distinction between plenum and void, in order to identify matter, as Leibniz said, with "completely naked extension and its modification."[3]

Thus each cosmological school admits in its explanations certain properties of matter which the next school refuses to take as real, for the latter regards them as mere words designating more deeply hidden realities without revealing them; it groups them, in short, with the occult qualities created in so much profusion by scholasticism.

It is hardly necessary to recall that all the cosmological schools other than the Aristotelian have agreed in attacking the latter for the arsenal of qualities which it stored in substantial form, an arsenal which added a new quality each time a new phenomenon had to be explained. But Aristotelian physics has not been the only one obliged to meet such criticisms.

The Newtonians who endow material elements with attractions and repulsions acting at a distance seem to the atomists and Cartesians to be adopting one of those purely verbal explanations usual with the old Scholasticism. Newton's *Principia* had hardly been published when his work excited the sarcasm of the atomistic clan grouped around Huygens. "So far as concerns the cause of the tides given by Mr. Newton," Huygens wrote Leibniz, "I am far from satisfied, nor do I feel happy about any of his other theories built on his principle of attraction, which to me appears absurd."[4]

If Descartes had been alive at that time, he would have used a language similar to that of Huygens. In fact, Father Mersenne had submitted to Descartes a work by Roberval[5] in which the author adopted a form of universal gravitation long before Newton. On April 20, 1646 Descartes expressed his opinion as follows:

"Nothing is more absurd than the assumption added to the foregoing; the author assumes that a certain property is inherent in each of the parts of the world's matter and that, by the force of this property, the parts are carried toward one another and attract each other. He also assumes that a like property inheres in each part of the earth considered in relation with the other parts of the earth, and that this property does not in any way disturb the preceding one. In order to understand this, we must not only assume that each material particle is animated, and even animated by a large number of diverse souls that do not disturb each other, but also that these souls of material particles are endowed with knowledge of a truly divine sort, so that they may know without any medium what takes place at very great distances and act accordingly."[6]

The Cartesians agree, then, with the atomists when it comes to condemning as an occult quality the action at a distance which Newtonians invoke in their theories; but turning next against the atomists, the Cartesians deal just as harshly with the hardness and indivisibility attributed to corpuscles by the atomists. The Cartesian Denis Papin wrote to the atomist Huygens: "Another thing that bothers me is . . . that you believe that perfect hardness is of the essence of bodies; it seems to me that you are there assuming an inherent quality which takes us beyond mathematical or mechanical principles."[7] The atomist Huygens, it is true, did not deal less harshly with Cartesian opinion: "Your other difficulty," he replied to Papin, "is that I assume hardness to be of the essence of bodies whereas you and Descartes admit only their extension. By which I see that you have not yet rid yourself of that opinion which I have for a long time judged very absurd."[8]

5. No Metaphysical System Suffices in Constructing a Physical Theory

Each of the metaphysical schools scolds its rivals for appealing in its explanations to notions which are themselves unexplained and are really occult qualities. Could not this criticism be nearly always applied to the scolding school itself?

In order for the philosophers belonging to a certain school to declare themselves completely satisfied with a theory constructed by the physicists of the same school, all the principles used in this theory would have to be deduced from the metaphysics professed by that school. If an appeal is made, in the course of the explanation of a physical phenomenon, to some law which that metaphysics is powerless to justify, then no explanation will be forthcoming and physical theory will have failed in its aim.

Now, no metaphysics gives instruction exact enough or detailed enough to make it possible to derive all the elements of a physical theory from it.

In fact, the instruction furnished by a metaphysical doctrine concerning the real nature of bodies consists most often of negations. The Aristotelians, like the Cartesians, deny the possibility of empty space; the Newtonians reject any quality which is not reducible to a force acting among material points; the atomists and Cartesians deny any action at a distance;

the Cartesians do not recognize among the diverse parts of matter any distinctions other than shape and motion.

All these negations are appropriately argued when it is a matter of condemning a theory proposed by an adverse school; but they appear singularly sterile when we wish to derive the principles of a physical theory.

Descartes, for example, denied that there is anything else in matter than extension in length, breadth, depth, and its diverse modes—that is to say, shapes and motions; but with these data alone, he could not even begin to sketch the explanation of a physical law.

At the very least, before attempting the construction of any theory, he would have had to know the general laws governing diverse motions. Hence, he proceeded from his metaphysical principles to attempt first of all to deduce a dynamics.

The perfection of God requires him to be immutable in his plans; from this immutability the following consequence is drawn: God preserves as constant the quantity of motion that he gave the world in the beginning.

But this constancy of the quantity of motion in the world is still not a sufficiently precise or definite principle to make it possible for us to write any equation of dynamics. We must state it in a quantitative form, and that means translating the hitherto very vague notion of "quantity of motion" into a completely determined algebraic expression.

What, then, will be the mathematical meaning to be attached by the physicist to the words "quantity of motion"?

According to Descartes, the quantity of motion of each material particle will be the product of its mass—or of its volume, which in Cartesian physics is identical with its mass—times the velocity with which it is animated, and the quantity of motion of all matter in its entirety will be the sum of the quantities of motion of its diverse parts. This sum should in any physical change retain a constant value.

Certainly the combination of algebraic magnitudes through which Descartes proposed to translate the notion of "quantity of motion" satisfies the requirements imposed in advance by our instinctive knowledge of such a translation. It is zero for a whole at rest, and always positive for a group of bodies agitated by a certain motion; its value increases when a determined mass increases the velocity of its movement; it increases again when a given velocity affects a larger mass. But an infinity of other expressions might just as well have satisfied these requirements: for the velocity we might notably have substituted the square of the velocity. The algebraic expression obtained would then have coincided with what Leibniz was to call "living force"; instead of drawing from divine immutability the constancy of the Cartesian quantity of motion in the world, we should have deduced the constancy of the Leibnizian living force.

Thus, the law which Descartes proposed to place at the base of dynamics undoubtedly agrees with the Cartesian metaphysics; but this agreement is not necessary. When Descartes reduced certain physical effects to mere consequences of such a law, he proved, it is true, that these effects do not contradict his principles of philosophy, but he did not give an explanation of the law by means of these principles.

What we have just said about Cartesianism can be repeated about any metaphysical doctrine which claims to terminate in a physical theory; in this theory there are always posited certain hypotheses which do *not* have as their grounds the principles of the metaphysical doctrine. Those who follow the thought of Boscovich admit that all the attractions or repulsions which are observable at a perceptible distance vary inversely with the square of the distance. It is this hypothesis which permits them to construct three systems of mechanics: celestial, electrical, and magnetic; but this form of law is dictated to them by the desire to have their explanations agree with the facts and not by the requirements of their philosophy. The atomists admit that a certain law governs the collisions of corpuscles; but this law is a singularly bold extension to the atomic world of another law which is permissible only when masses big enough to be observed are considered; it is not deduced from the Epicurean philosophy.

We cannot therefore derive from a metaphysical system all the elements necessary for the construction of a physical theory. The latter always appeals to propositions which the metaphysical system has not furnished and which consequently remain mysteries for the partisans of that system. At the root of the explanations it claims to give there always lies the unexplained.

. . . .

Chapter 2. Physical Theory and Natural Classification

1. What Is the True Nature of a Physical Theory and the Operations Constituting It?

While we regard a physical theory as a hypothetical explanation of material reality, we make it dependent on metaphysics. In that way, far from giving it a form to which the greatest number of minds can give their assent, we limit its acceptance to those who acknowledge the philosophy it insists on. But even they cannot be entirely satisfied with this theory since it does not draw all its principles from the metaphysical doctrine from which it is claimed to be derived.

These thoughts, discussed in the preceding chapter, lead us quite naturally to ask the following two questions:

Could we not assign an aim to physical theory that would render it *autonomous?* Based on principles which do not arise from any metaphysical doctrine, physical theory might be judged in its own terms without including the opinions of physicists who depend on the philosophical schools to which they may belong.

Could we not conceive a method which might be *sufficient* for the construction of a physical theory? Consistent with its own definition the theory would employ no principle and have no recourse to any procedure which it could not legitimately use.

We intend to concentrate on this aim and this method, and to study both.

Let us posit right now a definition of physical theory; the sequel of this book will clarify it and will develop its complete content: A physical theory is not an explanation. It is a system of mathematical propositions, deduced from a small number of principles, which aim to represent as simply, as completely, and as exactly as possible a set of experimental laws.

In order to start making this definition somewhat more precise, let us characterize the four successive operations through which a physical theory is formed:

1. Among the physical properties which we set ourselves to represent we select those we regard as simple properties, so that the others will supposedly be groupings or combinations of them. We make them correspond to a certain group of mathematical symbols, numbers, and magnitudes, through appropriate methods of measurement. These mathematical symbols have no connection of an intrinsic nature with the properties they represent; they bear to the latter only the relation of sign to thing signified. Through methods of measurement we can make each state of a physical property correspond to a value of the representative symbol, and vice versa.

2. We connect the different sorts of magnitudes, thus introduced, by means of a small number of propositions which will serve as principles in our deductions. These principles may be called "hypotheses" in the etymological sense of the word for they are truly the grounds on which the theory will be built; but they do not claim in any manner to state real relations among the real properties of bodies. These hypotheses may then be formulated in an arbitrary way. The only absolutely impassable barrier which limits this arbitrariness is logical contradiction either among the terms of the same hypothesis or among the various hypotheses of the same theory.

3. The diverse principles or hypotheses of a theory are combined together according to the rules of mathematical analysis. The requirements of algebraic logic are the only ones which the theorist has to satisfy in the course of this development. The magnitudes on which his calculations bear

are not claimed to be physical realities, and the principles he employs in his deductions are not given as stating real relations among those realities; therefore it matters little whether the operations he performs do or do not correspond to real or conceivable physical transformations. All that one has the right to demand of him is that his syllogisms be valid and his calculations accurate.

4. The various consequences thus drawn from the hypotheses may be translated into as many judgments bearing on the physical properties of the bodies. The methods appropriate for defining and measuring these physical properties are like the vocabulary and key permitting one to make this translation. These judgments are compared with the experimental laws which the theory is intended to represent. If they agree with these laws to the degree of approximation corresponding to the measuring procedures employed, the theory has attained its goal, and is said to be a good theory; if not, it is a bad theory, and it must be modified or rejected.

270

Thus a true theory is not a theory which gives an explanation of physical appearances in conformity with reality; it is a theory which represents in a satisfactory manner a group of experimental laws. A false theory is not an attempt at an explanation based on assumptions contrary to reality; it is a group of propositions which do not agree with the experimental laws. *Agreement with experiment is the sole criterion of truth for a physical theory.*

The definition we have just outlined distinguishes four fundamental operations in a physical theory: (1) the definition and measurement of physical magnitudes; (2) the selection of hypotheses; (3) the mathematical development of the theory; (4) the comparison of the theory with experiment.

Each one of these operations will occupy us in detail as we proceed with this book, for each of them presents difficulties calling for minute analysis. But right now it is possible for us to answer a few questions and to refute a few objections raised by the present definition of physical theory.

2. What Is the Utility of a Physical Theory? Theory Considered as an Economy of Thought

And first, of what use is such a theory?

Concerning the very nature of things, or the realities hidden under the phenomena we are studying, a theory conceived on the plan we have just drawn teaches us absolutely nothing, and does not claim to teach us anything. Of what use is it, then? What do physicists gain by replacing the laws which experimental method furnishes directly with a system of mathematical propositions representing those laws?

First of all, instead of a great number of laws offering themselves as independent of one another, each having to be learnt and remembered on its own account, physical theory substitutes a very small number of propositions, viz., fundamental hypotheses. The hypotheses once known, mathematical dedution permits us with complete confidence to call to mind all the physical laws without omission or repetition. Such condensing of a multitude of laws into a small number of principles affords enormous relief to the human mind, which might not be able without such an artifice to store up the new wealth it acquires daily.

The reduction of physical laws to theories thus contributes to that "intellectual economy" in which Ernst Mach sees the goal and directing principle of science.[9]

The experimental law itself already represented a first intellectual economy. The human mind had been facing an enormous number of concrete facts, each complicated by a multitude of details of all sorts; no man could have embraced and retained a knowledge of all these facts; none could have communicated this knowledge to his fellows. Abstraction entered the scene. It brought about the removal of everything private or individual from these facts, extracting from their total only what was general in them or common to them, and in place of this cumbersome mass of facts it has substituted a single proposition, occupying little of one's memory and easy to convey through instruction: it has formulated a physical law.

"Thus, instead of noting individual cases of light-refraction, we can mentally reconstruct all present and future cases, if we know that the incident ray, the refracted ray, and the perpendicular lie in the same plane and that $\sin i/\sin r = n$. Here, instead of the numberless cases of refraction in different combinations of matter and under all different angles of incidence, we have simply to note the rule above stated and the values of n—which is much easier. The economical purpose here is unmistakable."[10]

The economy achieved by the substitution of the law for the concrete facts is redoubled by the mind when it condenses experimental laws into theories. What the law of refraction is to the innumerable facts of refraction, optical theory is to the infinitely varied laws of light phenomena.

Among the effects of light only a very small number had been reduced to laws by the ancients; the only laws of optics they knew were the law of the rectilinear propagation of light and the laws of reflection. This meager contingent was reinforced in Descartes' time by the law of refraction. An optics so slim could do without theory; it was easy to study and teach each law by itself.

Today, on the contrary, how can a physicist who wishes to study optics, as we know it, acquire even a superficial knowledge of this enormous domain without the aid of a theory? Consider the effects of simple refraction, of double refraction by uniaxial or biaxial crystals, of reflection on isotropic or crystalline media, of interference, of diffraction, of polarization by reflection and by simple or double refraction, of chromatic polarization, of rotary polarization, etc. Each one of these large categories of phenomena may occasion the statement of a large number of experimental laws whose number and complication would frighten the most capable and retentive memory.

Optical theory supervenes, takes possession of these laws, and condenses them into a small number of principles. From these principles we can always, through regular and sure calculation, extract the law we wish to use. It is no longer necessary, therefore, to keep watch over the knowledge of all these laws; the knowledge of the principles on which they rest is sufficient.

This example enables us to take firm hold of the way the physical sciences progress. The experimenter constantly brings to light facts hitherto unsuspected and formulates new laws, and the theorist constantly makes it possible to store up these acquisitions by imagining more condensed representations, more economical systems. The development of physics incites a continual struggle between "nature that does not tire of providing" and reason that does not wish "to tire of conceiving."

3. Theory Considered as Classification

Theory is not solely an economical representation of experimental laws; it is also a *classification* of these laws.

Experimental physics supplies us with laws all lumped together and, so to speak, on the same plane, without partitioning them into groups of laws united by a kind of family tie. Very often quite accidental causes or rather superficial analogies have led observers in their research to bring together different laws. Newton put into the same work the laws of the dispersion of light crossing a prism and the laws of the colors adorning a soap bubble, simply because of the colors that strike the eye in these two sorts of phenomena.

On the other hand, theory, by developing the numerous ramifications of the deductive reasoning which connects principles to experimental laws, establishes an order and a classification among these laws. It brings some laws together, closely arranged in the same group; it separates some of the others by placing them in two groups very far apart. Theory gives, so to speak, the table of contents and the chapter headings under which the science to be studied will be methodically divided, and it indicates the laws which are to be arranged under each of these chapters.

Thus, alongside the laws which govern the spectrum formed by a prism it arranges the laws governing the colors of the rainbow; but the laws according to which the colors of Newton's rings are ordered go elsewhere to join the laws of fringes discovered by Young and Fresnel; still in another category, the elegant coloration analyzed by Grimaldi is considered related to the diffraction spectra produced by Fraunhofer. The laws of all these phenomena, whose striking colors lead to their confusion in the eyes of the simple observer, are, thanks to the efforts of the theorist, classified and ordered.

These classifications make knowledge convenient to use and safe to apply. Consider those utility cabinets where tools for the same purpose lie side by side, and where partitions logically separate instruments not designed for the same task: the worker's hand quickly grasps, without fumbling or mistake, the tool needed. Thanks to theory, the physicist finds with certitude, and without omitting anything useful or using anything superfluous, the laws which may help him solve a given problem.

Order, wherever it reigns, brings beauty with it. Theory not only renders the group of physical laws it represents easier to handle, more convenient, and more useful, but also more beautiful.

It is impossible to follow the march of one of the great theories of physics, to see it unroll majestically its regular deductions starting from initial hypotheses, to see its consequences represent a multitude of experimental laws down to the smallest detail, without being charmed by the beauty of such a construction, without feeling keenly that such a creation of the human mind is truly a work of art.

4. A Theory Tends to Be Transformed into a Natural Classification[11]

This esthetic emotion is not the only reaction that is produced by a theory arriving at a high degree of perfection. It persuades us also to see a natural classification in a theory.

Now first, what is a natural classification? For example, what does a naturalist mean in proposing a natural classification of vertebrates?

The classification he has imagined is a group of intellectual operations not referring to concrete individuals but to abstractions, species; these species are arranged in groups, the more particular under the more general. In order to form these groups the naturalist considers the diverse organs—vetebral column, cranium, heart, digestive tube, lungs, swim-bladder—not in the particular and concrete forms they assume in each individual, but in the abstract, general, schematic forms which fit all the species of the same group. Among these organs thus transfigured by abstraction he establishes comparisons, and notes analogies and differences; for example, he declares the swim-bladder of fish analogous to the lung of vertebrates. These homologies are purely ideal connections, not referring to real organs but to generalized and simplified conceptions formed in the mind of the naturalist; the classification is only a synoptic table which summarizes all these comparisons.

When the zoologist asserts that such a classification is natural, he means that those ideal connections established by his reason among abstract conceptions correspond to real relations among the associated creatures brought together and embodied in his abstractions. For example, he means that the more or less striking resemblances which he has noted among various species are the index of a more or less close blood-relationship, properly speaking, among the individuals composing these species; that the cascades through which he translates the subordination of classes, of orders, of families, and of genera reproduce the genealogical tree in which the various vertebrates are branched out from the same trunk and root. These relations of real family affiliation can be established only by comparative anatomy; to grasp them in themselves and put them in evidence is the business of physiology and of paleontology. However, when he contemplates the order which his methods of comparison introduce into the confused multitude of animals, the anatomist cannot assert these relations, the proof of which transcends his methods. And if physiology and paleontology should someday demonstrate to him that the relationship imagined by him cannot be, that the evolutionist hypothesis is controverted, he would continue to believe that the plan drawn by his classification depicts real relations among animals; he would admit being deceived about the nature of these relations but not about their existence.

The neat way in which each experimental law finds its place in the classification created by the physicist and the brilliant clarity imparted to this group of laws so perfectly ordered persuade us in an overwhelming manner that such a classification is not purely artificial, that such an order does not result from a purely arbitrary grouping imposed on laws by an ingenious organizer. Without being able to explain our conviction, but also without being able to get rid of it, we see in the exact ordering of this system the mark by which a natural classification is recognized. Without claiming to explain the reality hiding under the phenomena whose laws we group, we feel that the groupings established by our theory correspond to real affinities among the things themselves.

The physicist who sees in every theory an explanation is convinced that he has grasped in light vibration the proper and intimate basis of the quality which our senses reveal in the form of light and color; he believes in an ether, a body whose parts are excited by this vibration into a rapid to-and-fro motion.

Of course, we do not share these illusions. When, in the course of an optical theory, we talk about luminous vibration, we no longer think of a real to-and-fro motion of a real body; we imagine only an abstract magnitude, i.e., a pure, geometrical expression. It is a periodically variable length which helps us state the hypotheses of optics, and to regain by regular calculations the experimental laws governing light. This vibration is to our mind a *representation,* and not an *explanation.*

But when, after much groping, we succeed in formulating with the aid of this vibration a body of fundamental hypotheses, when we see in the plan drawn by these hypotheses a vast domain of optics, hitherto encumbered by so many details in so confused a way, become ordered and organized; it is impossible for us to believe that this order and this organization are not the reflected image of a real order and organization; that the phenomena which are brought together by the theory, e.g., interference bands and colorations of thin layers, are not in truth slightly different manifestations of the same property of light; and that phenomena separated by the theory, e.g., the spectra of diffraction and of dispersion, do not have good reasons for being in fact essentially different.

Thus, physical theory never gives us the explanation of experimental laws; it never reveals realities hiding under the sensible appearances; but the more complete it becomes, the more we apprehend that the logical order in which theory orders experimental laws is the reflection of an ontological order, the more we suspect that the relations it establishes among the data of observation correspond to real relations among things,[12] and the more we feel that theory tends to be a natural classification.

The physicist cannot take account of this conviction. The method at his disposal is limited to the data of observation. It therefore cannot prove that the order established among experimental laws reflects an order transcending experience; which is all the more reason why his method cannot suspect the nature of the real relations corresponding to the relations established by theory.

But while the physicist is powerless to justify this conviction, he is nonetheless powerless to rid his reason of it. In vain is he filled with the idea that his theories have no power to grasp reality, and that they serve only to give experimental laws a summary and classificatory representation. He cannot compel himself to believe that a system capable of ordering so simply and so easily a vast number of laws, so disparate at first encounter, should be a purely artificial system. Yielding to an intuition which Pascal would have recognized as one of those reasons of the heart "that reason does not know," he asserts his faith in a real order reflected in his theories more clearly and more faithfully as time goes on.

Thus the analysis of the methods by which physical theories are constructed proves to us with complete evidence that these theories cannot be offered as explanations of experimental laws; and, on the other hand, an act of faith, as incapable of being justified by this analysis as of being frustrated by it, assures us that these theories are not a purely artificial system, but a natural classification. And so, we may here apply that profound thought of Pascal: "We have an impotence to prove, which cannot be conquered by any dogmatism; we have an idea of truth which cannot be conquered by any Pyrrhonian skepticism."

5. Theory Anticipating Experiment

276

There is one circumstance which shows with particular clarity our belief in the natural character of a theoretical classification; this circumstance is present when we ask of a theory that it tell us the results of an experiment before it has occurred, when we give it the bold injunction: "Be a prophet for us."

A considerable group of experimental laws had been established by investigators; the theorist has proposed to condense the laws into a very small number of hypotheses, and has succeeded in doing so, each one of the experimental laws is correctly represented by a consequence of these hypotheses.

But the consequences that can be drawn from these hypotheses are unlimited in number; we can, then, draw some consequences which do not correspond to any of the experimental laws previously known, and which simply represent possible experimental laws.

Among these consequences, some refer to circumstances realizable in practice, and these are particularly interesting, for they can be submitted to test by facts. If they represent exactly the experimental laws governing these facts, the value of the theory will be augmented, and the domain governed by the theory will annex new laws. If, on the contrary, there is among these consequences one which is sharply in disagreement with the facts whose law was to be represented by the theory, the latter will have to be more or less modified, or perhaps completely rejected.

Now, on the occasion when we confront the predictions of the theory with reality, suppose we have to bet for or against the theory; on which side shall we lay our wager?

If the theory is a purely artificial system, if we see in the hypotheses on which it rests statements skillfully worked out so that they represent the experimental laws already known, but if the theory fails to hint at any reflection of the real relations among the invisible realities, we shall think that such a theory will fail to confirm a new law. That, in the space left free among the drawers adjusted for other laws, the hitherto unknown law should find a drawer already made into which it may be fitted exactly would be a marvelous feat of chance. It would be folly for us to risk a bet on this sort of expectation.

If, on the contrary, we recognize in the theory a natural classification, if we feel that its principles express profound and real relations among things, we shall not be surprised to see its consequences anticipating experience and stimulating the discovery of new laws; we shall bet fearlessly in its favor.

The highest test, therefore, of our holding a classification as a natural one is to ask it to indicate in advance things which the future alone will reveal. And when the experiment is made and confirms the predictions obtained from our theory, we feel strengthened in our conviction that the relations established by our reason among abstract notions truly correspond to relations among things.

Thus, modern chemical symbolism, by making use of developed formulas, establishes a classification in which diverse compounds are ordered. The wonderful order this classification brings about in the tremendous arsenal of chemistry already assures us that the classification is not a purely artificial system. The relations of analogy and derivation by substitution it establishes among diverse compounds have meaning only in our mind; yet, we are convinced that they correspond to kindred relations among substances themselves, whose nature remains deeply hidden but whose reality does not seem doubtful. Nevertheless, for this conviction to change into overwhelming certainty, we must see the theory write in advance the formulas of a multitude of bodies and, yielding to these indications, synthesis must bring to light a large number of substances whose composition and several properties we should know even before they exist.

Just as the syntheses announced in advance sanction chemical notation as a natural classification, so physical theory will prove that it is the reflection of a real order by anticipating observation.

Now the history of physics provides us with many examples of this clairvoyant guesswork; many a time has a theory forecast laws not yet observed, even laws which appear improbable, stimulating the experimenter to discover them and guiding him toward that discovery.

The Académie des Sciences had set, as the subject for the physics prize that was to be awarded in the public meeting of March 1819, the general examination of the phenomena of the diffraction of light. Two memoirs were presented, and one by Fresnel was awarded the prize, the commission of judges consisting of Biot, Arago, Laplace, Gay-Lussac, and Poisson.

From the principles put forward by Fresnel, Poisson deduced through an elegant analysis the following strange consequence: If a small, opaque, and circular screen intercepts the rays emitted by a point source of light, there should exist behind the screen, on the very axis of this screen, points which are not only bright, but which shine exactly as though the screen were not interposed between them and the source of light.

Such a corollary, so contrary, it seems, to the most obvious experimental certainties, appeared to be a very good ground for rejecting the theory of diffraction proposed by Fresnel. Arago had confidence in the natural character arising from the clairvoyance of this theory. He tested it, and observation gave results which agreed absolutely with the improbable predictions from calculation.[13]

Thus physical theory, as we have defined it, gives to a vast group of experimental laws a condensed representation, favorable to intellectual economy.

It classifies these laws and, by classifying, renders them more easily and safely utilizable. At the same time, putting order into the whole, it adds to their beauty.

It assumes, while being completed, the characteristics of a natural classification. The groups it establishes permit hints as to the real affinities of things.

This characteristic of natural classification is marked, above all, by the fruitfulness of the theory which anticipates experimental laws not yet observed, and promotes their discovery.

That sufficiently justifies the search for physical theories, which cannot be called a vain and idle task even though it does not pursue the explanation of phenomena.

. . . .

Appendix. Physics of a Believer

. . . .

Consider someone . . . who would take physical theory just as we have it, in the year of grace 1905, presented by the majority of those who teach it. Anyone who would listen closely to the talk in classes and to the gossip of the laboratories without looking back or caring for what used to be taught, would hear physicists constantly employing in their theories molecules, atoms, and electrons, counting these small bodies and determining their size, their mass, their charge. By the almost universal assent favoring these theories, by the enthusiasm they raise, and by the discoveries they incite or attribute to them, they would undoubtedly be regarded as prophetic forerunners of the theory destined to triumph in the future. He would judge that they reveal a first draft of the ideal form which physics will resemble more each day; and as the analogy between these theories and the cosmology of the atomists strikes him as obvious, he would obtain an eminently favorable presumption for this cosmology.

How different his judgment will be if he is not content with knowing physics through the gossip of the moment, if he studies deeply all its branches, not only those in vogue but also those that an unjust oblivion has let be neglected, and especially if the study of history by recalling the errors of past centuries puts him on his guard against the unreasoned exaggerations of the present time!

Well, he will see that the attempts at explanation based on atomism have accompanied physical theory for the longest time; whereas in physical theory he will recognize a work produced by the power of abstraction, these attempts at explanation will show themselves to him as the efforts of the mind that wishes to imagine what ought to be merely conceived; he will see them constantly being reborn, but constantly aborted; each time the fortunate daring of an experimenter will have discovered a new set of experimental laws, he will see the atomists, with feverish haste, take possession of this scarcely explored domain and construct a mechanism approximately representing these new findings. Then, as the experimenter's discoveries become more numerous and detailed, he will see the atomist's combinations get complicated, disturbed, overburdened with arbitrary complications without succeeding, however, in rendering a precise account of the new laws or in connecting them solidly to the old laws; and during this period he will see abstract theory, matured through patient labor, take possession of the new lands the experimenters have explored, organize these conquests, annex them to its old domains, and make a perfectly coordinated empire of their union. It will appear clearly to him that the physics of atomism, condemned to perpetual fresh starts, does not tend by continued progress to the ideal form of physical theory; whereas he will surmise the gradually complete realization of this ideal when he contemplates the development which abstract theory has undergone from Scholasticism to Galileo and Descartes; from Huygens, Leibniz and Newton to D'Alembert, Euler, Laplace, and Lagrange; from Sadi Carnot and Clausius to Gibbs and Helmholtz.

NOTES

1. Nicolaus Cabeus, S. J., *Philosophia magnetica, in qua magnetis natura penitus explicatur et omnium quae hoc lapide cernuntur causae propriae afferuntur, multa quoque dicuntur de electricis et aliis attractionibus, et eorum causis* (Cologne: Joannem Kinckium, 1629).

2. P. Rogerio Josepho Boscovich, S. J., *Theoria philosophiae naturalis redacta ad unicam legem virium in natura existentium* (Vienna, 1758).

3. G. W. Leibniz, *Oeuvres,* ed. Gerhardt, IV, 464. [Translator's note: See Leibniz, *Selections* (Charles Scribner's Sons, 1951), pp. 100ff.].

4. Christian Huygens to G. W. Leibniz, Nov. 18, 1690, *Oeuvres complètes de Huygens,*

Correspondance, 10 vols. (The Hague, 1638–1695), IX, 52. [Translator's note: The complete edition of Huygens' Collected Works was published in twenty-two volumes by the Holland Society of Sciences (Haarlem, 1950).]

5. *Aristarchi Samii "De mundi systemate, partibus et motibus ejusdem, liber singularis"* (Paris, 1643). This work was reproduced in 1647 in Volume III of the *Cogitata physico-mathematica* by Marin Mersenne.

6. R. Descartes, *Correspondance,* ed. P. Tannery and C. Adam, Vol. IV (Paris, 1893), Letter CLXXX, p. 396.

7. Denis Papin to Christian Huygens, June 18, 1690, *Oeuvres complètes de Huygens . . . ,* IX, 429.

8. Christian Huygens to Denis Papin, Sept. 2, 1690, *ibid.,* IX, 484.

9. E. Mach, "Die ökonomische Natur der physikalischen Forschung," *Populärwissenschaftliche Vorlesungen* (3rd ed.; Leipzig, 1903), Ch. XIII, p. 215. [Translator's note: Translated by T. J. McCormack, "The Economical Nature of Physical Research," Mach's *Popular Scientific Lectures* (3rd ed.; La Salle, Ill.: Open Court, 1907), Ch. XIII.]

See also E. Mach, *La Mécanique; exposé historique et critique de son développement* (Paris, 1904), Ch. IV, Sec. 4: "La Science comme économie de la pensée," p. 449. [Translator's note: Translated from the German 2nd ed. by T. J. McCormack, *The Science of Mechanics: a Critical and Historical Account of Its Development* (Open Court, 1902), Ch. IV, Sec. iv: "The Economy of Science," pp. 481–494.]

10. E. Mach, *La Mécanique . . . ,* p. 453. [Translator's note: Translated in *The Science of Mechanics . . . ,* p. 485.]

11. We have already noted natural classification as the ideal form toward which physical theory tends in "L'Ecole anglaise et les théories physiques," Art. 6, *Revue des questions scientifiques,* October 1893.

12. Cf. H. Poincaré, *La Science et l'Hypothèse* (Paris, 1903), p. 190. [Translator's note: Translated by Bruce Halsted, "Science and Hypothesis" in *Foundations of Science* (Lancaster, Pa.: Science Press, 1905).]

13. *Oeuvres complètes d'Augustin Fresnel,* 3 vols. (Paris, 1866–1870), I, 236, 365, 368.

16

Van Fraassen's Antirealism

From

The Scientific Image / Bas C. van Fraassen

Chapter 2. Arguments Concerning Scientific Realism

The rigour of science requires that we distinguish well the undraped figure of nature itself from the gay-coloured vesture with which we clothe it at our pleasure.

—Heinrich Hertz, quoted by Ludwig Boltzmann, letter to *Nature*, 28 February 1895

In our century, the first dominant philosophy of science was developed as part of logical positivism. Even today, such an expression as 'the received view of theories' refers to the views developed by the logical positivists, although their heyday preceded the Second World War.

In this chapter I shall examine, and criticize, the main arguments that have been offered for scientific realism. These arguments occurred frequently as part of a critique of logical positivism. But it is surely fair to discuss them in isolation, for even if scientific realism is most easily understood as a reaction against positivism, it should be able to stand alone. The alternative view which I advocate—for lack of a traditional name I shall call it *constructive empiricism*—is equally at odds with positivist doctrine.

Reprinted from *The Scientific Image* (Oxford: Clarendon Press, 1980), 6–23. Reprinted by permission of Oxford University Press.

§1. Scientific Realism and Constructive Empiricism

In philosophy of science, the term 'scientific realism' denotes a precise position on the question of how a scientific theory is to be understood, and what scientific activity really is. I shall attempt to define this position, and to canvass its possible alternatives. Then I shall indicate, roughly and briefly, the specific alternative which I shall advocate and develop in later chapters.

§1.1 Statement of Scientific Realism

What exactly is scientific realism? A naïve statement of the position would be this: the picture which science gives us of the world is a true one, faithful in its details, and the entities postulated in science really exist: the advances of science are discoveries, not inventions. That statement is too naïve; it attributes to the scientific realist the belief that today's theories are correct. It would mean that the philosophical position of an earlier scientific realist such as C. S. Peirce had been refuted by empirical findings. I do not suppose that scientific realists wish to be committed, as such, even to the claim that science will arrive in due time at theories true in all respects—for the growth of science might be an endless self-correction; or worse, Armageddon might occur too soon.

But the naïve statement has the right flavour. It answers two main questions: it characterizes a scientific theory as a story about what there really is, and scientific activity as an enterprise of discovery, as opposed to invention. The two questions of what a scientific theory is, and what a scientific theory does, must be answered by any philosophy of science. The task we have at this point is to find a statement of scientific realism that shares these features with the naïve statement, but does not saddle the realists with unacceptably strong consequences. It is especially important to make the statement as weak as possible if we wish to argue against it, so as not to charge at windmills.

As clues I shall cite some passages most of which will also be examined below in the contexts of the authors' arguments. A statement of Wilfrid Sellars is this:

> to have good reason for holding a theory is *ipso facto* to have good reason for holding that the entities postulated by the theory exist.[1]

This addresses a question of epistemology, but also throws some indirect light on what it is, in Sellars's opinion, to hold a theory. Brian Ellis, who calls himself a scientific entity realist rather than a scientific realist, appears to agree with that statement of Sellars, but gives the following formulation of a stronger view:

> I understand scientific realism to be the view that the theoretical statements of science are, or purport to be, true generalized descriptions of reality.[2]

This formulation has two advantages: It focuses on the understanding of the theories without reference to reasons for belief, and it avoids the suggestion that to be a realist you must believe current scientific theories to be true. But it gains the latter advantage by use of the word 'purport', which may generate its own puzzles.

Hilary Putnam . . . gives a formulation which he says he learned from Michael Dummett:

> A realist (with respect to a given theory or discourse) holds that (1) the sentences of that theory are true or false; and (2) that what makes them true or false is something external—that is to say, it is not (in general) our sense data, actual or potential, or the structure of our minds, or our language, etc.[3]

He follows this soon afterwards with a further formulation which he credits to Richard Boyd:

> That terms in mature scientific theories typically refer (this formulation is due to Richard Boyd), that the theories accepted in a mature science are typically approximately true, that the same term can refer to the same thing even when it occurs in different theories—these statements are viewed by the scientific realist . . . as part of any adequate scientific description of science and its relations to its objects.[4]

None of these were intended as definitions. But they show I think that truth must play an important role in the formulation of the basic realist position. They also show that the formulation must incorporate an answer to the question what it is to *accept* or *hold* a theory. I shall now propose such a formulation, which seems to me to make sense of the above remarks, and also renders intelligible the reasoning by realists which I shall examine below—without burdening them with more than the minimum required for this.

Science aims to give us, in its theories, a literally true story of what the world is like; and acceptance of a scientific theory involves the belief that it is true. This is the correct statement of scientific realism.

Let me defend this formulation by showing that it is quite minimal, and can be agreed to by anyone who considers himself a scientific realist. The naïve statement said that science tells a true story; the correct statement says only that it is the aim of science to do so. The aim of science is of course not to be identified with individual scientists' motives. The aim of the game of chess is to checkmate your opponent; but the motive for playing may be fame, gold, and glory. What the aim is determines what counts as success in the enterprise as such; and this aim may be pursued for any number of reasons. Also, in calling something *the* aim, I do not deny that there are other subsidiary aims which may or may not be means to that end: everyone will readily agree that simplicity, informativeness, predictive power, explanation are (also) virtues. Perhaps my formulation can even be accepted by any philosopher who considers the most important aim of science to be something which only *requires* the finding of true theories—given that I wish to give the weakest formulation of the doctrine that is generally acceptable.

I have added 'literally' to rule out as realist such positions as imply that science is true if 'properly understood' but literally false or meaningless. For that would be consistent with conventionalism, logical positivism, and instrumentalism. I will say more about this below. . . .

The second part of the statement touches on epistemology. But it only equates acceptance of a theory with belief in its truth.[5] It does not imply that anyone is ever rationally warranted in forming such a belief. We have to make room for the epistemological position, today the subject of considerable debate, that a rational person never assigns personal probability 1 to any proposition except a tautology. It would, I think, be rare for a scientific realist to take this stand in epistemology, but it is certainly possible.[6]

To understand qualified acceptance we must first understand acceptance *tout court*. If acceptance of a theory involves the belief that it is true, then tentative acceptance involves the tentative adoption of the belief that it is true. If belief comes in degrees, so does acceptance, and we may then speak of a degree of acceptance involving a certain degree of belief that the theory is true. This must of course be distinguished from belief that the theory is approximately true, which seems to mean belief that some member of a class centering on the mentioned theory is (exactly) true. In this way the proposed formulation of realism can be used regardless of one's epistemological persuasion.

§1.2 Alternatives to Realism

Scientific realism is the position that scientific theory construction aims to give us a literally true story of what the world is like, and that acceptance of a scientific theory involves the belief that it is true. Accordingly, antirealism is a position according to which the aim of science can well be served without giving such a literally true story, and acceptance of a theory may properly involve something less (or other) than belief that it is true.

What does a scientist do then, according to these different positions? According to the realist, when someone proposes a theory, he is asserting it to be true. But according to the anti-realist, the proposer does not assert the theory; *he displays it,* and claims certain virtues for it. These virtues may fall short of truth: empirical adequacy, perhaps; comprehensiveness, acceptability for various purposes. This will have to be spelt out, for the details here are not determined by the denial of realism. For now we must concentrate on the key notions that allow the generic division.

The idea of a literally true account has two aspects: the language is to be literally construed; and so construed, the account is true. This divides the anti-realists into two sorts. The first sort holds that science is or aims to be true, properly (but not literally) construed. The second holds that the language of science should be literally construed, but its theories need not be true to be good. The anti-realism I shall advocate belongs to the second sort.

It is not so easy to say what is meant by a literal construal. The idea comes perhaps from theology, where fundamentalists construe the Bible literally, and liberals have a variety of allegorical, metaphorical, and analogical interpretations, which 'demythologize'. The problem of explicating 'literal construal' belongs to the philosophy of language. . . . The term 'literal' is well enough understood for general philosophical use, but if we try to explicate it we find ourselves in the midst of the problem of giving an adequate account of natural language. It would be bad tactics to link an inquiry into science to a commitment to some solution to that problem. The following remarks . . . should fix the usage of 'literal' sufficiently for present purposes.

The decision to rule out all but literal construals of the language of science, rules out those forms of anti-realism known as *positivism* and *instrumentalism.* First, on a literal construal, the apparent statements of science really are statements, *capable of* being true or false. Secondly, although a literal construal can elaborate, it cannot change logical relationships. (It is possible to elaborate, for instance, by identifying what the terms designate. The 'reduction' of the language of phenomenological thermodynamics to that of statistical mechanics is like that: bodies of gas are identified as aggregates of molecules, temperature as mean kinetic energy, and so on.) On the positivists' interpretation of science, theoretical terms have meaning only through their connection with the observable. Hence they hold that two theories may in fact *say the same thing* although in form they contradict each other. (Perhaps the one says that all matter consists of atoms, while the other postulates instead a universal continuous medium; they will say the same thing nevertheless if they agree in their observable consequences, according to the positivists.) But two theories which contradict each other in such a way can 'really' be saying the same thing only if they are not literally construed. Most specifically, if a theory says that something exists, then a literal construal may elaborate on what that something is, but will not remove the implication of existence.

There have been many critiques of positivist interpretations of science, and there is no need to repeat them. . . .

§1.3 Constructive Empiricism

To insist on a literal construal of the language of science is to rule out the construal of a theory as a metaphor or simile, or as intelligible only after it is 'demythologized' or subjected to some other sort of 'translation' that does not preserve logical form. If the theory's statements include 'There are electrons', then the theory says that there are electrons. If in addition they include 'Electrons are not planets', then the theory says, in part, that there are entities other than planets.

But this does not settle very much. It is often not at all obvious whether a theoretical term refers to a concrete entity or a mathematical entity. Perhaps one tenable interpretation of classical physics is that there are no concrete entities which are forces—that 'there are forces such that . . .' can always be understood as a mathematical statement asserting the existence of certain functions. That is debatable.

Not every philosophical position concerning science which insists on a literal construal of the language of science is a realist position. For this insistence relates not at all to our epistemic attitudes toward theories, nor to the aim we pursue in constructing theories, but only to the correct understanding of *what a theory says.* (The fundamentalist theist, the agnostic, and the atheist presumably agree with each other (though not with liberal theologians) in their understanding of the statement that God, or gods, or angels exist.) After deciding that the language of science must be literally understood, we can still say that there is no need to believe good theories to be true, nor to believe *ipso facto* that the entities they postulate are real.

Science aims to give us theories which are empirically adequate; and acceptance of a theory involves as belief only that it is empirically adequate. This is the statement of the anti-realist position I advocate; I shall call it *constructive empiricism.*

This formulation is subject to the same qualifying remarks as that of scientific realism in Section 1.1 above. In addition it requires an explication of 'empirically adequate'. For now, I shall leave that with the preliminary explication that a theory is empirically adequate exactly if what it says about the observable things and events in this world, is true—exactly if it 'saves the phenomena'. A little more precisely: such a theory has at least one model that all the actual phenomena fit inside. I must emphasize that this refers to *all* the phenomena; these are not exhausted by those actually observed, nor even by those observed at some time, whether past, present, or future. . . .

The distinction I have drawn between realism and anti-realism, in so far as it pertains to acceptance, concerns only how much belief is involved therein. Acceptance of theories (whether full, tentative, to a degree, etc.) is a phenomenon of scientific activity which clearly involves more than belief. One main reason for this is that we are never confronted with a complete theory. So if a scientist accepts a theory, he thereby involves himself in a certain sort of research programme. That programme could well be different from the one acceptance of another theory would have given him, even if those two (very incomplete) theories are equivalent to each other with respect to everything that is observable—in so far as they go.

Thus acceptance involves not only belief but a certain commitment. Even for those of us who are not working scientists, the acceptance involves a commitment to confront any future phenomena by means of the conceptual resources of this theory. It determines the terms in which we shall seek explanations. If the acceptance is at all strong, it is exhibited in the person's assumption of the role of explainer, in his willingness to answer questions *ex cathedra.* Even if you do not accept a theory, you can engage in discourse in a context in which language use is guided by that theory—but acceptance produces such contexts. There are similarities in all of this to ideological commitment. A commitment is of course not true or false: The confidence exhibited is that it will be *vindicated.*

This is a preliminary sketch of the *pragmatic* dimension of theory acceptance. Unlike the epistemic dimension, it does not figure overtly in the disagreement between realist and anti-realist. But because the amount of belief involved in acceptance is typically less according to anti-realists, they will tend to make more of the pragmatic aspects. It is as well to note here the important difference. Belief that a theory is true, or that it is empirically adequate, does not imply, and is not implied by, belief that full acceptance of the theory will be vindicated. To see this, you need only consider here a person who has quite definite beliefs about the future of the human race, or about the scientific community and the influences thereon and practical limitations we have. It might well be, for instance, that a theory which is empirically adequate will not combine easily with some other theories which we have accepted in fact, or that Armageddon will occur before we succeed. Whether belief that a theory is true, or that it is empirically adequate, can be equated with belief that acceptance of it would, under ideal research conditions, be vindicated in the long run, is another question. It seems to me an irrelevant question within philosophy of science, because an affirmative answer would not obliterate the distinction we have already established by the preceding remarks. (The question may also assume that counterfactual statements are objectively true or false, which I would deny.)

Although it seems to me that realists and anti-realists need not disagree about the pragmatic aspects of theory acceptance, I have mentioned it here because I think that typically they do. We shall find ourselves returning time and again, for example, to requests for explanation to which realists typically attach an objective validity which anti-realists cannot grant.

§2. The Theory/Observation 'Dichotomy'

For good reasons, logical positivism dominated the philosophy of science for thirty years. In 1960, the first volume of *Minnesota Studies in the Philosophy of Science* published Rudolf Carnap's 'The Methodological Status of Theoretical Concepts', which is, in many ways, the culmination of the positivist programme. It interprets science by relating it to an observation language (a postulated part of natural language which is devoid of theoretical terms). Two years later this article was followed in the same series by Grover Maxwell's 'The Ontological Status of Theoretical Entities', in title and theme a direct counter to Carnap's. This is the *locus classicus* for the new realists' contention that the theory/observation distinction cannot be drawn.

I shall examine some of Maxwell's points directly, but first a general remark about the issue. Such expressions as 'theoretical entity' and 'observable–theoretical dichotomy' are, on the face of it, examples of category mistakes. Terms or concepts are theoretical (introduced or adapted for the purposes of theory construction); entities are observable or unobservable. This may seem a little point, but it separates the discussion into two issues. Can we divide our language into a theoretical and non-theoretical part? On the other hand, can we classify objects and events into observable and unobservable ones?

Maxwell answers both questions in the negative, while not distinguishing them too carefully. On the first, where he can draw on well-known supportive essays by Wilfrid Sellars and Paul Feyerabend, I am in total agreement. All our language is thoroughly theory-infected. If we could cleanse our language of theory-laden terms, beginning with the recently introduced ones like 'VHF receiver', continuing through 'mass' and 'impulse' to 'element' and so on into the prehistory of language formation, we would end up with nothing useful. The way we talk, and scientists talk, is guided by the pictures provided by previously accepted theories. This is true also, as Duhem already emphasized, of experimental reports. Hygienic reconstructions of language such as the positivists envisaged are simply not on. . . .

But does this mean that we must be scientific realists? We surely have more tolerance of ambiguity than that. The fact that we let our language be guided by a given picture, at some point, does not show how much we believe about that picture. When we speak of the sun coming up in the morning and setting at night, we are guided by a picture now explicitly disavowed. When Milton wrote *Paradise Lost* he deliberately let the old geocentric astronomy guide his poem, although various remarks in passing clearly reveal his interest in the new astronomical discoveries and speculations of his time. These are extreme examples, but show that no immediate conclusions can be drawn from the theory-ladenness of our language.

However, Maxwell's main arguments are directed against the observable–unobservable distinction. Let us first be clear on what this distinction was supposed to be. The term 'observable' classifies putative entities (entities which may or may not exist). A flying horse is observable—that is why we are so sure that there aren't any—and the number seventeen is not. There is supposed to be a correlate classification of human acts: an unaided act of perception, for instance, is an observation. A calculation of the mass of a particle from the deflection of its trajectory in a known force field, is not an observation of that mass.

It is also important here not to confuse *observing* (an entity, such as a thing, event, or process) and *observing that* (something or other is the case). Suppose one of the Stone Age people recently found in the Philippines is shown a tennis ball or a car crash. From his behaviour, we see that he has noticed them; for example, he picks up the ball and throws it. But he has not seen *that* it is a tennis ball, or *that* some event is a car crash, for he does not even have those concepts. He cannot get that information through perception; he would first have to learn a great deal. To say that he does not see the same things and events as we do, however, is just silly; it is a pun which trades on the ambiguity between seeing and seeing that. (The truth-conditions for our statement '*x* observes *that* A' must be such that what concepts *x* has, presumably related to the language *x* speaks if he is human, enter as a variable into the correct truth definition, in some way. To say that *x* observed the tennis ball, therefore, does not imply at all that *x* observed that it was a tennis ball; that would require some conceptual awareness of the game of tennis.)

The arguments Maxwell gives about observability are of two sorts: one directed against the possibility of drawing such distinctions, the other against the importance that could attach to distinctions that can be drawn.

The first argument is from the continuum of cases that lie between direct observation and inference:

> there is, in principle, a continuous series beginning with looking through a vacuum and containing these as members: looking through a windowpane, looking through glasses, looking through binoculars, looking through a low-power microscope, looking through a high-power microscope, etc., in the order given. The important consequence is that, so far, we are left without criteria which would enable us to draw a non-arbitrary line between 'observation' and 'theory'.[7]

This continuous series of supposed acts of observation does not correspond directly to a continuum in what is supposed observable. For if something can be seen through a window, it can also be seen with the window raised. Similarly, the moons of Jupiter can be seen through a telescope; but they can also be seen without a telescope if you are close enough. That something is observable does not automatically imply that the conditions are right for observing it now. The principle is:

> X is observable if there are circumstances which are such that, if X is present to us under those circumstances, then we observe it.

This is not meant as a definition, but only as a rough guide to the avoidance of fallacies.

We may still be able to find a continuum in what is supposed detectable: perhaps some things can only be detected with the aid of an optical microscope, at least; perhaps some require an electron microscope, and so on. Maxwell's problem is: where shall we draw the line between what is observable and what is only detectable in some more roundabout way?

Granted that we cannot answer this question without arbitrariness, what follows? That 'observable' is a *vague predicate.* There are many puzzles about vague predicates, and many sophisms designed to show that, in the presence of vagueness, no distinction can be drawn at all. In Sextus Empiricus, we find the argument that incest is not immoral, for touching your mother's big toe with your little finger is not immoral, and all the rest differs only by degree. But predicates in natural language are almost all vague, and there is no problem in their use; only in formulating the logic that governs them.[8] A vague predicate is usable provided it has clear cases and clear counter-cases. Seeing with the unaided eye is a clear case of observation. Is Maxwell then perhaps challenging us to present a clear counter-case? Perhaps so, for he says 'I have been trying to support the thesis that any (non-logical) term is a *possible* candidate for an observation term.'

A look through a telescope at the moons of Jupiter seems to me a clear case of observation, since astronauts will no doubt be able to see them as well from close up. But the purported observation of micro-particles in a cloud chamber seems to me a clearly different case—if our theory about what happens there is right. The theory says that if a charged particle traverses a chamber filled with saturated vapour, some atoms in the neighbourhood of its path are ionized. If this vapour is decompressed, and hence becomes super-saturated, it condenses in droplets on the ions, thus marking the path of the particle. The resulting silver-grey line is similar (physically as well as in appearance) to the vapour trail left in the sky when a jet passes. Suppose I point to such a trail and say: 'Look, there is a jet!'; might you not say: 'I see the vapour trail, but where is the jet?' Then I would answer: 'Look just a bit ahead of the trail . . . there! Do you see it?' Now, in the case of the cloud chamber this response is not possible. So while the particle is detected by means of the cloud chamber, and the detection is based on observation, it is clearly not a case of the particle's being observed.

As second argument, Maxwell directs our attention to the 'can' in 'what is observable is what can be observed.' An object might of course be temporarily unobservable—in a rather different sense: it cannot be observed in the circumstances in which it actually is at the moment, but could be observed if the circumstances were more favourable. In just the same way, I might be temporarily invulnerable or invisible. So we should concentrate on 'observable' *tout court,* or on (as he prefers to say) 'unobservable in principle'. This Maxwell explains as meaning that the relevant scientific theory *entails* that the entities cannot be observed in any circumstances. But this never happens, he says, because the different circumstances could be ones in which we have different sense organs—electron-microscope eyes, for instance.

This strikes me as a trick, a change in the subject of discussion. I have a mortar and pestle made of copper and weighing about a kilo. Should I call it breakable because a giant could break it? Should I call the Empire State Building portable? Is there no distinction between a portable and a console record player? The human organism is, from the point of view of physics, a certain kind of measuring apparatus. As such it has certain inherent limitations—which will be described in detail in the final physics and biology. It is these limitations to which the 'able' in 'observable' refers—our limitations, *qua* human beings.

As I mentioned, however, Maxwell's article also contains a different sort of argument: even if there is a feasible observable/unobservable distinction, this distinction has no importance. The point at issue for the realist is, after all, the reality of the entities postulated in science. Suppose that these entities could be classified into observables and others; what relevance should that have to the question of their existence?

Logically, none. For the term 'observable' classifies putative entities, and has logically nothing to do with existence. But Maxwell must have more in mind when he says: 'I conclude that the drawing of the observational–

theoretical line at any given point is an accident and a function of our physiological make-up, . . . and, therefore, that it has no ontological significance whatever.'[9] No ontological significance if the question is only whether 'observable' and 'exists' imply each other—for they do not; but significance for the question of scientific realism?

Recall that I defined scientific realism in terms of the aim of science, and epistemic attitudes. The question is what aim scientific activity has, and how much we shall believe when we accept a scientific theory. What is the proper form of acceptance: belief that the theory, as a whole, is true; or something else? To this question, what is observable by us seems eminently relevant. Indeed, we may attempt an answer at this point: to accept a theory is (for us) to believe that it is empirically adequate—that what the theory says *about what is observable* (by us) is true.

It will be objected at once that, on this proposal, what the anti-realist decides to believe about the world will depend in part on what he believes to be his, or rather the epistemic community's, accessible range of evidence. At present, we count the human race as the epistemic community to which we belong; but this race may mutate, or that community may be increased by adding other animals (terrestrial or extra-terrestrial) through relevant ideological or moral decisions ('to count them as persons'). Hence the anti-realist would, on my proposal, have to accept conditions of the form

> If the epistemic community changes in fashion *Y*, then my beliefs about the world will change in manner *Z*.

To see this as an objection to anti-realism is to voice the requirement that our epistemic policies should give the same results independent of our beliefs about the range of evidence accessible to us. That requirement seems to me in no way rationally compelling; it could be honoured, I should think, only through a thorough-going scepticism or through a commitment to wholesale leaps of faith. But we cannot settle the major questions of epistemology *en passant* in philosophy of science; so I shall just conclude that it is, on the face of it, not irrational to commit oneself only to a search for theories that are empirically adequate, ones whose models fit the observable phenomena, while recognizing that what counts as an observable phenomenon is a function of what the epistemic community is (that *observable* is *observable-to-us*).

The notion of empirical adequacy in this answer will have to be spelt out very carefully if it is not to bite the dust among hackneyed objections. . . . But the point stands: even if observability has nothing to do with existence (is, indeed, too anthropocentric for that), it may still have much to do with the proper epistemic attitude to science.

§3. Inference to the Best Explanation

A view advanced in different ways by Wilfrid Sellars, J. J. C. Smart, and Gilbert Harman is that the canons of rational inference require scientific realism. If we are to follow the same patterns of inference with respect to this issue as we do in science itself, we shall find ourselves irrational unless we assert the truth of the scientific theories we accept. Thus Sellars says: 'As I see it, to have good reason for holding a theory is *ipso facto* to have good reason for holding that the entities postulated by the theory exist.'[10]

The main rule of inference invoked in arguments of this sort is the rule of *inference to the best explanation*. The idea is perhaps to be credited to C. S. Peirce,[11] but the main recent attempts to explain this rule and its uses have been made by Gilbert Harman.[12] I shall only present a simplified version. Let us suppose that we have evidence *E*, and are considering several hypotheses, say *H* and *H'*. The rule then says that we should infer *H* rather than *H'* exactly if *H* is a better explanation of *E* than *H'* is. (Various qualifications are necessary to avoid inconsistency: we should always try to move to the best over-all explanation of all available evidence.)

It is argued that we follow this rule in all 'ordinary' cases: and that if we follow it consistently everywhere, we shall be led to scientific realism, in the way Sellars's dictum suggests. And surely there are many telling 'ordinary' cases: I hear scratching in the wall, the patter of little feet at midnight, my cheese disappears—and I infer that a mouse has come to live with me. Not merely that these apparent signs of mousely presence will continue, not merely that all the observable phenomena will be as if there is a mouse; but that there really is a mouse.

Will this pattern of inference also lead us to belief in unobservable entities? Is the scientific realist simply someone who consistently follows the rules of inference that we all follow in more mundane contexts? I have two objections to the idea that this is so.

First of all, what is meant by saying that we all *follow* a certain rule of inference? One meaning might be that we deliberately and consciously 'apply' the rule, like a student doing a logic exercise. That meaning is much too literalistic and restrictive; surely all of mankind follows the rules of logic much of the time, while only a fraction can even formulate them. A second meaning is that we act in accordance with the rules in a sense that does not require conscious deliberation. That is not so easy to make precise, since each logical rule is a rule of permission (*modus ponens* allows you to infer *B* from *A* and (if *A then B*), but does not forbid you to infer (*B or A*) instead). However, we might say that a person behaved in accordance with a set of rules in that sense if every conclusion he drew could be reached from his premisses via those rules. But this meaning is much too loose; in this sense we always behave in accordance with the rule that any conclusion may be inferred from any premiss. So it seems that to be following a rule, I must be willing to believe all conclusions it allows, while definitely unwilling to believe conclusions at variance with the ones it allows—or else, change my willingness to believe the premisses in question.

Therefore the statement that we all follow a certain rule in certain cases, is a *psychological hypothesis* about what we are willing and unwilling to do. It is an empirical hypothesis, to be confronted with data, and with rival hypotheses. Here is a rival hypothesis: we are always willing to believe that the theory which best explains the evidence, is empirically adequate (that all the observable phenomena are as the theory says they are).

In this way I can certainly account for the many instances in which a scientist appears to argue for the acceptance of a theory or hypothesis, on the basis of its explanatory success. (A number of such instances are related by Thagard.[13]) For, remember: I equate the acceptance of a scientific theory with the belief that it is empirically adequate. We have therefore two rival hypotheses concerning these instances of scientific inference, and the one is apt in a realist account, the other in an anti-realist account.

Cases like the mouse in the wainscoting cannot provide telling evidence between those rival hypotheses. For the mouse *is* an observable thing; therefore 'there is a mouse in the wainscoting' and 'All observable phenomena are as if there is a mouse in the wainscoting' are totally equivalent; each implies the other (given what we know about mice).

It will be countered that it is less interesting to know whether people do follow a rule of inference than whether they ought to follow it. Granted; but the premiss that we all follow the rule of inference to the best explanation when it comes to mice and other mundane matters—that premiss is shown wanting. It is not warranted by the evidence, because that evidence is not telling *for* the premiss *as against* the alternative hypothesis I proposed, which is a relevant one in this context.

My second objection is that even if we were to grant the correctness (or worthiness) of the rule of inference to the best explanation, the realist needs some further premiss for his argument. For this rule is only one that dictates a choice when given a set of rival hypotheses. In other words, we need to be committed to belief in one of a range of hypotheses before the rule can be applied. Then, under favourable circumstances, it will tell us which of the hypotheses in that range to choose. The realist asks us to choose between different hypotheses that explain the regularities in certain ways; but his opponent always wishes to choose among hypotheses of the form 'theory T_i is empirically adequate'. So the realist will need his special extra premiss that every universal regularity in nature needs an explanation, before the rule will make realists of us all. And that is just the premiss that distinguishes the realist from his opponents. . . .

The logically minded may think that the extra premiss can be bypassed by logical *léger-de-main.* For suppose the data are that all facts observed so far accord with theory *T*; then *T* is one possible explanation of those data. A rival is *not-T* (that *T* is false). This rival is a very poor explanation of the data. So we *always* have a set of rival hypotheses, and the rule of inference to the best explanation leads us unerringly to the conclusion that *T* is true. Surely I am committed to the view that *T* is true or *T* is false?

This sort of epistemological rope-trick does not work of course. To begin, I am committed to the view that *T* is true or *T* is false, but not thereby committed to an inferential move to one of the two! The rule operates only if I have decided not to remain neutral between these two possibilities.

Secondly, it is not at all likely that the rule will be applicable to such logically concocted rivals. Harman lists various criteria to apply to the evaluation of hypotheses *qua* explanations.[14] Some are rather vague, like simplicity (but is simplicity not a reason to use a theory whether you believe it or not?). The precise ones come from statistical theory which has lately proved of wonderful use to epistemology:

> H is a better explanation than H' (*ceteris paribus*) of E, provided:
> (a) $P(H) > P(H')$—H has higher probability than H'
> (b) $P(E/H) > P(E/H')$—H bestows higher probability on E than H' does.

The use of 'initial' or *a priori* probabilities in (a)—the initial plausibility of the hypotheses themselves—is typical of the so-called *Bayesians.* More traditional statistical practice suggests only the use of (b). But even that supposes that H and H' bestow definite probabilities on E. If H' is simply the denial of H, that is not generally the case. (Imagine that H says that the probability of E equal ¾. The very most that *not-H* will entail is that the probability of E is some number other than ¾; and usually it will not even entail that much, since H will have other implications as well.)

Bayesians tend to cut through this 'unavailability of probabilities' problem by hypothesizing that everyone has a specific subjective probability (degree of belief) for every proposition he can formulate. In that case, no matter what E, H, H' are, all these probabilities really are (in principle) available. But they obtain this availability by making the probabilities thoroughly subjective. I do not think that scientific realists wish their conclusions to hinge on the subjectively established initial plausibility of there being unobservable entities, so I doubt that this sort of Bayesian move would help here. . . .

I have kept this discussion quite abstract. . . . It should at least be clear that there is no open-and-shut argument from common sense to the unobservable. Merely following the ordinary patterns of inference in science does not obviously and automatically make realists of us all.

. . . .

NOTES

1. [See n. 10.]

2. Brian Ellis, *Rational Belief Systems* (Oxford: Blackwell, 1979), p. 28.

3. Hilary Putnam, *Mathematics, Matter and Method* (Cambridge: Cambridge University Press, 1975), vol. 1, pp. 69f.

4. Putnam, *op. cit.*, p. 73, n. 29. The argument is reportedly developed at greater length in Boyd's forthcoming book *Realism and Scientific Epistemology* (Cambridge University Press).

5. Hartry Field has suggested that 'acceptance of a scientific theory involves the belief that it is true' be replaced by 'any reason to think that any part of a theory is not, or might not be, true, is reason not to accept it.' The drawback of this alternative is that it leaves open what epistemic attitude acceptance of a theory does involve. This question must also be answered, and as long as we are talking about full acceptance—as opposed to tentative or partial or otherwise qualified acceptance—I cannot see how a realist could do other than equate that attitude with full belief. (That theories believed to be false are used for practical problems, for example, classical mechanics for orbiting satellites, is of course a commonplace.) For if the aim is truth, and acceptance requires belief that the aim is served . . . I should also mention the statement of realism at the beginning of Richard Boyd, 'Realism, Underdetermination, and a Causal Theory of Evidence', *Noûs*, 7 (1973), 1–12. Except for some doubts about his use of the terms 'explanation' and 'causal relation' I intend my statement of realism to be entirely in accordance with his. Finally, see C. A. Hooker, 'Systematic Realism', *Synthese*, 26 (1974), 409–97; esp. pp. 409 and 426.

6. More typical of realism, it seems to me, is the sort of epistemology found in Clark Glymour's forthcoming book, *Theory and Evidence* (Princeton: Princeton University Press, 1980), except of course that there it is fully and carefully developed in one specific fashion. (See esp. his chapter 'Why I am not a Bayesian' for the present issue.) But I see no reason why a realist, as such, could not be a Bayesian of the type of Richard Jeffrey, even if the Bayesian position has in the past been linked with anti-realist and even instrumentalist views in philosophy of science.

7. G. Maxwell, 'The Ontological Status of Theoretical Entities', *Minnesota Studies in Philosophy of Science*, III (1962), p. 7.

8. There is a great deal of recent work on the logic of vague predicates; especially important, to my mind, is that of Kit Fine ('Vagueness, Truth, and Logic', *Synthese*, 30 (1975), 265–300) and Hans Kamp. The latter is currently working on a new theory of vagueness that does justice to the 'vagueness of vagueness' and the context-dependence of standards of applicability for predicates.

9. *Op. cit.*, p. 15.

10. *Science, Perception and Reality* (New York: Humanities Press, 1962); cf. the footnote

on p. 97. See also my review of his *Studies in Philosophy and its History*, in *Annals of Science*, January 1977.

11. Cf. P. Thagard, doctoral dissertation, University of Toronto, 1977, and 'The Best Explanation: Criteria for Theory Choice', *Journal of Philosophy*, 75 (1978), 76–92.

12. 'The Inference to the Best Explanation', *Philosophical Review*, 74 (1965), 88–95 and 'Knowledge, Inference, and Explanation', *American Philosophical Quarterly*, 5 (1968), 164–73. Harman's views were further developed in subsequent publications (*Noûs*, 1967; *Journal of Philosophy*, 1968; in M. Swain (ed.), *Induction*, 1970; in H.-N. Castañeda (ed.), *Action, Thought, and Reality*, 1975; and in his book *Thought*, Ch. 10). I shall not consider these further developments here.

13. [See n. 11.]

14. See esp. 'Knowledge, Inference, and Explanation', p. 169.

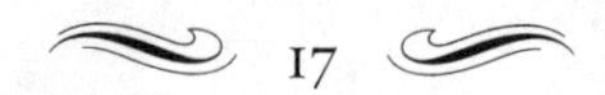

17

Perrin's Realism and Argument for Molecules

From

Atoms / Jean Perrin

Preface

Two kinds of intellectual activity, both equally instinctive, have played a prominent part in the progress of physical science.

One is already developed in a child who, while holding an object, knows what will happen if he relinquishes his grasp. He may possibly never have had hold of the particular object before, but he nevertheless recognises something in common between the muscular sensations it calls forth and those which he has already experienced when grasping other objects that fell to the ground when his grasp was relaxed. Men like Galileo and Carnot, who possessed this *power of perceiving analogies* to an extraordinary degree, have by an analogous process built up the doctrine of energy by successive generalisations, cautious as well as bold, from experimental relationships and objective realities.

In the first place they observed, or it would perhaps be better to say that everyone has observed, that not only does an object fall if it be dropped, but that once it has reached the ground it will not rise *of itself.* We have *to pay* before a lift can be made to ascend, and the more dearly the swifter and

Reprinted from *Atoms,* trans. D. Ll. Hammick (Woodbridge, Conn.: Ox Bow Press, 1990), v–viii, 83–94, 104–6.

higher it rises. Of course, the real price is not a sum of money, but the external compensation given for the work done by the lift (the fall of a mass of water, the combustion of coal, chemical change in a battery). The money is only the symbol of this compensation.

This once recognised, our attention naturally turns to the question of how small the payment can be. We know that by means of a wheel and axle we can raise 1,000 kilogrammes through 1 metre by allowing 100 kilogrammes to fall 10 metres; is it possible to devise a more economical mechanism that will allow 1,000 kilogrammes to be raised 1 metre for the same price (100 kilogrammes falling through 10 metres)?

Galileo held that it is possible to affirm that, under certain conditions, 200 kilogrammes could be raised 1 metre without external compensation, "for nothing." Seeing that we no longer believe that this is possible, we have to recognise *equivalence between mechanisms* that bring about the elevation of one weight by the lowering of another.

In the same way, if we cool mercury from 100° C. to 0° C. by melting ice, we always find (and the general expression of this fact is the basis of the whole of calorimetry) that 42 grammes of ice are melted for every kilogramme of mercury cooled, whether we work by direct contact, radiation, or any other method (provided always that we end with melted ice and mercury cooled from 100° C. to 0° C.). Even more interesting are those experiments in which, through the intermediary of friction, a heating effect is produced by the falling of weights (Joule). However widely we vary the mechanism through which we connect the two phenomena, we always find one large calory of heat produced for the fall of 428 kilogrammes through 1 metre.

Step by step, in this way the First Principle of Thermodynamics has been established. It may, in my opinion, be enunciated as follows:

If by means of a certain mechanism we are able to connect two phenomena in such a way that each may accurately compensate the other, then it can never happen, however the mechanism employed is varied, that we could obtain, as the external effect of one of the phenomena, first the other and then another phenomenon in addition, which would represent a gain.[1]

Without going so fully into detail, we may notice another similar result, established by Sadi Carnot, who, grasping the essential characteristic common to all heat engines, showed that the production of work is always accompanied "by the passage of caloric from a body at a higher temperature to another at a lower temperature." As we know, proper analysis of this statement leads to the Second Law of Thermodynamics.

Each of these principles has been reached by noting analogies and generalising the results of experience, and our lines of reasoning and statements of results have related only to objects that can be observed and to experiments that can be performed. Ostwald could therefore justly say that in the doctrine of energy there are no *hypotheses.* Certainly when a new machine is invented we at once assert that it cannot create work; but we can at once verify our statement, and we cannot call an assertion a hypothesis if, as soon as it is made, it can be checked by experiment.

Now, there are cases where hypothesis is, on the contrary, both necessary and fruitful. In studying a machine, we do not confine ourselves only to the consideration of its visible parts, which have objective reality for us only as far as we can dismount the machine. We certainly observe these visible pieces as closely as we can, but at the same time we seek to divine the *hidden* gears and parts that explain its apparent motions.

To divine in this way the existence and properties of objects that still lie outside our ken, *to explain the complications of the visible in terms of invisible simplicity,* is the function of the intuitive intelligence which, thanks to men such as Dalton and Boltzmann, has given us the doctrine of Atoms. This book aims at giving an exposition of that doctrine.

The use of the intuitive method has not, of course, been used only in the study of atoms, any more than the inductive method has found its sole application in energetics. A time may perhaps come when atoms, directly perceptible at last, will be as easy to observe as are microbes to-day. The true spirit of the atomists will then be found in those who have inherited the power to divine another universal structure lying hidden behind a vaster experimental reality than ours.

I shall not attempt, as too many have done, to decide between the merits of the two methods of research. Certainly during recent years intuition has gone ahead of induction in rejuvenating the doctrine of energy by the incorporation of statistical results borrowed from the atomists. But its greater fruitfulness may well be transient, and I can see no reason to doubt the possibility of further discovery that will dispense with the necessity of employing any unverifiable hypothesis.

. . . .

Chapter 3. The Brownian Movement—Emulsions

History and General Characteristics

50.—The Brownian Movement.—Direct perception of the molecules in agitation is not possible, for the same reason that the motion of the waves is not noticed by an observer at too great a distance from them. But if a ship comes in sight, he will be able to see that it is rocking, which will enable him to infer the existence of a possibly unsuspected motion of the sea's surface. Now may we not hope, in the case of microscopic particles suspended in a fluid, that the particles may, though large enough to be followed under the microscope, nevertheless be small enough to be noticeably agitated by the molecular impacts?

It is possible that an inquiry on the above lines might have led to the discovery of the extraordinary phenomenon which microscopical observation first brought within our ken and which has given us such a profound insight into the properties of the fluid state.

To our observations on the usual scale, all portions of a liquid in equilibrium appear to be at rest. On placing any denser object in the liquid it sinks, vertically if it is spherical, and we know, of course, that once it has got to the bottom of the containing vessel it will stay there and will not attempt to rise to the surface by itself.

Though these are quite familiar points, they nevertheless are valid only on the dimensional scale to which we are accustomed. We have only to examine under the microscope a collection of small particles suspended in water to notice at once that each one of them, instead of sinking steadily, is quickened by an extremely lively and wholly haphazard movement. Each particle spins hither and thither, rises, sinks, rises again, without ever tending to come to rest. This is the *Brownian movement,* so called after the English botanist Brown, who discovered it in 1827, just after the introduction of the first achromatic objectives.[2]

This remarkable discovery attracted little attention. Those physicists who mentioned the agitation likened it, I think, to the movements of the dust particles to be seen with the naked eye dancing in a sunbeam under the influence of air currents produced by small inequalities in pressure and temperature. But in this case neighbouring particles move in approximately the same direction as the air currents and roughly indicate the conformation of the latter. The Brownian movement, on the other hand, cannot be watched for any length of time without it becoming apparent that the movements of any two particles are completely independent, even when they approach one another to within a distance less than their diameter (Brown, Wiener, Gouy).

The agitation cannot, moreover, be due to vibration of the object glass carrying the drop under observation, for such vibration, when produced expressly, produces *general* currents which can be recognised without hesitation and which can be seen superimposed upon the irregular agitation of the grains. The Brownian movement, again, is produced on a firmly fixed support, at night and in the country, just as clearly as in the daytime, in town and on a table constantly shaken by the passage of heavy vehicles (Gouy). Again, it makes no difference whether great care is taken to ensure uniformity of temperature throughout the drop; all that is gained is the suppression of the general convection currents, which are quite easy to recognise and which have no connection whatever with the irregular agitation under observation (Wiener, Gouy). Great diminution in the intensity of the illuminating light or change in its colour is without effect (Gouy).

Of course, the phenomenon is not confined to suspensions in water, but takes place in all fluids, though more actively the less viscous the fluid.[3] Thus it is just perceptible in glycerine and extremely active, on the other hand, in gases (Bodoszewski, Zsygmondy).

Incidentally, I have been able to observe it with minute spheres of water supported by the "black spots" on soap bubbles. The spherules were 100 to 1,000 times thicker than the thin film which served to support them. They

thus bore to the black spots very nearly the same relationship that an orange bears to a sheet of paper. Their Brownian movement, which is negligible in the direction perpendicular to the pellicule, is very active in the plane of the latter (almost as active as if the spherules were in a gas).

In a given fluid the size of the grains is of great importance, the agitation being the more active the smaller the grains. This property was pointed out by Brown at the time of his original discovery. The nature of the grains appears to exert little influence, if any at all. In the same fluid two grains are agitated to the same degree if they are of the same size, whatever the substance of which they are composed and whatever their density (Jevons, Ramsay, Gouy). Incidentally, the absence of any influence exerted by the nature of the grains destroys any analogy with the displacements of large amplitude undergone by specks of camphor when thrown upon water; the moving fragments moreover finally come to rest (when the water has become saturated with camphor).

In fact—and this is perhaps its strangest and most truly novel feature—the Brownian movement never ceases. Inside a small closed cell (so that evaporation may be avoided) it may be observed over periods of days, months, and years. It is seen in the liquid inclusions that have remained shut up in quartz for thousands of years. *It is eternal and spontaneous.*

All these characteristics force us to conclude, with Wiener (1863), that "the agitation does not originate either in the particles themselves or in any cause external to the liquid, but must be attributed to internal movements, characteristic of the fluid state," movements which the grains follow more faithfully the smaller they are. *We are thus brought face to face with an essential property of what is called a fluid in equilibrium; its apparent repose is merely an illusion due to the imperfection of our senses and corresponds in reality to a permanent condition of unco-ordinated agitation.*

This view agrees completely with the requirements of the molecular hypotheses, which indeed find in the Brownian movement such confirmation as was looked for above. Every granule suspended in a fluid is being struck continually by the molecules in its neighbourhood and receives impulses from them that do not in general exactly counterbalance each other; consequently it is tossed hither and thither in an irregular fashion.

51.—The Brownian Movement and Carnot's Principle.—We have therefore to deal with an agitation that continues indefinitely and is without external cause. Clearly the agitation cannot go on in contradiction to the principle of the conservation of energy. This condition is satisfied if every increment of velocity acquired by a grain is accompanied by the cooling of the fluid in its immediate neighbourhood, and similarly if every diminution in velocity is accompanied by local heating. It merely becomes apparent that *thermal equilibrium is itself only a statistical equilibrium.*

But it must be remembered (Gouy, 1888) that the Brownian movement, which is a fact beyond dispute, provides an experimental proof of those conclusions (deduced from the molecular agitation hypothesis) by means of which Maxwell, Gibbs, and Boltzmann robbed *Carnot's principle* of its claim to rank as an absolute truth and reduced it to the mere expression of a very high probability.

The principle asserts, as we know, that in a medium in thermal equilibrium no contrivance can exist capable of transforming the calorific energy of the medium into work. Such a machine would, for example, allow of a ship being propelled by the cooling of the sea water; and because of the vastness of such a reserve of energy, this would be of practically the same advantage to us as a machine capable of "perpetual motion." That is to say, it would be doing work without taking anything in exchange and without external compensation. But this *perpetual motion of the second kind* is held to be impossible.

Now we have only to follow, in water in thermal equilibrium, a particle denser than water, to notice that at certain instants it rises spontaneously, thus transforming a part of the heat of the medium into work. If we were no bigger than bacteria, we should be able at such moments to fix the dust particle at the level reached in this way, without going to the trouble of lifting it and to build a house, for instance, without having to pay for the raising of the materials.

But the bulkier the particle to be raised, the smaller is the chance that molecular agitation will raise it to a given height. Imagine a brick weighing a kilogramme suspended in the air by a rope. It must have a Brownian movement, though it will certainly be very feeble. As a matter of fact we shall shortly be in a position to calculate the time we would have to wait before we had an even chance of seeing the brick rise to a second level by virtue of its Brownian movement. (That time[4] will be found to be such that by comparison the duration of geological epochs and perhaps of our universe itself will be quite negligible.) Common sense tells us, of course, that it would be foolish to rely upon the Brownian movement to raise the bricks necessary to build a house. Thus the practical importance of Carnot's principle *for magnitudes and lengths of time on our usual dimensional scale* is not affected; nevertheless we shall evidently gain a better understanding of the ultimate significance of that law of probability by stating it as follows:—

On the scale of magnitudes that are of practical interest to us, perpetual motion of the second kind is in general so insignificant that it would be foolish to take it into consideration.

It would, moreover, be incorrect to say that Carnot's principle is incompatible with the conception of molecular motions. On the contrary, it follows as a consequence of that motion, though in the form of a law of probability. In order to escape the restrictions imposed by that law and to transform at will all the energy of motion of the molecules in a fluid in thermal equilibrium into work, it must be possible to *co-ordinate,* or to make parallel, the velocities of all of them.

52.—Wiener's researches and conclusions might have exercised a considerable influence on the mechanical theory of heat, then in process of development; but, embarrassed by confused ideas as to the mutual actions of material atoms and "ether atoms," they remained unknown. Sir W. Ramsay (1876), and afterwards Professors Delsaulx and Carbonelle, arrived at a clearer understanding of the manner in which molecular motion is able to produce the Brownian movement. According to them, "the internal movements which constitute the heat content of fluids is well able to account for the facts." And, going more into detail, "in the case of large surfaces, molecular impacts, which cause pressure, will produce no displacement of the suspended body, because taken altogether they tend to urge the body in all directions at once. But, if the surface is smaller than the area necessary to ensure that all irregular motions will be compensated, we must expect pressures that are unequal and continually shifting from point to point. These pressures will not be made uniform by the law of aggregates and, their resultant being no longer zero, they will vary continuously in intensity and direction. . . ." (Delsaulx and Carbonelle).

The same conclusion was reached by Gouy, whose exposition of the question was particularly brilliant (1888), by Seidentopf (1900), and finally by Einstein (1905), who succeeded in formulating a quantitative theory of the phenomenon; I shall give an account of his work later.

However seductive the hypothesis may be that finds the origin of the Brownian movement in the agitation of the molecules, it is nevertheless a hypothesis only. As I shall explain later on, I have attempted (1908) to subject the question to a definite experimental test that will enable us to verify the molecular hypothesis as a whole.

If the agitation of the molecules is really the cause of the Brownian movement, and if that phenomenon constitutes an accessible connecting link between our dimensions and those of the molecules, we might expect to find therein some means for getting at these latter dimensions. This is indeed the case, and we have moreover a choice of methods we may employ. I shall discuss first the one that seems to me the most illuminating.

Statistical Equilibrium in Emulsions

53.—EXTENSION OF THE GAS LAWS TO DILUTE EMULSIONS.—We have seen . . . how the gas laws were extended by van't Hoff to dilute solutions, where *osmotic* pressure (exerted on a *semi-permeable membrane* which stops the passage of the dissolved substance but allows the solvent to pass through) takes the place of pressure in the gaseous state. At the same time . . . we saw that this law of van't Hoff's holds for all solutions that obey Raoult's laws.

Now Raoult's laws are applicable indiscriminately to all molecules, large or small, heavy or light. The sugar molecule, containing as many as 45 atoms, and the quinine sulphate molecule, containing more than 100, exert no greater or less effect than the active water molecule, which contains 3 atoms only.

Is it not conceivable, therefore, that there may be no limit to the size of the atomic assemblages that obey these laws? Is it not conceivable that even visible particles might still obey them accurately, so that a granule agitated by the Brownian movement would count neither more nor less than an ordinary molecule with respect to the effect of its impact upon a partition that stops it? In short, is it impossible to suppose that the laws of perfect gases may be applicable even to emulsions composed of visible particles?

I have sought in this direction for crucial experiments that should provide a solid experimental basis from which to attack or defend the Kinetic Theory. In the following paragraph I shall describe the one that appears to me to be the simplest.

54.—DISTRIBUTION OF EQUILIBRIUM IN A VERTICAL COLUMN OF GAS.—It is well known that the air is more rarefied in the mountains than at sea level and that, in general terms, any vertical column of gas is compressed under its own weight. The rarefaction has been given by Laplace (who obtained it when working out the connection between altitude and barometric indications).

In order to obtain his law, let us consider a thin horizontal cylindrical element, of unit cross-sectional area and of height h; slightly different pressures p and p' will be exerted on the two faces of the element. There would be no change in the condition of the element if it were to be enclosed between two pistons held in position by pressures equal to p and p'; the difference $(p - p')$ between them must balance the force gm due to gravity which tends to pull the mass m of the element downwards. This mass m, moreover, is to the gramme molecular mass M of the gas as its volume $(1 \times h)$ is to the volume v occupied by the gramme molecule under the same mean pressure, so that

$$p - p' = g \cdot \frac{M}{v} \cdot h.$$

And since the mean pressure differs very little from p, so that we may substitute (from the equation for perfect gases) RT/p for v, we may write

$$p - p' = \frac{M \cdot g \cdot h}{RT} \cdot p$$

or

$$p' = p\left(1 - \frac{M \cdot g \cdot h}{RT}\right).$$

Clearly, when the thickness h of the element is fixed, the ratio between the pressures on its two faces is fixed, whatever the level of the element. For example, in air, at the ordinary temperature, the pressure falls by the same relative amount as we mount each step on a staircase (by about 1/40,000 of its value if the step is 20 centimetres high). If p_o is the pressure at the foot of the stairs, the pressure after mounting the first step is

$$p_o\left(1 - \frac{M \cdot g \cdot h}{RT}\right);$$

it is again lowered in the same ratio after the second step and becomes

$$p_o\left(1 - \frac{M \cdot g \cdot h}{RT}\right)^2.$$

After the hundredth step it will be

$$p_o\left(1 - \frac{M \cdot g \cdot h}{RT}\right)^{100}$$

and so on.

Moreover, it does not matter from what level the staircase starts. Hence, since it is clear that when we rise to the same height starting from the same level the fall in pressure does not depend on the number of steps into which we divide that height, it appears that the pressure will fall in the same ratio

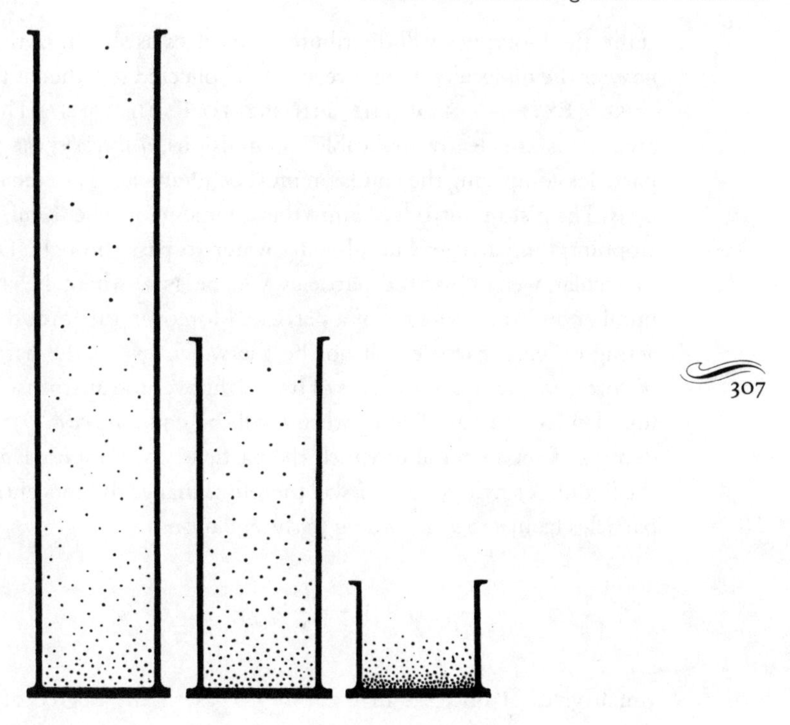

each time we rise through the height H, no matter from what level we start. In air (at the ordinary temperature) we find that the pressure becomes halved each time we rise through 6 kilometres. (In pure oxygen, at 0° C., 5 kilometres is sufficient to halve the pressure.)

Of course, since the pressure, being proportional to the density, is therefore proportional to the number of molecules in unit volume, the ratio p_o/p between pressures can be replaced by the ratio n/n_o between the numbers of molecules at the two levels considered.

But the elevation required to produce a given rarefaction varies with the nature of the gas. It is apparent from the formula that the ratio between the pressures does not change if the product M*h* remains constant. In other words, if the gramme molecular weight of a second gas is 16 times lighter than that of the first, the elevation required to produce the same rarefaction will be 16 times greater in the second gas than in the first. Since it is necessary to rise to a height of 5 kilometres in oxygen at 0° C. before its density is halved, a height 16 times greater (or 80 kilometres) will be necessary in hydrogen at 0° C. to produce the same result.

Above are represented three gigantic vertical gas jars (the largest being 300 kilometres high), containing the same number of molecules of hydrogen, helium, and oxygen respectively. Assuming the temperature to be constant, the molecules will distribute themselves as shown in the figure; the heavier the molecules, the more are they collected together at the bottom.

55.—EXTENSION OF THE THEORY TO EMULSIONS.—The preceding arguments are clearly applicable to emulsions, *if they obey the gas laws.* The particles composing the emulsion must be identical, as are the molecules of a gas. The pistons introduced into the argument must be "semi-permeable," stopping the particles but allowing water to pass through. The "gramme molecular weight" of the particles will be N*m,* where N is Avogadro's number and *m* is the mass of a particle. Moreover, the force due to gravity acting on each particle will not be the weight *mg* of the particle, but its

effective weight; that is, the excess of its weight over the up-thrust caused by its liquid surroundings. The up-thrust will be equal to $m(d/D)g$, if D is the density of the material of which the particles are composed, and *d* that of the liquid. A small elevation *h* will therefore change the concentration of the particles from *n* to *n′* according to the equation

$$\frac{n'}{n} = 1 - \frac{N}{RT} \cdot m \left(1 - \frac{d}{D}\right) . gh,$$

which gives at once, as in the case of gases,[6] the degree of rarefaction corresponding to any height H whatever. H may be considered to be subdivided like a flight of stairs into *q* small steps of height *h.*

Thus, once equilibrium has been reached between the opposing effects of gravity, which pulls the particles downwards, and of the Brownian movement, which tends to scatter them, equal elevations in the liquid will be accompanied by equal rarefactions. But if we find that we have only to rise 1⁄20 of a millimetre, that is, 100,000,000 times less than in oxygen, before the concentration of the particles becomes halved, we must conclude that the effective weight of each particle is 100,000,000 times greater than that of an oxygen molecule. *We shall thus be able to use the weight of the particle, which is measureable, as an intermediary or connecting link between masses on our usual scale of magnitude and the masses of the molecules.*

. . . .

65.—A DECISIVE PROOF.—Let us consider grains of such a kind that an elevation of 6μ is sufficient to halve their concentration. To reach the same degree of rarefaction in air, we have seen that a distance of 6 kilometres, which is nearly 10,000 million times as great, is necessary. If our theory is correct, the weight of an air molecule should therefore be one ten thousand-millionth of the weight, in water, of one of the grains. The weight of the hydrogen atom may be obtained in the same way, and it now only remains to be seen whether numbers obtained by this method are the same as those deduced from the kinetic theory.[7]

It was with the liveliest emotion that I found, at the first attempt, the very numbers that had been obtained from the widely different point of view of the kinetic theory. In addition, I have varied widely the conditions of experiment. The volumes of the grains have had values distributed between limits which were to each other as 1 is to 50. I have also varied the nature of the grains (with the aid of M. Dabrowski), using mastic instead of gamboge. I have varied the intergranular liquid (with the help of M. Niels Bjerrum) and studied gamboge grains suspended in glycerine containing 12 per cent. of water, the mixture being 125 times more viscous than water.[8] I have varied the apparent density of the grains, in ratios varying from 1 to 5; in glycerine it becomes *negative* (in which case the influence of the changed sign of their weight accumulated the grains in the *upper* layers of the emulsion). Finally, M. Bruhat has, under my direction, studied the influence of temperature and observed the grains first in *super-cooled* water (−9° C.) and then in hot water (60° C.); the viscosity in the latter case was half what it was at 20° C., so that the viscosity varied in the ratio of 1 to 250.

In spite of all these variations, the value found for Avogadro's number N remains approximately constant, varying irregularly between 65×10^{22} and 72×10^{22}. Even if no other information were available as to the molecular magnitudes, *such constant results would justify the very suggestive hypotheses that have guided us,* and we should certainly accept as extremely probable the values obtained with such concordance for the masses of the molecules and atoms.

But the number found agrees with that (62×10^{22}) given by the kinetic theory from the consideration of the viscosity of gases. *Such decisive agreement can leave no doubt as to the origin of the Brownian movement.* To appreciate how particularly striking the agreement is, it must be remembered that before these experiments were carried out we should certainly not have been in a position either to deny that the fall in concentration through the minute height of a few microns would be negligible, in which case an infinitely small value for N would be indicated, or, on the other hand, to assert that all the grains do not ultimately collect in the immediate vicinity of the bottom, which would indicate an infinitely large value for N. It cannot be supposed that, out of the enormous number of values *a priori* possible, values so near to the predicted number have been obtained by chance for every emulsion and under the most varied experimental conditions.

The objective reality of the molecules therefore becomes hard to deny. At the same time, molecular movement has not been made visible. The Brownian movement is a faithful reflection of it, or, better, it is a molecular movement in itself, in the same sense that the infra-red is still light. From the point of view of agitation, there is no distinction between nitrogen molecules and the visible molecules realised in the grains of an emulsion,[9] which have a gramme molecule of the order of 100,000 tons.

Thus, as we might have supposed, an emulsion is actually a miniature ponderable atmosphere; or, rather, it is an atmosphere of colossal molecules, which are actually visible. The rarefaction of this atmosphere varies with enormous rapidity, but it may nevertheless be perceived. In a world with such an atmosphere, Alpine heights might be represented by a few microns, in which case individual atmospheric molecules would be as high as hills.

. . . .

NOTES

1. At least, the other phenomenon could only be one of those which we know can occur without external compensation (such as isothermal change of volume of a gaseous mass, according to a law discovered by Joule). In that case the gain may still be looked upon as non-existent.

2. Buffon and Spallanzani knew of the phenomenon but, possibly owing to the lack of good microscopes, they did not grasp its nature and regarded the "dancing particles" as rudimentary animalculæ (Ramsay: Bristol Naturalists' Society, 1881).

3. The addition of impurities (such as acids, bases, and salts) has no influence *whatever* on the phenomenon (Gouy, Svedberg). That the contrary has been maintained, after a superficial examination, is due to the fact that impurities cause the small particles to stick to the glass when they happen to touch the sides of the containing vessel; the movement of the remainder, however, is unaffected. We might as well say that the motion of the waves is stopped when we fasten a wave-tossed plank against a quay.

4. Considerably more than $10^{10^{10}}$ years.

5. If the staircase had q steps, the ratio p/p_0 between the pressures at the top and at the bottom would be

$$\frac{p}{p_0} = \left(1 - \frac{M \,.\, g \,.\, h}{RT}\right) q.$$

The calculation is simplified by taking logarithms of the two sides, which gives (using ordinary logarithms to base 10) by a simple transformation

$$2.3 \log \frac{p_0}{p} = \frac{M \,.\, g \,.\, H}{RT},$$

where H is the distance between the higher and lower levels and is regarded as being divided into a very large number q of steps each of height h.

6. As with columns of gases, the calculation may be simplified by using logarithms, which gives the following form to the equation for the distribution of the particles:

$$2.3 \log \frac{n_o}{n} = \frac{N}{RT} \cdot m \cdot \left(1 - \frac{d}{D}\right) gH,$$

or, if we wish to introduce the volume V of a particle:

$$2.3 \log \frac{n_o}{n} = \frac{N}{RT} \cdot V \cdot \left(D - d\right) gH.$$

7. The calculations are simplified if the distribution equation given in the note to para. 55 is used.

8. The Brownian movement, though much abated, is nevertheless perceptible; several days are required before a permanent equilibrium is reached. I should have liked to study the distribution in an even more viscous medium, but, when less than 5 per cent. of water was added to the glycerine (very feebly acid), the grains collect upon the sides and permanent equilibrium could no longer be observed. I have subsequently made use of this circumstance in extending the gas laws to these viscous emulsions. . . .

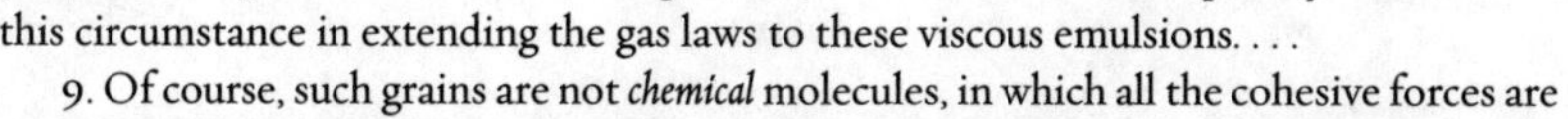

9. Of course, such grains are not *chemical* molecules, in which all the cohesive forces are of the nature of those uniting the carbon to the four hydrogen atoms in methane.

18

Salmon's Empirical Defense of Realism

From

Scientific Explanation and the Causal Structure of the World / Wesley C. Salmon

The Common Cause Principle and Molecular Reality

One great focus of interest in twentieth-century philosophy of science has been the controversy over instrumentalism and scientific realism. Major contributions to the discussion have been made by Mach, Russell, Duhem, Carnap, Reichenbach, Popper, Nagel, and Hempel—to name only a few. Moreover, the climate of opinion on this issue has changed radically. During the earlier decades of this century, various forms of instrumentalism seemed in general to hold sway; when Reichenbach and Popper defended realism in the thirties, they were departing drastically from the mainstream. In recent years, various forms of scientific realism have gained ascendancy, and instrumentalism has few important defenders. Although I had not joined the debate in print before 1978, I felt the tension between these two viewpoints; I felt my own sentiments moving away from instrumentalism toward scientific realism. At the same time, I have wondered whether the general shift in opinion has been more than merely a change in philosophi-

---

cal fashion, for the arguments that were offered in support of scientific realism have struck me as quite unconvincing. Bas van Fraassen has been the sharpest recent critic of the realist position, and I agree almost entirely with his criticisms of the prorealist arguments. I began to suspect that what moved me (and many other philosophers as well, perhaps) was not philosophical argument, but, rather—to use Russell's apt phrase—growth of a more "robust sense of reality." It is, of course, disquieting to find that one's main philosophical commitments arise—not from reasoned argument—but from purely psychological considerations.

In an effort to alleviate this intellectual discomfort, I decided to try an empirical approach to the philosophical problem. Since it seemed unlikely that scientists would have been moved by the kinds of arguments supplied by philosophers, I felt that some insight might be gained if we were to consider the evidence and arguments that convinced scientists of the reality of unobservable entities. Although scientists, by and large, seem committed to the existence of a variety of unobservable entities—for example, the planetary orbits of Copernicus and Maxwell's electromagnetic waves (outside of the range of visible light)—the existence of atoms and molecules, as the microphysical constituents of ordinary matter, is the most clear and compelling example. It is fortunate that a superb account of this historical case is provided by Mary Jo Nye in *Molecular Reality* (1972). It is also fortunate that the French physical chemist Jean Perrin, who played a central role in this historical development, provided a lucid account in his semi-popular book, *Les Atomes* (1913). Examination of these works reveals a clear-cut form of argument.

As everyone knows, a primitive form of atomism was propounded in antiquity by Democritus, Lucretius, and others, but no very strong empirical evidence in support of such a theory was available before the beginning of the nineteenth century. However, from the time of Dalton's work (early in that century) until the end of the century, the atomic hypothesis was the subject of considerable discussion and debate. During this period, responsible, well-informed scientists could reasonably adopt divergent viewpoints on the question. Michael Gardner (1979) has given an enlightening discussion of these nineteenth-century developments, relating them to the realism/instrumentalism issue. Within a dozen years after the turn of the century, however, the issue was scientifically settled in favor of the atomic/molecular hypothesis, and the scientific community—with the notable exception of Ernst Mach—appears to have been in agreement about it.

The decisive achievement was the determination of Avogadro's number *N*, the number of molecules in a mole of any substance. This number is *the* link between the macrocosm and the microcosm; once it is known a variety of microquantities are directly ascertainable. Loschmidt gave what seems to be the first crude estimate of N in 1865; at the turn of the century, rough approximations based upon the kinetic theory of gases were available. Perrin's chief experimental efforts were devoted to the precise ascertainment of N through the study of Brownian movement—the random agitation of microscopic particles (e.g., pollen or smoke) suspended in fluid. The difficulties that stood in the way of success in such a venture were awesome, and Perrin's achievement is a milestone in the emergence of modern microphysics. Since Nye's excellent account is available, I shall not attempt to provide a historical treatment of this topic, but I shall give a brief chronology of some major developments in the first few years of the twentieth century.

Perrin, who was born in 1870, recognized early in his career the crucial importance of providing a scientifically convincing demonstration of the reality of atoms and molecules. The achievement of this aim constituted the major motivation for his research, at least until 1913. His doctoral dissertation was on X rays and cathode rays—topics that were at the forefront of microphysics just before the turn of the century. Shortly after receiving his doctorate in 1897, he was put in charge of a course in physical chemistry at the Sorbonne. In organizing the materials for this course, he became interested in various aspects of thermodynamics and the microstructure of matter that eventually led to his work on Brownian movement. In 1901, in commenting upon the then-current debate between the 'partisans of the plenum' and the 'partisans of the void' (atomists), he explicitly noted the common view that the controversy was a matter of philosophical speculation that could never be settled by experimental evidence. At the same time, he pointed to the possibility of verifying physical statements about objects whose dimensions fall far below 0.25 microns, the limit of resolution of an ordinary microscope (Nye, 1972, p. 83).

A good deal of Perrin's early research concerned the nature of colloids—suspensions of particles larger than molecules but still too small to be discerned under an ordinary microscope. In 1903, Siedentopf and Zsigmondy published an account of the ultramicroscope, which made possible the observation of particles with diameters as small as 5×10^{-3} microns. With this amount of magnification, the Brownian movement of colloidal particles could be viewed. Perrin's earliest determination of a precise value for Avogadro's number was based upon observations of the vertical distribution of particles in colloidal suspensions.

In 1905–1906, Einstein and Smoluchowski provided the first adequate theoretical accounts of the phenomenon that Robert Brown had observed early in the nineteenth century. At about the same time (1906), Perrin maintained that particles undergoing Brownian motion reveal *qualitatively* the nature of the fluid in which they are suspended (Nye, p. 85). The

Einstein-Smoluchowski theory had little, if any, influence upon the experimental work of Perrin that culminated in his first precise determination of Avogadro's number (Nye, p. 111), but by the time these results were published (1908), Perrin realized that they constituted *quantitative* experimental confirmation for the Einstein-Smoluchowski theory. Einstein was pleased, but also astonished, by Perrin's achievement, for he had held the opinion that the practical difficulties standing in the way of such experiments would be insuperable (Nye, p. 135).

Perrin's experimental achievement was, indeed, monumental. The experiments required the preparation of tiny spheres—less than 1 micron in diameter—of gamboge, a bright yellow resin. To serve adequately in the experiments, these spheres had to be uniform in size and density, and the size had to be measured with great precision. Vast numbers of painstaking observations were needed. The results were presented in four papers that were published in the *Comptes rendus* of the Académie des Sciences in 1908. It is of the greatest importance to *our* story to note that these papers included not only the precise value of Avogadro's number ascertained on the basis of his study of Brownian movement, but also a comparison of that value with the results of several other determinations based upon entirely different methods, including Rutherford's study of radioactivity and Planck's work on blackbody radiation (Nye, pp. 109–111).

Perrin's 1908 results had an immediate impact in some quarters. Ostwald, who was one of the most prominent and staunch opponents of the atomic/molecular hypothesis, had stated categorically in 1906 that "atoms are only hypothetical things" (quoted by Nye, p. 151). However, in the fourth edition of his *Grundriss der physikalischen Chemie* (1908), Ostwald did an about-face, writing:

> I have satisfied myself that we arrived a short time ago at the possession of experimental proof for the discrete or particulate nature of matter—proof which the atomic hypothesis has vainly sought for a hundred years, even a thousand years. . . . [The results] entitle even the cautious scientist to speak of an experimental proof for the atomistic constitution of space-filled matter. (Quoted by Nye, *ibid.*)

Ostwald is referring here to the work of J. J. Thomson on the kinetic theory of gases, as well as to Perrin's work on Brownian movement.

Perrin's work, up to this time, had dealt with translational Brownian movement. Recognizing that the principle of equipartition implies that thermal energy also produces rotations, he made further determinations of N on the basis of the rotational motions of Brownian particles. In order to carry out this experiment, he had to create imperfect spheres of gamboge, for the rotational motion of a perfect sphere is impossible to observe. With characteristic ingenuity, he accomplished this feat. In a 1909 paper, Perrin reported not only his own ascertainments of Avogadro's number by three distinct methods—namely, the vertical distribution of Brownian particles, the translational motion of Brownian particles, and the rotational motion of Brownian particles—but also eight other distinct determinations of *N* by other investigators (Nye, pp. 133–135). The results were all in striking agreement with one another.

In 1913, Perrin published *Les Atomes,* in which he recounted his experimental efforts and summarized the evidence for the reality of molecules. Instead of focusing upon one or two beautiful experimental determinations of precise values of Avogadro's number, he lays great stress upon the fact that *N* has been ascertained in a considerable number of independent ways—in fact, he lists thirteen distinct methods (Perrin, 1913, p. 206; Nye, p. 161). In her careful historical analysis of the situation, Nye also places great emphasis upon the variety of independent methods of ascertaining *N* (pp. 110–111, 133–135). At a 1912 conference, Poincaré likewise emphasized the variety of independent determinations (Nye, p. 157). Since I shall be arguing that this variety plays a decisive role in the argument, let me give a somewhat oversimplified indication of its range. Emulating the venerable tradition established by St. Thomas, I shall offer five ways.

1. *Brownian movement.* If one applies the principle of equipartition to the Brownian particles, then it follows immediately that the average kinetic energy of the Brownian particles is equal to the average kinetic energy of the molecules of the fluid in which they are suspended. This is a straightforward interpretation of the claim that the system is in thermal equilibrium. Molecular velocities are, in principle, directly measurable; moreover, they can be derived from macroscopic quantities via Herapath's formula

$$P = \tfrac{1}{3} \rho v^2$$

where P is the pressure and ρ the density of a sample of gas. In addition, it is possible in principle to measure both the mass and the velocities of Brownian particles. From these quantities it is easy to calculate the mass of a molecule of the fluid, and hence, Avogadro's number, which is the molecular weight divided by the mass of a single molecule.

In fact, this simple approach is impractical; because the Brownian particle changes its direction of motion with extremely great frequency, its velocity cannot be measured directly. Perrin found it necessary to take a less direct approach. In his first experiments, he examined the vertical distribution of Brownian particles in suspension under equilibrium conditions.

Such particles are subject to two opposing forces. Given that the particles are denser than water, the gravitational force makes them tend to sink to the bottom. A force due to osmotic pressure (a diffusion process) tends to raise them. By carefully ascertaining the exponential distribution according to height, Perrin was able to calculate Avogadro's number. This result is, in principle, similar to the simpler idea outlined in the preceding paragraph, for it leads to values for the average kinetic energy of the Brownian particles and of the molecules of the fluid. (For details see Nye, 1972, pp. 102–111.)

2. *Alpha decay.* During the first decade of the present century, it was discovered that alpha particles are helium nuclei. Alpha particles were detected by scintillation techniques; it was, consequently, straightforward to ascertain the rate at which a given substance—radium, for example—produced helium nuclei. Rutherford recognized that the alpha particles quickly pick up electrons, transforming them into helium atoms, and that the helium generated by a given quantity of radium in a fixed period of time could be captured and weighed. In this way, it was possible to count the number of atoms required to make up a given mass of helium. Avogadro's number comes directly from such data.

3. *X-ray diffraction.* M. von Laue had the idea that the arrangement of atoms in a crystal could serve as a diffraction grating for X rays. Given a knowledge of the wavelengths of the X rays, it is possible to examine the diffraction patterns and calculate the spacing between the atoms. With this information, one can ascertain the number of atoms in a given crystal. Experiments of this sort were actually conducted in 1912 by W. Friedrich and P. Knipping. Avogadro's number follows directly.

4. *Blackbody radiation.* In the closing days of the nineteenth century (December 14, 1900), Max Planck presented his derivation of the law of blackbody radiation, from which he derived the relation

$$\lambda_{max} \times kT/c = 0.2014 \times h$$

where k is Boltzmann's constant, T the temperature of the blackbody, λ the wavelength at which the maximum energy is being radiated, c the speed of light, and h Planck's constant. From this relation, if h is known, we can derive the value of Boltzmann's constant, since all of the other quantities can be measured macroscopically. The speed of light was well known by the end of the nineteenth century.

There are various ways of ascertaining a value for Planck's constant; one obvious method comes from Einstein's theory of the photoelectric effect. If we let E stand for the total energy that a photon can transfer to an electron—that is, the maximum kinetic energy of photoelectrons plus the work function of the metal upon which the photons impinge—we have h as a constant of proportionality between E and ν, the frequency of the incident light.

During the nineteenth century, the ideal gas law

$$PV = nRT$$

was derived from the kinetic theory of gases, and the universal gas constant R was measured empirically. Since n represents the number of moles of the ideal gas, R is the gas constant *per mole.* Boltzmann's constant k is the gas constant *per molecule.* Hence $R/k = N$.

5. *Electrochemistry.* The faraday F is the amount of charge (96,500 coulombs) required to deposit by electrolysis one mole of a monovalent metal—for example, silver—on an electrode. It takes the charge of one electron to deposit each ion, that is,

$$F/e = N.$$

The experimental determination of e, the charge of the electron, by Robert A. Millikan in 1911, thus furnished another way of ascertaining Avogadro's number.

The foregoing list of ways to establish the value of Avogadro's number does not coincide with Perrin's list, but it gives a fair indication of the variety of methods available in the first few years of the present century.[1] As I have already remarked, the variety of these approaches is remarkable and striking, and it was fully appreciated at the time. At the conclusion of a 1912 conference at which the atomic/molecular hypothesis was a main topic of discussion, Poincaré remarked:

> A first reflection is sure to strike all the listeners; the long-standing mechanistic and atomistic hypotheses have recently taken on enough consistency to cease almost appearing to us as hypotheses; atoms are no longer a useful fiction; things seem to us in favour of saying that we see them since we know how to count them. . . . The brilliant determinations of the number of atoms by M. Perrin have completed this triumph of atomism. . . . In the procedures deriving from Brownian movement or in those where the law of radiation is invoked . . . in . . . the blue of the sky . . . when radium is dealt with. . . . The atom of the chemist is now a reality. (Quoted by Nye, 1972, p. 157)

In *Les Atomes,* published in the next year, Perrin also places considerable weight upon the variety of methods. Immediately following his table of thirteen distinct experimental approaches to Avogadro's number, he writes:

> Our wonder is aroused at the very remarkable agreement found between values derived from the consideration of such widely different phenomena. Seeing that not only is the same magnitude obtained by each method when the conditions under which it is applied are varied as much as possible, but that the numbers thus established also agree among themselves, without discrepancy, for all the methods employed, the real existence of the molecule is given a probability bordering on certainty. (Perrin, 1913, pp. 215–216)

It seems clear that Perrin would not have been satisfied with the determination of Avogadro's number by any single method, no matter how carefully applied and no matter how precise the results. It is the "remarkable agreement" among the results of many diverse methods that supports his conclusion about the reality of molecules.

If there were no such micro-entities as atoms, molecules, and ions, then these different experiments designed to ascertain Avogadro's number would be genuinely independent experiments, and the striking numerical agreement in their results would constitute an utterly astonishing coincidence. To those who were in doubt about the existence of such micro-entities, the "remarkable agreement" constitutes strong *evidence* that these experiments are not fully independent—that they reveal the existence of such entities. To those of us who believe in the existence of these sorts of micro-entities, their very existence and characteristics—as specified by various theories—*explain* this "remarkable agreement."

The claim I should like to make about the argument hinted by Poincaré, and stated somewhat more explicitly by Perrin, is that it relies upon the principle of the common cause—indeed, that it appeals to a conjunctive fork. We can say, very schematically, that the coincidence to be explained is the "remarkable agreement" among the values of N that result from independent determinations. The situation is, I think, quite analogous to the testimony of independent witnesses. Suppose that a murder has been committed, and that one crucial factor in the investigation concerns whether a particular suspect, John Doe, was in the vicinity of the crime at about the time it was committed. Suppose that several different witnesses testify, not only to his whereabouts, but also to a variety of details—such as how he was dressed, what he was carrying, and the fact that he arrived in a taxi. Indeed, suppose that all of these witnesses were able to give the license number of the taxi. If collusion can be ruled out, and if we can be confident that they were not coached, then the agreement of the witnesses on details of the foregoing sorts would constitute strong evidence that they were reporting facts that they had observed. Moreover, even if none of the witnesses could be considered particularly reliable, the agreement among their reports would greatly enhance our confidence in the veracity of their account. It would be too improbable a coincidence for all of them to have fabricated their stories independently, and for these stories to exhibit such strong agreement in precise detail.

For the sake of argument, let us suppose that there are five witnesses, each of whom claims to have observed the arrival of John Doe at the scene of the crime. Each claims to remember the license. Assume that any license consists of two letters and four digits. Since there are over six million different combinations, the probability that five independent witnesses would choose the same combination by chance is too small to consider seriously. We might think about the question of filling the blank in the following statement: The number of the license on the taxi is ———. If each of the witnesses fills the blank in the same way, we do not believe it can be due to mere chance coincidence. If the witnesses had not been present to see the suspect arrive in the taxi, or if they had been present but no taxi had been involved, we could hardly expect such agreement in testimony.

The situation pertaining to Avogadro's number can be viewed in much the same way. Suppose we have five scientists. One of them is studying Brownian movement (1), another is studying alpha decay (2), another is doing X-ray diffraction experiments on crystals (3), still another is working on the spectrum of blackbody radiation (4), while the remaining one is doing an experiment in electrolysis (5). Notice what a wide variety of substances are involved and how diverse are the phenomena being observed: (1) random movements of tiny spheres of gamboge suspended in water are viewed under a powerful microscope; (2) scintillations on a screen exposed to alpha radiation are carefully counted, and the quantities of helium generated in the presence of a radioactive substance are carefully measured; (3) spatial relations among the light and dark areas on a photographic film that has been exposed to X-radiation are measured; (4) the light from the mouth of a blast furnace is spectroscopically separated on the basis of wavelength for the measurement of intensity; and (5) metallic silver is observed to collect upon an electrode placed in an electrolytic solution. These experiments seem on the surface to have nothing to do with one another. Nevertheless, we ask each scientist to fill in the blank in this statement: On the basis of my experiments, assuming matter to be composed of molecules, I calculate the number of molecules in a mole of any given substance to be ———. When we find that all of them write numbers that, within the accuracy of their experiments, agree with 6×10^{23}, we are impressed by the "remarkable agreement" as were Perrin and Poincaré. Certainly, these five hypothetical scientists have been counting entities that are objectively real.

Let us consider the alternative hypothesis. At the turn of the century, the main hypothesis that was opposed to the atomic theory—one which rejected the kinetic-molecular theory of gases—was a phenomenological view. According to this theory, often known as energeticism, matter that appears under close scrutiny to be continuous is in fact continuous; it is not composed of myriad tiny corpuscles and (in the case of gases, at least) a great deal of empty space. Proponents of the phenomenological view maintained that the laws of phenomenological thermodynamics are strictly true down to the microscopic and submicroscopic levels. Ostwald, in particular, was greatly impressed by the fact that energy is a conserved quantity, and that the thermodynamic equations governing the behavior of energy are differential equations. He took these considerations to be a basis for doubting the hypothesis that matter is composed of discrete corpuscles, and as support for the view that energy, which is continuous, constitutes the underlying reality. Arguments of the foregoing sort were frequently coupled, among opponents of atomism, with extreme positivistic arguments about the unknowability of the fine structure of matter. Proponents of the kinetic-molecular theory, in contrast, were forced to regard at least some of the basic laws of thermodynamics—especially the second law—as statistical, and consequently, to admit that the phenomenological laws were subject to violation at the molecular and microscopic levels. The dispute about 'molecular reality' clearly involved a strange mixture of factual and philosophical components. One is reminded of the man, accused of drunken behavior, who protests, "Everyone knows I am a teetotaler, and besides, I only had two drinks."

Now suppose, for instance, that carbon is not composed of atoms. We take various forms of carbon—a piece of coal and diamond, for example—and try to count the constituent particles. We pulverize the coal and the diamond in whatever ways we can devise. It seems plausible to suppose that the results of such endeavors would *not* yield constant numbers for 12 grams of coal if we tried it many times, or for 12 grams of diamonds. It would be *most* unreasonable to expect numerical agreement between the results of counting coal particles and those of counting diamond fragments. Likewise, if water and alcohol were continuous fluids on every scale, no matter how fine, then we should not expect consistent results if we attempt to create fine sprays and count the number of droplets. These suggestions are, of course, egregiously crude from an experimental standpoint. Nevertheless, it seems to me, regardless of how carefully and ingeniously they were refined, we would not get the kinds of consistent numerical results that have actually emerged in the various ascertainments of Avogadro's number unless matter really does possess an atomic/molecular structure. If matter were ultimately continuous, the number of particles would depend solely upon the method by which the substance is broken into pieces, for the substance itself is not in any straightforward sense *composed of* constituent particles. What we find, in fact, is that when we try, in any of a variety of suitable ways, to count the number of atoms in 12 grams of carbon—whether it be in the form of coal, graphite, lampblack, or diamond—the number is always the same. Moreover, that number is the same as the number of water molecules in 18 grams of water, or the number of helium atoms in 4 grams of helium. Still other substances yield the same number of molecules per mole. These results apply to matter in the gaseous, liquid, and solid states, and with respect both to physical and to chemical properties. Such numerical consistency would be an unthinkably improbable coincidence if molecules, atoms, and ions did not exist.

I maintained previously that the argument by which the existence of atoms and molecules was established to the satisfaction of virtually all physical scientists in the first dozen years of the present century has the form of a common cause argument, and that it relies upon the structure of the conjunctive fork. We should recall explicitly at this point that conjunctive forks differ in a fundamental physical way from interactive forks. In the interactive fork, we have two or more causal processes that intersect spatiotemporally with one another. In the conjunctive fork, by contrast, we find causal processes that are physically independent of one another—that do not even intersect—but that arise out of common background conditions. The statistical dependency among the effects is a result of common background conditions. Remember, for instance, the victims of mushroom poisoning; their common illness arose from the fact that each of them consumed food from a common pot. Similarly, I think, the agreement in values arising from different ascertainments of Avogadro's number results from the fact that in each of the physical procedures mentioned, the experimenter was dealing with substances composed of atoms and molecules—in accordance with the theory of the constitution of matter that we have all come to accept. The historical argument that convinced scientists of the reality of atoms and molecules is, I believe, philosophically impeccable.

In presenting the idea of a conjunctive common cause, I emphasized the fact that there must be causal processes leading from the causal background to the correlated effects, and that a set of statistical relations . . . must be satisfied. There is little difficulty, I think, in seeing that the required causal processes do connect the existence and behavior of the micro-particles with the experimental results that furnish the basis for the calculation of Avogadro's number. In the case of X-ray diffraction, for example, electromagnetic radiation of known wavelength is emitted from an X-ray source, it travels a spatiotemporally continuous path from the source to the crystal, where it interacts with a grating in a way that is well established in optics. The diffracted radiation then travels from the crystal to a photographic plate, where another causal interaction occurs, yielding an interference pattern when the plate is developed. In the case of Brownian movement, the observed random motion of the Brownian particle is caused by a very large number of collisions (causal interactions) with the molecules of the fluid in which the particle is suspended. The other approaches to Avogadro's number patently involve similar sorts of causal processes and interactions. Indeed, it is by virtue of the causal properties of atoms, molecules, and ions that the various experimental ascertainments of *N* are in principle possible. The causal processes that lead from the microstructure of matter to the observed phenomena that yield values of N are quite complex, but our physical theories provide a straightforward account of them.[2]

Let us now try to see how the statistical relations used to define conjunctive forks apply to the ascertainment of *N;* for ease of reference, these relations are here repeated:

$$(1)\ P(A.B|C) = P(A|C) \times P(B|C)$$
$$(2)\ P(A.B|\overline{C}) = P(A|\overline{C}) \times P(B|\overline{C})$$
$$(3)\ P(A|C) > P(A|\overline{C})$$
$$(4)\ P(B|C) > P(B|\overline{C})$$

In order to apply these formulas directly, we shall confine our consideration to two methods—alpha radiation and Brownian motion—though the formulas could obviously be generalized to apply to five or thirteen or any desired number of methods. We need to exercise a bit of care in assigning meanings to the variables *A, B,* and C. Let us suppose that *A* and *B* stand for classes of individual experiments involving alpha radiation and Brownian motion, respectively, that yield sufficiently accurate values for Avogadro's number. In Perrin's previously mentioned table of thirteen methods, the values of N range from 6×10^{23} to 7.5×10^{23}; the currently accepted value is 6.022×10^{23}. Let us say that values lying between 4×10^{23} and 8×10^{23} are acceptable. Given the difficulty of the experiments, we should by no means expect uniformly successful results. Next, let us understand C to represent experiments actually conducted under the initial conditions supposed to obtain in the case at hand. In the experiments on alpha radiation and helium production, for example, C would involve specification of the correct atomic weight for helium and the correct decay rate for the radioactive source of alpha particles. A mistake on either of these scores would provide us with an instance of $\overline{C}$. Similarly, if alpha particles did not come from the disintegration of atoms and if helium did not have an atomic structure, we would not have an instance of C. In the experiments on Brownian motion, if we suppose that the particles of gamboge are suspended in ordinary water, they must not be suspended in alcohol or heavy water. Likewise, we must have accurate values for the size and density of the Brownian particles. Given an accurate value for the molecular weight of the suspending fluid, and accurate values of the parameters characterizing the spheres of gamboge, we may say that C is satisfied. Of course, C would fail radically to be fulfilled if suspending fluids were not composed of molecules, but were strictly continuous media.

With this general understanding of the interpretation of formulas (1)–(4), it is easy to see that these relations are satisfied. Equation (1) holds because the experiments on Brownian motion are physically separate and distinct from those on alpha radiation and helium production. The most famous ones were conducted in different countries. On the assumption that the initial conditions for these experiments have been satisfied, the probability of a successful outcome of an experiment of the one type is uninfluenced by successes or failures in experiments of the other type. Even if Rutherford and his associates in England had been inept experimentalists, that would not have had any influence upon the results of experiments conducted by Perrin in France, and conversely. Similarly, on the assumption that experiments of either type are conducted under incorrectly assigned initial conditions, it would seem reasonable to suppose that a successful outcome would simply be fortuitous, and that a lucky correct result on an experiment of one type would have no influence upon the chance of lucky correct results on experiments of the other. Thus equation (2) appears also to be satisfied.

There is, however, an important ground for uneasiness about relation (2). We are especially interested in one particular sort of mistake about initial conditions—namely, the mistake of supposing that material objects are composed of molecules and atoms when, in fact, they are not. Since material objects *are* composed of molecules, experiments cannot actually be conducted under these conditions. The reference class of such experiments is empty; hence the probabilities that occur in equation (2) *might seem* not to be well defined if we confine our attention to experimental conditions that fail to be instances of C for that reason. This consequence does not necessarily follow. If we can provide theoretical reasons for asserting what the outcome would be if the kinetic-molecular theory were mistaken, then, it seem to me, we can counterfactually assign values to the probabilities in (2).[3] Such assignments will clearly depend heavily upon the alternative hypotheses concerning the structure of matter that are available and regarded as serious contenders.

Consider Brownian motion. Perhaps water is a continuous medium that is subject to internal microscopic vibrations related to temperature. Suspended particles might behave in just the way Brown and others observed. Moreover, these vibrations might bear some relationship to the chemical properties of different substances that we (mistakenly) identify with molecular weight. Conceivably, such vibratory behavior might lead to precisely the results of experiments on Brownian motion that Perrin found, leading us (mistakenly) to think that we had successfully ascertained the number of molecules in a mole of the suspending fluid. It is for this reason that no *single* experimental method of determining Avogadro's number, no matter how ingeniously and beautifully executed, could serve to establish decisively the existence of molecules. The opponent could plausibly respond that, with respect to Brownian motion, fluids behave merely *as if* they were composed of molecules, and, indeed, as if they contained a specific number of molecules.

When we turn to helium production by alpha radiation, we must find another alternative explanation. Assuming that radium is *not* composed of atoms, we may still suppose that tiny globs of stuff are ejected from a mass of radium at a statistically stable rate. We may further claim that, for some reason, all of these globs are the same size. Moreover, we may assert that the globs are bits of helium that can be collected and weighed. We find, consequently, that it takes a given number of little globs of helium to make up 4 grams. Even given all of these assumptions, it is extremely implausible to suppose that the number of globs in 4 grams of helium should equal the number that Perrin (mistakenly) inferred to be the number of molecules in 18 grams of water. It is difficult—to say the least—to see any connection between the vibratory motions of water that account for Brownian motion on the noncorpuscular hypothesis and the size of the globs of helium that pop out of chunks of radium. And even if—in some fashion that is not clear to me—it were possible to provide a noncorpuscular account of the substances involved in experiments of these two types, we would still face the problem of extending this noncorpuscular theory to deal, not just qualitatively, but also quantitatively with the results of the many other types of experiments to which we have made reference. To scientists like Ostwald, the task must have seemed insuperable.

The claim that inequalities (3) and (4) are satisfied is quite straightforward. The chances of obtaining the appropriate results if the initial conditions are satisfied must surely be greater than the chances of obtaining them fortuitously under experimental conditions from which they should not have been expected to arise. It seems to me that, in a rough and schematic way at least, we have justified the assertion that the overall structure of the argument that was taken by most scientists early in the present century to establish conclusively the existence of 'molecular reality' is that of the conjunctive common cause.

Van Fraassen (1980, p. 123) has criticized my earlier appeal to the common cause principle for support of theoretical realism (in Salmon, 1978) on the ground that the effects that are correlated—the values of Avogadro's number derived from different experiments—cannot be traced back to any *event* that serves as a common cause. The fact that such results arise out of an experiment on Brownian motion and an experiment on electrolysis shows, at best, that there is some common feature of the conditions under which these experiments are performed. Quite so. In my previous exposition of this argument for theoretical realism, I was not clear on this crucial distinction between interactive forks and conjunctive forks; consequently, van Fraassen's criticism of that discussion is entirely well-founded. . . . I am extremely grateful to van Fraassen for forcing me to see that there are these two distinct kinds of causal forks, and for making me recognize that conjunctive forks involve physically distinct (nonintersecting) processes arising out of common background conditions. These considerations strengthen and clarify the use of common cause arguments on behalf of scientific realism.

. . . . .

NOTES

1. An excellent discussion of the significance of Avogadro's number *N*, and the way in which it serves to link the macrocosm and the microcosm, can be found in (Wichmann, 1967, chap. 1). Further details regarding several of the previously mentioned methods used to ascertain the value *N* are presented there.

2. It goes almost without saying, of course, that none of the methods used to ascertain the value of Avogadro's number does so exclusively on the basis of directly observable quantities without any appeal to auxiliary hypotheses. As Clark Glymour argues in detail in (Glymour, 1980, chap. 5), this fact does not undermine the validity of the argument.

3. In (Reichenbach, 1954, Appendix) a way is proposed for assigning well-defined probability values in certain kinds of cases in which the reference classes are empty.

REFERENCES

Gardner, Michael. 1979. "Realism and Instrumentalism in 19th Century Atomism." *Philosophy of Science* 46:1–34.

Glymour, Clark. 1980. *Theory and Evidence.* Princeton: Princeton University Press.

Nye, Mary Jo. 1972. *Molecular Reality.* London: Macdonald.

Perrin, Jean. 1913. *Les Atomes.* Paris: Alcan.

Reichenbach, Hans. 1954. *Nomological Statements and Admissible Operations.* Amsterdam: North-Holland. Reprinted, with a new Foreword by Wesley C. Salmon, as (Reichenbach, 1976).

———. 1976. *Laws, Modalities, and Counterfactuals.* Berkeley and Los Angeles: University of California Press. Reprint of (Reichenbach, 1954).

Salmon, Wesley C. 1978. "Why ask, 'Why?'?" *Proceedings and Addresses of the American Philosophical Association* 51:683–705. Reprinted in (Salmon, 1979).

———. 1979. *Hans Reichenbach: Logical Empiricist.* Dordrecht: D. Reidel.

Van Fraassen, Bas C. 1980. *The Scientific Image.* Oxford: Clarendon Press.

Wichmann, Eyvind H. 1967. *Quantum Physics* (Berkeley Physics Course, vol. 4). New York: McGraw-Hill.

19

Realism and Perrin's Argument for Molecules

Is There a Valid Experimental Argument for Scientific Realism? / Peter Achinstein

Long before atoms could be detected individually, scientists deduced their existence from the way dust motes danced in droplets of liquid; atoms making up the liquid were colliding with and jostling the dust.

—*New York Times*[1]

Wesley C. Salmon[2] claims that there is a valid experimental argument for scientific realism, and that one of the best examples is that provided by Jean Perrin[3] early in the twentieth century. In 1908, Perrin conducted a series of experiments on Brownian motion on the basis of which he claimed that Avogadro's number N, the number of molecules in a substance whose weight in grams equals its molecular weight, is approximately 6×10^{23}. Perrin drew the conclusion that unobservable molecules exist (*ibid.*, pp. 213–27).

By "scientific realism," Salmon means a doctrine committed at least to the claim that unobservable entities exist. (What else, if anything, scientific realism does or should entail is controversial; the question will be taken up in sections V–VII.) In seeking an argument to establish the claim that unobservables exist, Salmon writes:

Reprinted from *Journal of Philosophy* 99 (Sept. 2002): 470–95.
In memory of Wesley C. Salmon.

> ... I decided to try an empirical approach to the philosophical problem [of scientific realism]. Since it seemed unlikely that scientists would have been moved by the kinds of arguments supplied by philosophers, I felt that some insight might be gained if we were to consider the evidence and arguments that convinced scientists of the reality of unobservable entities. Although scientists, by and large, seem committed to the existence of a variety of unobservable entities ..., the existence of atoms and molecules, as the microphysical constituents of ordinary matter, is the most clear and compelling example. (*Op. cit.*, pp. 213–14)

Salmon proceeds to reconstruct Perrin's argument and to claim that it establishes the existence of molecules experimentally. It is a simple step to scientific realism:

> Molecules exist.
> Molecules are unobservable entities.
> Therefore, unobservable entities exist.

Since the first step is itself the conclusion of an argument based on Perrin's experiments with Brownian motion, we seem to have an experimental argument for scientific realism. Because this argument convinced at least some antirealist scientists of the reality of molecules,[4] Salmon challenges antirealist philosophers to say why they should not follow suit.

In what follows, I want to see how antirealist philosophers respond, or could respond, to Salmon's challenge to accept an experimental argument for scientific realism. First, however, I turn to Perrin's argument itself.

I. Perrin's Experimental Argument

A discovery was made in 1827 by the English botanist Robert Brown that small microscopic particles suspended in a liquid do not sink but exhibit rapid, haphazard motion—so-called Brownian motion. In 1908, Perrin conducted a series of experiments involving microscopic particles of gamboge (a gum resin extracted from certain Asiatic trees) in a dilute emulsion. These particles exhibited Brownian motion which was visible using a microscope. The emulsion was contained in a cylinder of known height h. Perrin determined the density D of the material making up the particles, the density d of the liquid, the mass m of the particles (all of them prepared by him to be the same weight), the temperature T of the liquid, and (with microscopes) the number of suspended particles per unit volume at various heights. He performed experiments with different emulsions, particles of different sizes and mass, different liquids, and different temperatures.

Using an argument that I shall note presently, he derived the following equation, which relates the quantities just cited:[5]

$$(1)\ \frac{n'}{n} = 1 - \frac{Nmg(1 - d/D)h}{RT}$$

Employing different experimental values obtained for all the quantities in equation (1) except for *N*, Perrin used this equation to determine whether Avogadro's number *N* is really a constant, and if so, what its value is.

He arrived at equation (1) by assuming that the motions of the visible Brownian particles are caused by collisions with the molecules making up the dilute liquid in which these visible particles are suspended. Accordingly, he also assumed that the visible particles will mirror the behavior of the invisible molecules. Finally, he assumed that molecules in a dilute solution of the sort in question will behave like molecules in a gas with respect to their vertical distribution. He then derived the following formula (the law of atmospheres) that governs a volume of gas contained in a thin cylinder of unit cross-sectional area and very small elevation *h*:[6]

$$(2)\ \frac{p'}{p} = 1 - \frac{Mgh}{RT}$$

The pressure of a gas is proportional to its density, and hence, Perrin assumed, to the number of molecules per unit volume. So he replaced the ratio of pressures p'/p by the ratio n'/n, where n' and n are the number of molecules per unit volume at the upper and lower levels. He also replaced *M* by *Nm*, where *m* is a mass of a molecule of gas and *N* is Avogadro's number. With these substitutions, Perrin obtained

$$(3)\ \frac{n'}{n} = 1 - \frac{Nmgh}{RT}$$

He now transformed equation (3) into equation (1) by letting n' and n represent the number of Brownian particles per unit volume at the upper and lower levels; and in (3) substituting the expresion $mg(1 - d/D)$—the "effective weight" of a Brownian particle—for *mg*, the weight of a molecule.[7] Strictly speaking, in equation (1), *N* represents a number for Brownian particles: any quantity of these particles equal to their molecular weight will contain the same number *N* of particles. Perrin assumed that this number will be the same as Avogadro's number for molecules.

On the basis of various experiments involving different values for the observable quantities n', *n*, *m*, *h*, and *T*, Perrin used equation (1) to determine a value for Avogadro's number *N*, and discovered that *N* is indeed a constant, whose approximate value is 6×10^{23}. He concluded that molecules exist:

> Even if no other information were available as to the molecular magnitudes, *such constant results would justify the very suggestive hypotheses that have guided us* [including that molecules exist], and we should certainly accept as extremely probable the values obtained with such concordance for the masses of the molecules and atoms. . . . The objective reality of the molecules therefore becomes hard to deny.[8]

Now for antirealist responses. There are two general kinds I shall consider: first, that Perrin's argument for molecules is invalid; second, that even if valid, it does not suffice to establish scientific realism.

II. The Circularity Objection

The first charge is that at the outset Perrin assumes that molecules exist while arguing that they exist. He derives equation (1), which he uses to obtain a value for Avogadro's number N and to see whether N is a constant, from equation (3), which presupposes the existence of molecules. In (3), n', n, and m are quantities associated with invisible molecules in a gas. He then assumes that a modified version of (3), namely, (1), can be applied to much larger visible Brownian particles suspended not in a gas but in a dilute fluid.

Reply.[9] The argument presented above is only part of Perrin's reasoning, and indeed not the first part. In his book *Atoms,* long before he gets to this argument from Brownian motion, he devotes eighty-two pages to a development of atomic theory, giving chemical arguments in favor of the existence of molecules. His 1909 article begins with a qualitative description of Brownian motion followed by qualitative arguments that Brownian motion is caused by collisions with molecules, and hence that molecules exist. In this article, he writes:

> It was established by the work of M. Gouy (1888), not only that the hypothesis of molecular agitation gave an admissible explanation of the Brownian movement, but that no other cause of the movement could be imagined, which especially increased the significance of the hypothesis.[10]

Gouy performed experiments in which he examined possible external causes of Brownian motion. These included vibrations transmitted to the fluid by passage of heavy vehicles in the street; convection currents produced when thermal equilibrium was not yet attained; and artificial illumination of the fluid. When these and other external sources of motion were reduced or eliminated, the Brownian motion was not altered. Perrin concludes: "these [Brownian] particles simply serve to reveal an internal agitation of the fluid . . . " (*ibid.*, p. 511).

Perrin offers a second argument from Brownian motion to molecules (*ibid.*, pp. 514–15). Since the Brownian particles are continually accelerating and decelerating, they must be subject to forces exerted upon them, in such a way as to satisfy conservation of momentum. These forces are not present in the particles themselves, nor, as he argued earlier, are they produced by forces external to the fluid. Accordingly, he concludes, they must be produced by the perpetual motions of unobservable particles in the fluid itself.

These arguments for the existence of molecules are presented before Perrin's argument from the law of atmospheres given in section I above. Both involve eliminative-causal reasoning of the following sort:

Eliminative-causal argument:

(1) Given what is known, the possible causes of effect *E* (for example, Brownian motion) are *C*, $C_1, \ldots, C_n$ (for example, the motion of molecules, external vibrations, heat convection currents). (In probabilistic terms, given what is known, the probability is high that *E* is caused by one of the *C*s cited.)

(2) $C_1, \ldots, C_n$ do not cause *E* (since *E* continues when these factors are absent or altered).

So probably,

(3) *C* causes *E*.

Later, I shall consider whether such an argument is valid for molecules (or anything else). At the moment, I am claiming only that an argument of this type is employed by Perrin, and that it is employed prior to, and in addition to, the argument of section I. Perrin offers additional arguments for molecules which are also independent of the law-of-atmospheres argument.[11]

Accordingly, independently of his own law-of-atmospheres experiments on Brownian motion, Perrin presents both experimental and theoretical arguments for the following claim:

H: Chemical substances are composed of molecules, the number *N* of which in a gram molecular weight of any substance is approximately 6×10^{23}.

Let b represent the information, other than the law-of-atmospheres considerations, on the basis of which Perrin infers H. He is claiming that the probability of H is high, given b, that is,

$$(4)\ p(H/b) > k$$

where k is some threshold of high probability, say 1/2.[12]

Now, on the basis of his law-of-atmospheres experiments, using equation (1), Perrin claims that:

H': The calculation of N done by means of Perrin's experiments on Brownian particles, using equation (1), is 6×10^{23}, and this number remains constant even when various values in equation (1) are varied.

In H', the number N represents a number for Brownian particles. Perrin can be understood as assuming that, given both H and b, the probability of H' is (substantially) increased over what it is given b alone. That is, the probability that Perrin's experiments will yield result H' for Brownian particles is greater given the assumption that molecules exist, and that their number in a gram molecular weight of a substance is constant and approximately equal to 6×10^{23}, than it is without this assumption. That is,

$$p(H'/H \& b) > p(H'/b)$$

It follows that

$$(5)\ p(H/H' \& b) > p(H/b)$$

assuming that neither $p(H)$ nor $p(H'/b)$ is zero. In short, the probability of the molecular hypothesis H is (substantially) increased by H', the results of Perrin's experiments with Brownian motion. From (4) and (5) we can conclude that the molecular hypothesis H is highly probable given H' and b, that is,

$$p(H/H' \& b) > k$$

Perrin's argument, so represented, does not involve circularity. Even though Perrin derives the law of atmospheres (3) involving molecules from the assumption that molecules exist, and even though he derives H' from the law of atmospheres for molecules, H' itself does not state or presuppose that molecules exist or that Avogadro's number for molecules is 6×10^{23}. H' is established experimentally. On Perrin's view, given b, H' bestows substantial probability on H (which does state that molecules exist and that Avogadro's number for molecules is a constant).

III. The "Multiplicity" Objection

A second objection is that, given Perrin's experimental results, various hypotheses can be shown to receive just as much support as the particular molecular one that he accepts. And these alternatives need not be committed to the existence of molecules. The multiplicity objection comes in two forms.

A. Parallel antirealist argument. The idea, a generalization of one offered by Bas van Fraassen,[13] is that, for each argument used by Perrin whose conclusion is that molecules exist, another argument, at least as good, can be constructed from the same premises which does not conclude that molecules exist, but only that the molecular theory "saves the phenomena." Indeed, an argument of the latter sort is simpler and stronger than the former, since it commits one to much less than does any argument of the former sort. It commits one only to the "empirical adequacy" of the molecular theory, not to its truth, whereas an argument of the former sort is committed to both.

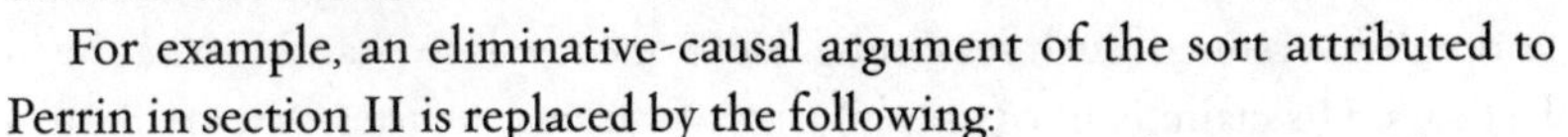

For example, an eliminative-causal argument of the sort attributed to Perrin in section II is replaced by the following:

Antirealist eliminative-causal argument:

(1) Given what is known, of the theories that attribute a cause for effect E (for example, Brownian motion), the possible candidates for theories that save the phenomena are ones that invoke causes $C, C_1, \ldots, C_n$. (In probabilistic terms, given what is known, the probability is high that at least one of the theories cited saves the phenomena.)

(2) Theories that invoke causes $C_1, \ldots, C_n$ do not save the phenomena (since these theories predict that, under certain observable conditions, effect E will continue, which new information shows to be false).

So probably,

(3) The theory that invokes cause C (for example, molecules) saves the phenomena.

The conclusion is not that the theory that claims that C causes E is true, or that the entities it postulates (for example, molecules) exist or cause E, but simply that it is empirically adequate.

A similar claim is made for any argument that concludes with the truth of a theory postulating the existence of unobservables. For example, with respect to parallel "inference-to-the-best-explanation" arguments, van Fraassen offers two objections (*ibid.*, pp. 20–21). The first is that the claim that scientists reason in accordance with realist versions of such arguments (or analogous eliminative-causal ones) is an *empirical* hypothesis "to be confronted with data, and with rival hypotheses" (*ibid.*, p. 20). Van Fraassen proposes a rival hypothesis, namely, that scientists reason in accordance with the antirealist version of inference-to-the-best-explanation (or an analogous version of the eliminative-causal argument given above). Both hypotheses, he seems to be saying, are compatible with the empirical fact that scientists employ explanatory (or eliminative-causal) reasoning. Van Fraassen's second objection is that the realist needs an extra premise for his argument, one that van Fraassen regards as false. For inference-to-the-best-explanation, the extra premise is that "every universal regularity in nature needs an explanation" (*ibid.*, p. 21). For causal reasoning, the extra premise would be that every phenomenon in nature has a cause.

Reply. (1) To be sure, the fact that Perrin invokes eliminative-causal reasoning does not by itself establish that he is employing a realist or an antirealist version of such reasoning. Here one needs to examine what he in fact says. His claim is not (simply) that of various theories that attribute a cause for Brownian motion, the ones he cites are the possible candidates for saving the phenomena; it is the stronger claim that these theories are the possible candidates for truth. Nor is his conclusion (simply) that the theory invoking molecules as causes in fact saves the phenomena. It is the stronger claim that such a theory is true, that molecules exist ("the objective reality of the molecules therefore becomes hard to deny"), and that molecules cause Brownian motion. If, as van Fraassen insists, it is an empirical question what form of reasoning a scientist in fact is using, then the best empirical method to determine this is to look at what he actually says.

(2) Perrin does not need an extra premise asserting that every phenomenon in nature has a cause (or that every universal regularity needs an explanation). All he needs is the assumption that Brownian motion has a cause. To be sure, he does not explicitly defend this assumption, although he clearly makes it.[14] Presumably, if required to defend this assumption Perrin would have appealed to the idea that Brownian motion involves accelerations of bodies—both changes in speed and direction—which, in the classical physics that he was assuming—require causes. Perrin had no need to introduce some very general assumption that every universal regularity in nature has a cause or needs an explanation (an assumption not even made in classical mechanics). He simply needed to assume that the Brownian motion of the observable particles of gamboge has a cause.

B. Multiplicity-of-causes objection. The idea behind this second form of the multiplicity objection is to attack Perrin's particular eliminative-causal argument, which assumes that, given that known observable causes of Brownian motion are eliminated, one can infer an unobservable cause. The problem is that not all possible observable causes have been considered. This objection could be raised as a general criticism of eliminative-causal reasoning ('In general, how do you know you have cited all possible causes of a phenomenon?'). Or it could be raised as a specific one against Perrin ('What reason was there for supposing that Perrin had considered all possible causes of Brownian motion?').

Reply. (1) The *general* criticism is based on the assumption that one is justified in employing an eliminative-causal argument only if all but one of the possible causes of the phenomenon have been listed and eliminated. But this is too demanding. One can be justified in employing an eliminative-causal argument if, *given one's background information,* one has considered and eliminated all but one of the possible causes, or at least, all but one of the causes that (on the basis of the background information) have any significant probability of causing the phenomenon in question. The claim that the possible causes cited probably include the actual one can be defended by appeal to the fact that the phenomenon in question is of a certain type that, experience has shown, in other cases is caused by one or the other of the causes cited.

(2) The *specific* criticism is that, even if eliminative-causal reasoning can be reasonable, if properly employed, Perrin did not properly employ it. He did not cite all the (plausible) causal candidates for Brownian motion, given information available to him. To be sure, he cites several candidates and argues that experiments eliminate these. But why suppose that these are the only possible or plausible causes permitted by his background information? And even if they are the only possible *observable* causes, why suppose that the motion of particles is the only possible *unobservable* cause?

Here, I suggest, the burden of proof is on the critic. Perrin cites various known causes of motion in a fluid and argues that experiments, particularly those of Christian Wiener in 1863 and Gouy in 1888 and 1889, show that Brownian motion continues unabated when these causes are altered or eliminated.[15] Perrin, quoting Wiener, concludes that the motion "does not originate either in the particles themselves or in any cause external to the liquid, but must be attributed to internal movements. . . ."[16] Since Perrin cited and eliminated various possible causes, it is, I think, up to the critic to say what other possible causes he should have eliminated, given his information (and why these should be assumed to be observable).

(3) Both the general and the specific criticisms cited above, if valid, could equally well be used against the antirealist version of the eliminative-causal

argument. That version cites theories that invoke causes C, $C_1, \ldots, C_n$ as the possible candidates for saving the phenomena and then eliminates all but one of these theories. But, the critic can ask, why suppose that these are the only possible causal theories that will save the phenomena? And the critic can ask Perrin the specific form of this question for Brownian motion. So an antirealist version of eliminative-causal reasoning is no better off than the realist version. My response is simply to reject the critic's claims for the reasons given in replies 1 and 2 above.

IV. The "Limits of Experience" Objection

The general idea here is simple and powerful. By observing nature, we can only make inferences about what is observable in nature, not what is unobservable. Here are claims of two prominent antirealists. First, Pierre Duhem[17]:

> Now these two questions—Does there exist a material reality distinct from sensible appearances? and What is the nature of this reality?—do not have their source in experimental method, which is acquainted only with sensible appearances and can discover nothing beyond them. (*Ibid.*, p. 10)

Second, van Fraassen[18]:

> I explicate the general limits [of experience] as follows: *experience can give us information only about what is both observable and actual.* (*Ibid.*, p. 253)

The claim, then, is that any argument from what is observed to the truth of claims about what is unobservable is unjustified. Since Perrin claimed truth for his theory about unobservable molecules, and since his argument is based on observed results of experiments that he and others conducted, his argument is unjustified.

What is the basis for the claim that by observing nature one can make inferences only about what is observable? A simple defense, one indeed suggested by Duhem, is that this is the essence of empiricism; inferring beyond the observable is metaphysics, which is not empirical. Indeed, immediately following the passage quoted above, Duhem claims that the resolution of questions as to what lies beyond the "sensible appearances" "transcends the methods used by physics; it is the object of metaphysics" (*op. cit.*, p. 10). But this begs the issue, which in this case is that of characterizing empiricism. Perrin, no less than Duhem, regarded himself as engaged in empirical science, not metaphysics. He believed that empirical arguments could be given for the truth of a theory about unobservable molecules. It will not do simply to claim that his argument is not (entirely) empirical, or that he has gone beyond what empiricism allows.

Van Fraassen regards as extreme a policy that permits inferences beyond the "range of possible additional evidence," that is, inferences to the truth of claims about unobservables.[19] If only evidence can justify a belief, he goes on to say, then belief in the truth of a theory is "supererogatory," since we can have evidence for truth only via evidence for empirical adequacy (*ibid.*, p. 255). Van Fraassen's first claim presupposes, or at least strongly suggests, that inferences to the truth of claims about unobservables are inferences beyond what the evidence allows. His second claim is that one can have evidence only for the empirical adequacy, not the truth, of a theory postulating unobservables, or that evidence for truth only amounts to evidence for empirical adequacy. Both claims, I suggest, require justification. Perrin, for example, clearly believed that he had evidence for the truth of propositions about unobservable molecules, not simply for their empirical adequacy; he had evidence that molecules exist, not simply evidence that the molecular theory saves the phenomena. Why was he wrong in this belief?

Let me suggest an argument an antirealist might offer in the spirit of the two passages quoted above. Suppose that having observed a great many *A*s and found them all to be *B*, one infers that all *A*s are *B*. For example, from the fact that all observed bodies have mass one infers that all bodies, including any unobservable ones, do too.[20] (In Perrin's case, we might consider an inference from 'All observed accelerating bodies in contact with other bodies exert forces on them' to 'All accelerating bodies, including molecules (if any exist), in contact with other bodies exert forces on them'.) Is an inference of this sort legitimate? Not necessarily. The sample observed may be unrepresentative. Even if we do not know that it is unrepresentative, we may have no positive empirical reason to think it is not biased. Suppose that all the *A*s chosen for observation satisfy a condition C. Then unless we can argue that C is irrelevant—that C does not bias the sample with respect to *B*—we should infer not that all *A*s are *B*, but, at best, that all *A*s that satisfy C are *B*.

Now, says the antirealist, let C be the condition of *being observable*. All observed *A*s satisfy this condition. Indeed, one can never observe an *A* that fails to satisfy it. So, unless one can demonstrate that this does not bias the sample of *A*s observed with respect to *B*, from the fact that all *A*s observed have been *B*, all we can legitimately infer is that all observable *A*s are *B*. The only way of demonstrating that "observability" does not bias the sample of *A*s observed with respect to *B* is by collecting a suitable sample of unobservable *A*s and showing that in this sample the unobservable *A*s are all *B* (or perhaps by showing that in other cases, samples of unobservables match observables with respect to some property). But, of course, one cannot do that! One cannot observe unobservables. So in order to exclude potential bias, we are restricted to making (inductive and causal) inferences to what is observable. This argument can be extended to cover all types of nondemonstrative inferences. That is why Duhem is right in claiming that we can discover nothing beyond appearances, and why van Fraassen is justified in claiming that experience can give us information only about what is observable.

Reply. An antirealist who argues in the previous manner looks at the situation as one involving so-called *stratified sampling.* The population of *A*s is divided into two classes or strata: the observables and the unobservables (if any). To make inferences about the entire class of *A*s with respect to a property *B,* the antirealist is claiming, one needs to select randomly members from both strata for observation. Since one cannot select unobservable members for observation, one cannot legitimately, without potential bias, make inferences about this stratum, but only about the observable stratum.

If this argument is derived from the general principle that to make an inductive inference about a class, samples must be taken from all strata of that class, then no inference from any observed sample will be possible. Let the stratifying condition C be 'has been observed' (or 'has been observed prior to 2025'). All the *A*s observed satisfy condition C. We cannot (now) take a sample from the subclass satisfying not-C. Accordingly, if the general principle of stratification for any condition C is correct, then no generalizations are possible.

Assuming that the antirealist does not accept this general principle with respect to all stratifying conditions C, what reason can he offer for supposing that the specific stratifying condition 'being observable' is a biasing condition? Does he have any *empirical* reason for supposing that, in general, when considering a class of *A*s, the unobservable stratum of *A*s, if any exist, is different with respect to *B* from the observable stratum? No, he does not, since, by his own admission, unobservables cannot be sampled. Nor does he have some a priori argument showing that the two strata are different with respect to *B.* All he has is the weak a priori claim that the strata *may* be different—a claim about a mere logical possibility. But this is not sufficient to justify the methodological injunction that to make a legitimate inductive inference about the entire class consisting of observables and unobservables one must sample both strata. Indeed, if it were sufficient, the antirealist would be prevented from making any generalizations from what has been observed to what is observable but never observed. If one stratifies the class of observable *A*s into those which are or will be observed and those which never will be observed, then it is logically possible that these strata are different with respect to *B.*

The realist, who wants to make inferences about unobservables as well as observables, must also reject the general principle of stratification above. Can he do so without rejecting the idea that sampling requires variation? Can the realist offer an argument that provides support for his claim that "observability" is not, in general, a biasing condition?

I suggest that here the realist has an advantage over the antirealist. The realist can offer an empirical argument that provides support for his claim. There is a type of variation that can be produced for the purpose of meeting the antirealist challenge. One can vary conditions or properties in virtue of which something is observable (or unobservable). For example, items can be observable (or unobservable) in virtue of their size, their distance from us in space or time, their duration, their interactions (or lack of them) with other items, and so on. Suppose that all physical bodies that have been observed have mass. Have we biased the observed sample with respect to mass by observing only bodies that are observable? Observed physical bodies are necessarily observable. Therefore, we cannot vary the physical bodies observed by observing unobservable as well as observable ones. But we can vary properties of physical bodies in virtue of which they are observable (or unobservable).

Suppose we do the latter and find that bodies have mass even when we observe bodies that have different sizes, different distances from us, different durations, and different numbers and kinds of interactions with other bodies. In the absence of any contrary empirical information, we can then infer that size, distance, duration, and numbers and kinds of interactions do not alter the situation: bodies observed with different sizes, distances, durations, and interactions all have mass. So we infer that differences in these properties—differences that make some bodies observable and others not—make no difference as far as having mass is concerned.[21] If we vary the conditions in virtue of which bodies are observable and find no differences in whether bodies have mass, and if we have no contrary empirical information, then we have offered an empirical argument to support the claim that the fact that all observed bodies are observable does not bias the observed sample with respect to the property of having mass. In so doing, we have provided empirical grounds for inferring that all bodies have mass, whatever their size, distance, duration, and so on, and hence, whether or not they are observable.[22]

An antirealist may vehemently repudiate this argument by insisting that the only way to vary a condition C that is satisfied by all observed As is to observe As that satisfy C and As that do not. The antirealist will then reject the claim that the types of variations above (which do not do this) suffice. But the claim that this is the only type of variation that can eliminate bias needs an argument, especially since the antirealist should allow variations in size, distance, and so on to count as varying the conditions when making inferences about observables.

For example, suppose an antirealist makes an inference he considers legitimate from the fact that all observed bodies have mass to all *observable* bodies have mass. In doing so, in order to preclude bias, one might vary the size of the bodies examined, their distance from other bodies, and a host of other properties. More generally, some variations involve changing "qualitative" properties by examining *A*s that have such properties and *A*s that do not. But there are also variations that involve changing "quantitative" properties by examining *A*s with varying amounts of such properties. If the antirealist rejects this second type of variation for eliminating bias, he needs an argument for doing so. If he accepts it, but only for inferences to observables, again he needs an argument for this restriction. The claim that unobservable bodies *may* be different from the observed ones with respect to having mass—in the sense of logical possibility—cuts no ice, since exactly the same could be said for the observable (but not yet observed) bodies. Nor, again, does the antirealist have either an empirical or an a priori argument to show that the observed bodies are unrepresentative of the unobservable ones with respect to having mass.[23]

V. Is This Scientific Realism?

The final objection, possibly the most important, is that whatever Perrin proved in his argument, even if by means of his experiments he did prove that molecules exist, this is not scientific realism. An objection of this sort will be raised by anyone, realist or antirealist, with a more demanding view of what scientific realism requires. Various candidates for scientific realism have been suggested in recent years by proponents as well as critics; three prominent ones will be noted.

(1) Scientific realism is a view about truth and reference in scientific theories generally. For example, Richard Boyd[24] and Stathis Psillos,[25] defenders of scientific realism, and Larry Laudan,[26] a severe critic, all claim that scientific realism involves a number of central theses, including these:

(A) Scientific theories (at least in the "mature" sciences) are typically approximately true, and more recent theories are closer to the truth than other theories in the same domain.

(B) The observational and theoretical terms within the theories of a mature science genuinely refer (roughly, there are substances in the world which correspond to the ontologies presumed by our best theories).

(C) Successive theories in any mature science will be such that they "preserve" the theoretical relations and the apparent referents of earlier theories (that is, earlier theories will be "limiting cases" of later theories).

(D) Acceptable new theories do and should explain why their predecessors were successful insofar as they were successful.

This is Laudan's formulation (*op. cit.,* pp. 219–20). To these theses Boyd adds:

(E) The reality which scientific theories describe is largely independent of our thoughts or theoretical commitments (*op. cit.,* p. 42).

Thesis (E) relates scientific realism to a core idea of what is sometimes called *metaphysical realism,* namely, that there is a mind- and theory-independent way the world is; to this core idea it adds the claim that scientific theories describe such an independent world.[27]

Even if Perrin proved that molecules exist, he did not prove theses (A)–(E), or even make them probable. Nothing Perrin did establishes or makes probable claims about scientific theories generally, or even about ones in mature sciences. Nor did Perrin establish that there is a mind- and theory-independent world or that theories in the mature sciences describe such a world approximately correctly.

(2) Scientific realism is a view about the *aim* of science. Here is van Fraassen's formulation:

> *Science aims to give us, in its theories, a literally true story of what the world is like; and acceptance of a scientific theory involves the belief that it is true.* This is the correct statement of scientific realism.[28]

Van Fraassen distinguishes the aim of science, as a type of activity, from the aims of particular scientists, which may include fame and fortune.[29] Now, proving that molecules exist does not establish that the aim of science is to provide a literally true story of the world (or to satisfy any other broad conditions, such as describing natural kinds and dependencies), not even if this were Perrin's aim in the case of molecules. One would need to show more than this to establish scientific realism in the present sense. (I shall return to this in section VII.)

(3) The scientific realism of interest to philosophers is not itself an *internal* scientific question, to be settled by scientific reasoning, but an *external* one concerning the adequacy of the scientific representation of the world. It cannot be established by empirical means. Both Rudolf Carnap[30]

and Arthur Fine[31] have defended a distinction between internal and external questions. Their views about internal questions are somewhat similar, although they take very different positions on external questions.

For Carnap, the question of whether molecules exist can be approached in two ways. First, it can be treated as an empirical question within what Carnap calls a "linguistic framework," which contains rules of language and of inference, including rules governing what counts as evidence for what. Considered as an internal question within a framework permitting "theoretical terms" for unobservable entities, the answer to whether molecules exist can be determined empirically by the sorts of arguments from experiments Perrin provided. But this is not what philosophers usually have in mind when they assert (or deny) scientific realism with respect to theoretical entities such as molecules. For these philosophers, the question is an external one concerning the adequacy of the "theoretical entity" framework within which molecular claims are made. For Carnap, the claim that the framework is adequate is not a claim about its relationship to a framework-independent world. It is a pragmatic claim about the employability of the framework based on features such as simplicity, ease of use, and familiarity. Accordingly, different frameworks can be adopted, some of which have no terms for unobservables such as molecules. Some frameworks may be more user-friendly than others, but none is "correct" or "incorrect."

Carnap's internal questions are part of what Fine calls the *natural ontological attitude* (NOA). So, Fine can agree, Perrin established the existence of molecules by means of his experiments. In this internal sense, an antirealist can agree with a realist about what Perrin accomplished scientifically. What a realist does, according to Fine, is to step outside of scientific activity and claim that theories correspond to reality. Fine does not endorse Carnap's line that such a claim is pragmatic. Rather, he says simply that what the realist is doing when he steps outside of scientific activity is tantamount to pounding the table and saying 'Molecules exist, really!'. This adds only emphasis to the internal claim. Fine, then, advocates NOA, which he regards as different from both realism and antirealism but as something that realists as well as antirealists can accept.

Both Carnap and Fine could agree that Perrin proved empirically that molecules exist, where the latter claim is understood as one internal to science (as part of NOA). For Carnap and Fine, Perrin did not prove, nor could he, that a framework containing terms for molecules corresponds to reality in some external sense. So if, as is typical, scientific realism is construed as an external doctrine, Perrin did not establish scientific realism. For Fine, neither Perrin nor anyone else could establish realism in an external sense. For Carnap, Perrin showed, by using the framework he did, at best that it was useful; he did not show that it is the only or the most useful one for dealing with questions involving Brownian motion.

These three views about scientific realism represent a spectrum of positions. Perrin's arguments do not establish scientific realism in a sense of that term advocated by any of the philosophers noted. So is Salmon mistaken in his claim that Perrin provided a valid experimental argument for scientific realism? To answer, some history is relevant.

VI. Historical Reasons for Rejecting Atomism

We need to consider reasons that certain scientists rejected atomic-molecular theory until the first decade of the twentieth century. These reasons were known to Perrin, who responded to them. Critics of atomic theory included Duhem, Ernst Mach, Friedrich Wilhelm Ostwald, Henri Poincaré, and Max Planck. Grounds cited for rejecting the theory were in part scientific.[32] But they also included important philosophical or methodological reasons. One was the claim that physicists such as James C. Maxwell and Ludwig Boltzmann, who proposed atomic-molecular theories, were employing an illegitimate "method of hypothesis." From unproved hypotheses about atoms and molecules, they were deducing observable consequences and claiming that from the truth of the latter, one could infer the truth of the hypotheses themselves.[33] While Mach, for example, did not repudiate the use of hypotheses, he did reject the idea that one could infer their truth or probability from true observed consequences. He regarded the hypotheses of the atomic theory as provisional "mental artifices" for summarizing known observational facts and facilitating observational predictions.[34] Once such hypotheses have served their organizational and predictive purposes, they are to be discarded, not accepted as true or probable. Indeed, Mach championed an antirealist "sensationalist" view about observable matter, which he regarded as complexes of sensations.[35]

A related philosophical claim made by some of the critics of atomic theory (whether or not they were Machian reductionists) was that hypotheses about unobservables can never be established as true or probable by empirical means; such hypotheses can never be known to be true. For example, Poincaré[36] speaks of the atomic hypothesis as "indifferent," meaning that although it may be useful, it can never be empirically established or shown to be more probable than rival hypotheses that assert the continuity of matter (*ibid.*, pp. 152–53). For Duhem,[37] as noted earlier, science can know only "sensible appearances and can discover nothing beyond them" (*ibid.*, p. 304). Unlike Poincaré and Mach, Duhem does not even regard atomic hypotheses as useful devices for summarizing and predicting observable phenomena.

Like some of the critics of the atomic theory, and no doubt because of their criticism, Perrin offers his own general philosophical/methodological reflections on how to proceed with scientific investigations. He distinguishes two scientific methods, which he calls the *inductive method* and the *intuitive method.*[38] The former, associated with the critics of atomic theory noted above, begins with what is observed and reasons only to statements about "objects that can be observed and to experiments that can be performed" (*ibid.,* p. vii). The second method infers the existence of an unobservable mechanism:

> In studying a machine, we do not confine ourselves only to the consideration of its visible parts, which have objective reality for us only as far as we can dismount the machine. We certainly observe these visible pieces as clearly as we can, but at the same time we seek to divine the *hidden* gears and parts that explain its apparent motions.
>
> To divine in this way the existence and properties of objects that still lie outside our ken, *to explain the complications of the visible in terms of invisible simplicity,* is the function of intuitive intelligence which, thanks to men such as Dalton and Boltzmann, has given us the doctrine of Atoms. This book aims at giving an exposition of that doctrine. (*Ibid.,* p. vii)

Perrin claims that in the times in which he is writing, the method of intuition has gone ahead of induction in rejuvenating the doctrine of energy by the incorporation of statistical results borrowed from atomists (*ibid.,* p. viii).

In these introductory passages, Perrin does not describe the intuitive method in detail other than to say that it is a method for inferring an invisible underlying reality from visible things and events in such a way that the former explains the latter. I suggest that we take his own arguments for molecules from experiments on Brownian motion as representing his use of the "intuitive method." If so, then that method is not simply a matter of speculating about an invisible realm. Nor is it a Machian provisional "mental artifice" for representing the facts, to be discarded once it has served its purpose. Nor is it a method simply for inferring the truth of claims about what can be observed (that is what he calls the inductive method). Nor, by contrast with Mach, does Perrin say or imply that assertions about ordinary matter are to be understood as claims about sensations; or (with later logical positivists) that assertions about an invisible realm of molecules are to be understood as claims about a visible realm such as that of Brownian motion; or that when he infers that molecules are real and that certain claims about them are true, by 'real' and 'true' he means 'useful' or 'saves the phenomena'.

Finally, Perrin's intuitive method is not the method of hypothesis, which infers the truth or probability of an hypothesis simply from the fact that the hypothesis explains and predicts observable phenomena. To be sure, Perrin's full argument involves citing explanatory chemical reasons from combinations of elements and compounds. But in addition, it involves eliminative-causal reasoning of a sort indicated in section II above. It involves an appeal to similar determinations of Avogadro's number N from a variety of experiments on phenomena other than Brownian motion. And, perhaps most importantly, it involves a calculation of N from Perrin's own experiments on Brownian motion and a demonstration that this number remains constant (proposition H' in section II). The former arguments, according to Perrin, show that the molecular hypothesis is highly probable. The latter sustains and even increases this probability on the basis of new, precise, experimental results.

Now we are in a position to respond to Salmon's challenge.

VII. Did Perrin Provide an Experimental Argument for Scientific Realism?

Or did he merely furnish an answer to what Carnap calls an internal question? That is, is Perrin's argument more than simply an empirical argument for the existence of a particular natural kind, namely, molecules—entities which had been previously postulated by scientists but not shown to exist? It is this and a good deal more. It is an argument to the existence of something that was regarded as dubious or objectionable on philosophical or methodological grounds. These grounds included the idea that from what is observed, one can make valid inferences only to what is observable, not to what is unobservable. Moreover, they are grounds for rejecting unobservables generally, not just molecules. (Duhem, for example, rejected a range of physical theories postulating unobservables.)

Perrin did not respond to these critics by making molecules observable. Rather, he claimed that there is a reasonable method or mode of reasoning (the intuitive method) that starts with observed experimental results and can be used to infer the existence of things that are unobservable (or 'invisible,' or 'hidden', to use Perrin's terms). He showed in detail how this method could be used to infer the existence of invisible molecules from experimental observations on Brownian motion. And he clearly regarded this intuitive method as a general one, not restricted simply to molecules.[39] This is not just scientific business as usual. It is not simply an argument establishing (or rejecting) the existence of just one more type of physical entity. Nor did Perrin or his antirealist opponents take this as scientific business as usual.

The scientific realism implicit in Perrin's arguments can be put like this:

(1) There are unobservables (for example, molecules).
(2) Their existence and their properties can be inferred (only) on empirical grounds, in some cases from experiments, so that a claim to know they exist and have these properties is justified.[40]
(3) A legitimate mode of reasoning that can be used for this purpose involves two important components:
   (a) causal-eliminative reasoning to the existence of the postulated entity, and to certain claims about its properties, from other experimental results;
   (b) an argument to the conclusion that the particular experimental results obtained are very probable given the existence of the postulated entity and properties.

Part (a) in (3) is intended to establish the high probability for the claim that the entity exists and has certain properties. Part (b) can be shown to sustain and possibly increase that probability on the basis of the new experiments. Reasoning involving (a) and (b) can be used to infer the existence of entities whether observable or unobservable.

The scientific realism reflected in points (1)–(3) does not say that theories about unobservables in mature sciences are (approximately) true and that the unobservables postulated exist, or that the aim of science is to obtain true theories about unobservables as well as observables. Nor does it adopt either some metaphysical viewpoint external to science, saying that there is a mind- and theory-independent way the world is and/or emphatically endorsing the adequacy of a scientific representation of this world, or a pragmatic viewpoint stressing the usefulness of scientific theorizing. It is not scientific realism *in these senses.*

Nor is it a restricted internal realism that says simply that molecules (electrons, or whatever) exist. It is much more general than this, and it has a methodological as well as ontological component. It claims that there are unobservables (Perrin's realm of "invisible simplicity"), and that valid arguments can be used to infer their existence and properties; these are entities which, on philosophical and/or methodological grounds, antirealists reject or understand in some nonrealistic way. It says, however, the arguments for these entities and claims about their properties are empirical ones, so that what entities exist in this realm and what properties they have are empirical questions. The particular empirical argument for a given unobservable will depend on, and vary with, the unobservable postulated. No *general* empirical argument can be given for all unobservables postulated. Nor can the issue of the existence of unobservables be settled a priori. Nevertheless, there is an a priori assumption that is essential to scientific realism so understood, namely, that a valid mode of reasoning can be employed (such as (a) and (b) in point (3) above) which can justify a belief in the truth of propositions about the unobservable entities inferred.

Does this deserve the name "scientific realism"? Fine may answer "no," claiming that it does not go beyond what his neutral NOA permits. Fine says that NOA, and hence realism and antirealism, "accept(s) the results of scientific investigations as 'true' on par with more homely truths."[41] Hence, he will insist, since the results of scientific investigation include empirically based inferences to the existence of unobservables such as molecules, NOA, together with realism and antirealism, accepts both (1) and (2). Moreover, he may add that NOA, and hence both realism and antirealism, accepts point (3), since, he says, NOA sanctions "ordinary relations of confirmation and evidential support, subject to the usual scientific canons" (*ibid.*, p. 98) (although he does not formulate any such canons, or indicate whether they include the type of inference in point (3)). But if NOA does endorse points (1)–(3), then, I suggest, *it endorses a form of scientific realism rejected by scientists of the sort Perrin was opposing who are generally classified as antirealists.*[42] These scientists, including Duhem, Mach, Ostwald, and Poincaré, claim either that point (1) is false or else that, if it is true, it is unknowable by empirical means. (They do not regard propositions about unobservables on a "par with more homely truths.") And they reject points (2) and (3), since they raise general methodological objections to inferences from what is observed to what is unobservable.

Moreover, if an antirealist says either that unobservables do not exist or that, if they do, they are empirically unknowable, and if he is an empiricist about science, then, as in the case of Duhem and van Fraassen, he is likely to deny that the aim of science is to provide true theories about unobservables. This is because he holds that such an aim cannot be satisfied at all or cannot be satisfied by empirical means. Although this position does not follow deductively from a denial of points (1)–(3), it is a natural claim to make if one wants to retain empiricism in science and hold that the aim of science can be achieved. Similarly, if one is a realist and states that unobservables do exist and that claims to know this in particular cases can be empirically justified, then it is natural to assert that the aim of science, or at least one of its aims, is to provide true theories about such unobservables.

The most important reason that antirealists such as Duhem and van Fraassen have for saying that the aim of science is (a) to provide theories that save the phenomena, rather than (b) to provide theories that are true, is that they regard (a) but not (b) as doable and empirically justifiable. By contrast, realists such as Perrin and Newton regard both (a) and (b) as doable and empirically justifiable. Since they do, and since they also regard unobservables (such as molecules and universal gravity, which Newton considered a force extended to bodies "beyond the range of the senses") as causally responsible for observable phenomena (such as Brownian motion and motions of the planets), and since they regard particular causal claims of the latter sort as justifiable empirically, it would be natural for them to hold that:

(4) One of the aims of science is to provide (approximately) true theories of what the world is like

where truth applies to unobservables as well as observables. Point (4) is indeed strongly suggested by the methodological remarks of Newton (for example, in his Rules 1 and 4)[43] and Perrin (in his intuitive method). Accordingly, since it is not entailed by realist points (1)–(3), it is reasonable to consider it part of the realist position. Its truth is not demonstrated by Perrin's empirical argument for molecules, but it is plausible to say that Perrin presupposed its truth in conducting his investigation into the cause of Brownian motion.

The same cannot be said, however, for the doctrine of realism defended by Boyd and Psillos and attacked by Laudan. Points (1)–(4) make no claims about whether, in general, scientific theories in the mature sciences are true, contain terms that refer to objects that exist, and describe a mind- and theory-independent world; such a realism is much stronger than (1)–(4). Nor were any such general claims about theories in the mature sciences made or presupposed by Perrin in his investigation (or by Newton in his argument for the existence of a universal gravitational force).

For scientific realism *of the sort that was supported by scientists such as Perrin and rejected by his opponents* points (1)–(4) suffice. These scientists claimed (or denied) that there is an empirically knowable realm of unobservables responsible for observable effects, and since there is, one of the aims of science is to provide true theories about this realm. Whether any particular theory is true is to be determined by empirical considerations specific to that theory. From the fact that the theory is part of a mature science one cannot infer that it is true or probable. Indeed, as Laudan has argued, historically many such theories have been empirically refuted. For scientific realism of the kind under attack by Duhem, Mach, and other scientists (as well as by van Fraassen), and defended by Perrin (and by Salmon), what is important is the idea that there is a realm of unobservables, claims about which can be empirically justified as true. Whether unobservables do exist, and if so which ones, and what properties they have, are issues to be determined by empirical arguments of the sort Perrin provided for molecules. Accordingly, I regard Salmon's conclusion as justified. In an historically and conceptually important sense of "scientific realism" (though not in every sense assigned to that term), Perrin's experimental argument for molecules provides an empirical basis for scientific realism.

NOTES

For very helpful questions and suggestions,
I am indebted to Joseph Berkovitz, Sean Greenberg, Gregory Morgan,
and Michael Williams.

1. March 29, 2001, p. A19.

2. *Scientific Explanation and the Causal Structure of the World* (Princeton: University Press, 1984), pp. 213–27.

3. "Brownian Motion and Molecular Reality," reprinted in Mary Jo Nye, ed., *The Question of the Atom* (Los Angeles: Tomash, 1984), pp. 507–601 (see *Annales de Chimie et de Physique* [1909]; and *Atoms* (Woodbridge, CT: Ox Bow, 1990).

4. For example, in 1909, Friedrich Wilhelm Ostwald, who had previously rejected atomic theory on antirealist grounds, wrote: "*I have convinced myself that we have recently come into possession of experimental proof of the discrete or grainy nature of matter, for which the atomic hypothesis had vainly been sought for centuries, even millenia.* The isolation and counting of gas ions on the one hand . . . and the agreement of Brownian movements with the predictions of the kinetic hypothesis on the other hand, which has been shown by a series of researchers, most completely by J. Perrin—this evidence now justifies even the most cautious scientist in speaking of the *experimental* proof of the atomistic nature of space-filling matter"—quoted in Stephen G. Brush, "A History of Random Processes," *Archive for History of the Exact Sciences,* v (1968): 1–36.

5. In addition to the quantities mentioned above, in equation (1), n' and n represent the number of Brownian particles per unit volume at the upper and lower levels; mg is the weight of a particle; R is the gas constant; and N is Avogadro's number.

6. In (2), p' and p are the pressures of the gas at the top and bottom of the cylinder; M is the gram molecular weight of the gas (the mass in grams equal to the molecular weight of the gas); g is the constant of gravitation; R is the gas constant; and T is absolute temperature.

7. The effective weight of a Brownian particle is the excess of its weight over the upward thrust caused by the liquid in which it is suspended.

8. *Atoms,* p. 105.

9. For an expanded discussion of issues in this reply, see my *The Book of Evidence* (New York: Oxford, 2001), chapter 12.

10. "Brownian Motion and Molecular Reality," pp. 510–11.

11. These involve chemical considerations and also the fact that experiments on various phenomena other than Brownian motion yield approximately 6×10^{23} for Avogadro's number, although Perrin claims that for the most part these other methods do not yield the same precision as the determination from Brownian motion. Salmon emphasizes this second part of Perrin's reasoning, which he takes to be an example of "common-cause reasoning." For a discussion of Salmon's particular analysis, see my *The Book of Evidence,* chapter 12.

12. For arguments that *k* is greater than or equal to 1/2, see my *The Book of Evidence.*

13. *The Scientific Image* (New York: Oxford, 1980).

14. See, for example, *Atoms,* pp. 85–86.

15. Wiener demonstrated that Brownian motion is not caused by infusoria (one-celled animals found in exposed bodies of water), by electrical forces, by temperature differences, or by evaporation of the fluid. In addition to these, Gouy's experiments eliminated causes pertaining to the size, composition, and density of the Brownian particles. For a discussion of various observable causes of Brownian motion postulated by scientists opposed to molecular theory, see Nye, *Molecular Reality* (London: Macdonald, 1972), pp. 22–27. Nye concludes (p. 27) that in his experiments, "Gouy [in 1889] refuted, point by point, all previous theories of Brownian movement other than the [molecular-] kinetic."

16. *Atoms,* p. 86.

17. *The Aim and Structure of Physical Theory* (Princeton: University Press, 1991).

18. *Images of Science,* Paul M. Churchland and Clifford A. Hooker, eds. (Chicago: University Press, 1985).

19. *Images of Science,* p. 254.

20. The points to be made below will apply as well to causal reasoning from all observed *A*s (for example, accelerations) are caused by *B*s (for example, forces) to all *A*s are caused by *B*s. Realists generally extend inductive generalizations and causal reasoning to unobservables. For example, Isaac Newton writes, in his discussion of his inductive Rule 3: "The extension of bodies is known to us only through our senses, and yet there are bodies beyond the range of these senses; but because extension is found in all sensible bodies, it is ascribed to all bodies universally"—*Principia,* I. B. Cohen and Anne Whitman, eds. (Berkeley: California UP, 1999), Book 3, p. 795. Other such properties Newton mentions in this discussion are hardness, impenetrability, mobility, inertia, and gravitational attraction. Newton also extends his causal Rule 2 to apply to reasoning from effects to unobservable causes. Antirealists such as Duhem and van Fraassen do not deny the existence of unobservables. Their claim, contrary to Newton's, is that one cannot extend inductive or causal reasoning to unobservables by claiming truth for the conclusions of such arguments.

21. This conclusion is subject to the restriction noted above, namely, "in the absence of any contrary information." With mass, there is no such contrary information. But with many properties, there are; for example, observed bodies of various sizes, distances from us, durations, and interactions have temperature. But we have other empirical information, including an empirically supported theory about the nature of heat, that prevents the inference that bodies of all sizes, and so on, have temperature (for example, individual molecules). In such cases, however, it is not the unobservability of such bodies that prevents the inference that all bodies have temperatures, but information about the nature of heat. More generally, with the addition of quantum-mechanical information, unknown of course to Perrin in 1908, various conclusions from observations made on observed macrobodies cannot be drawn to bodies of much smaller size such as electrons. But this is not because such bodies are unobservable but because bodies of that size exhibit quantum-mechanical properties.

22. This type of reasoning is analogous to what Philip Kitcher, in a recent perceptive article on realism, calls "the Galilean strategy"—"Real Realism: The Galilean Strategy," *The Philosophical Review,* cx (April 2001): 151–97, particularly pp. 173–80.

23. To be sure, there are variations that can be made to show that mass is not a

constant but varies with velocity. But this fact about mass has nothing to do with the observability or unobservability of the bodies with mass.

24. "The Current Status of Scientific Realism," in Jarrett Leplin, ed., *Scientific Realism* (Berkeley: California UP, 1984), pp. 41–82.

25. *Scientific Realism* (New York: Routledge, 1999), p. xix.

26. "A Confutation of Convergent Realism," in Leplin, pp. 218–49.

27. Psillos divides scientific realism into a "metaphysical" part that "asserts that the world has a definite and mind-independent natural-kind classification"; a "semantical" part that says that scientific theories are capable of truth values and that theoretical terms are capable of denoting unobservable entities in the world; and an "epistemic" part that claims that "mature and predictively successful scientific theories [are] well-confirmed and approximately true of the world" and that the theoretical terms they employ denote entities that exist.

28. *The Scientific Image,* p. 8. Van Fraassen's contrasting antirealism is this: "Science aims to give us theories which are empirically adequate, and acceptance of a theory involves as belief only that it is empirically adequate." Note that such an antirealism is compatible with the truth of theses (A), (B), (C), and (E) of the Boyd-Psillos-Laudan formulation of realism, though, van Fraassen would emphasize, it is not the aim of science to produce theories that satisfy these theses but only to produce theories that are empirically adequate.

29. Some realists insist on broader aims than the one van Fraassen assigns to realists. For example, Kitcher defends an account of realism that combines certain elements of van Fraassen's definition with that of the Boyd-Psillos-Laudan account—*The Advancement of Science* (New York: Oxford, 1993), chapter 5. Kitcher's realism, as expressed in this book, includes the idea that the aim of science is to recognize natural kinds and furnish a set of explanatory schemata that pick out dependencies in the world; it also includes the claim that various parts of the sciences achieve these aims. For his more recent account, see "Real Realism: The Galilean Strategy."

30. "Empiricism, Semantics, and Ontology," in Carnap, *Meaning and Necessity* (Chicago: University Press, 1956), pp. 205–21.

31. "The Natural Ontological Attitude," in Leplin, pp. 83–107.

32. For example, atomic theories, as proposed, were purely mechanical theories that should entail reversible processes; but the latter are incompatible with observed thermodynamic phenomena.

33. Whether Maxwell and Boltzmann were in fact doing this is questionable. For a discussion of Maxwell's reasoning, see my *Particles and Waves.*

34. He writes: "However well fitted atomic theories may be to reproduce certain groups of facts, the physical inquirer who has laid to heart Newton's rules will only admit those theories as *provisional* helps, and will strive to attain, in some more natural way, a satisfactory substitute. The atomic theory plays a part in physics similar to that of certain auxiliary concepts in mathematics; it is a mathematical *model* for facilitating the mental reproduction of facts"—*The Science of Mechanics* (LaSalle, IL: Open Court, 1960), p. 589.

35. ". . . ordinary 'matter' must be regarded merely as a highly natural, unconsciously constructed mental symbol for a complex of sensuous elements . . ."—*Contributions to the Analysis of the Sensations* (LaSalle, IL: Open Court, 1890), p. 152.

36. *Science and Hypothesis* (New York: Dover, 1952).

37. "Physics of a Believer," in *Aim and Structure of Physical Theory.*

38. *Atoms,* p. vii.

39. "The use of the intuitive method has not, of course, been used only in the study of atoms, any more than the inductive method has found its sole application in energetics"—*Atoms,* p. vii.

40. Statement (2) is close to what Leplin calls "minimal epistemic realism," a doctrine he defends in *A Novel Defense of Scientific Realism* (New York: Oxford, 1997).

41. Fine, "The Natural Ontological Attitude," p. 96.

42. Both Alan Musgrave—"NOA's Ark: Fine for Realism," *Philosophical Quarterly,* xxxix (1989): 383–98—and Psillos—*Scientific Realism,* chapter 10—make the general argument that Fine's NOA is not neutral, but a realist position that is incompatible with certain standard antirealist views.

43. Rule 1 requires that causes postulated be "both true and sufficient to explain their phenomena"; Rule 4 requires that "propositions gathered from phenomena by induction should be considered either exactly or very nearly true . . . until yet other phenomena make such propositions either more exact or liable to exceptions." Newton applies both rules to unobservables as well as observables.

Galileo's Tower Argument & Rejections of Universal Rules of Method

The methodologies discussed in parts I–III of this volume, whether empiricist or a priori, make a common assumption, namely, that there are universal rules to be followed in establishing or defending scientific propositions. Descartes proposes twenty-one such rules, which depend on the use of intuition and deduction. Both Newton and Mill propose rules involving causal and inductive reasoning. Whewell's hypothetico-deductivism invokes predictivity, consilience, and coherence as universal tests of a theory's truth. Even Popper, who believes that theories cannot be established, only refuted, proposes a form of hypothetico-deductive reasoning as one that is universally valid in science.

Feyerabend's Rejection of Universal Rules

Paul Feyerabend (1924–94), a student as well as critic of Karl Popper who became one of the most controversial and widely discussed philosophers of science during the second half of the twentieth century, emphatically rejects the idea that there are universal rules of method that always have been or ought to be followed by scientists. His essay "Against Method: An Anarchistic Theory of Knowledge," parts of which are included in this volume, was first published in 1970 and was later expanded into a book, *Against Method,* in 1975. The views Feyerabend expresses in these works, particularly his notorious slogan "anything goes," excited many readers while they enraged others. One of his aims was to excite and enrage. Doing so, he thought, is the best way to question otherwise entrenched positions. Like Popper, he thought that an important goal of science is to detect error. Unlike Popper, he denied the existence of universal rules of method for doing so. He believes that all such rules have been, and should have been, violated on various occasions. Otherwise scientific progress is impossible. He writes:

> The idea of a method that contains firm, unchanging, and absolutely binding principles for conducting the business of science gets into considerable difficulty when confronted with the results of historical research. We find, then, that there is not a single rule, however plausible, and however firmly grounded in epistemology, that is not violated at some time or other. It becomes evident that such violations are not

accidental events, they are not the results of insufficient knowledge or of inattention which might have been avoided. On the contrary, we see that they are necessary for progress. ("Against Method," sec. 1)

Feyerabend's positive proposal is what he calls the "principle of proliferation." The idea is that scientists should be "introducing, elaborating, and propagating hypotheses which are inconsistent either with well-established *theories* or with well-established *facts*" (sec. 2). Suppose that, in accordance with some widely accepted scientific method, say Newton's inductivism in the eighteenth century, a hypothesis, say the law of gravity, is well established by astronomical phenomena of the sort Newton invokes. In order to make progress in science Feyerabend rejects Newton's idea (expressed in his rule 4) simply to look for new phenomena that will either add to the inductive support of the law or show that it is "liable to exceptions." Rather, one needs to introduce an entirely different hypothesis that is inconsistent with Newton's law or with phenomena from which Newton derives his law, even if such a contrary hypothesis has no evidence whatever in its favor. This is the surest and best way to detect error in established positions and at the same time advance science by opening up new possibilities for consideration.

Galileo's Tower Argument

Although Feyerabend claims that many great scientists have proceeded in the above manner, the one example he discusses in detail is the "tower argument" introduced by Galileo in his defense of the Copernican idea that the earth turns on its axis as it rotates around the sun.

Galileo published his most famous work, *Dialogue on the Two Chief World Systems*, in March 1632. At the title indicates, the book consists of a dialogue, which involves three characters: Simplicio, a defender of the Ptolemaic theory that the planets and the sun revolve around the stationary earth; Salviati, a defender of the Copernican idea that the planets, including the earth, revolve around the sun and turn on their own axes; and Sagredo, a neutral observer whom the others are trying to convince. At the time of publication, the idea of an earth-centered universe was the official doctrine of the Catholic church. Although the dialogue form allowed Galileo the appearance of not contradicting Church doctrine, within a few months of the book's appearance the printer was ordered to stop selling it, and Galileo was summoned to Rome to face the Inquisition for discussing a heretical view. As a result of the trial, which began in April 1633, Galileo was placed under permanent house arrest, and his book was officially banned by being put on the Index of Prohibited Books, where it remained for more than 250 years.

The tower argument, the part of the *Dialogue* that Feyerabend focuses upon, is introduced by Salviati as representing the strongest argument that defenders of the stationary-earth (Ptolemaic) view invoke against the (Copernican) idea that the earth rotates on its axis as it moves around the sun. For the sake of argument, says the Ptolemaic astronomer, suppose the earth does rotate. Now drop a rock from the top of a tower. If the earth rotates, then, while the rock is in the air, the tower, which is connected to the earth, should travel a given distance, and the rock when it lands should be that distance from the base of the tower. But the rock does not do this at all. When dropped from the tower it falls straight down and lands at the base of the tower. Hence, concludes the Ptolemaic astronomer, the earth does not move.

In response, Salviati, defending the moving-earth view, reasons as follows. If the earth does move, and if the rock falls along the side of the tower and lands next to the base, then the motion of the rock would really be a compound of two motions: a motion in the direction of the earth's motion and a motion down toward the center of the earth. The resulting motion of the rock would then be a "slanting" one, not one parallel to the tower and perpendicular to the earth. So, Salviati concludes, just from seeing the motion of the falling rock and from the observed fact that it lands at the base of the tower, you could not infer that it falls straight down unless you assumed to begin with that the earth is stationary. What you see is compatible with both theories. It does not refute the moving-earth theory.

Now for Feyerabend's take on this. To begin with, according to Feyerabend, Galileo (in the guise of Salviati) is employing the principle of proliferation: he is introducing a hypothesis, namely, that the motion of the falling rock is a slanting one that is the result of the motion of the earth and the motion of the rock toward the earth's center. This hypothesis is incompatible with an established theory (that the earth is stationary), and it is incompatible with a fact established by observation (that the rock falls with a straight, not a slanting, motion). Accordingly, for Feyerabend, Galileo is proceeding "counterinductively." He is violating inductive rules of the sort espoused by Newton, among others.

What argument does Galileo use? None, says Feyerabend, or at least none that succeeds:

> How does he proceed? How does he manage to introduce absurd and counterinductive assertions such as the assertion that the earth moves, and how does he manage to get them a just and attentive hearing? One may anticipate that arguments will not suffice—an interesting, and highly important, limitation of rationalism [Feyerabend's term for arguing in accordance with general rules]—and Galileo's utterances are indeed arguments in appearance only. For Galileo uses *propaganda.* He uses *psychological tricks* in addition to whatever intellectual reasons he has to offer. These tricks are very successful; they lead him to victory. But they . . . obscure the fact that the experience on which Galileo wants to base the Copernican view is nothing but the result of his own fertile imagination, that it has been *invented.* ("Against Method," sec. 7)

So, according to Feyerabend, Galileo not only proceeds "counterinductively" by suggesting a theory that is inconsistent with established theories and facts, but he does so by *inventing* the experience (the "slanted" motion) on which he is basing the moving-earth theory. This, for Feyerabend, is not a criticism of Galileo but a reaffirmation of the way great scientists frequently do and ought to proceed.

Selections from Galileo's *Dialogue* in which the tower argument is introduced and discussed are included in the present volume. This part of the volume also includes an essay by the editor that criticizes Feyerabend's views about proliferation and his claim that Galileo was employing proliferation in the manner Feyerabend suggests.

Kuhn's Rejection of Universal Rules

In 1962 Thomas Kuhn, a physicist by training, published *The Structure of Scientific Revolutions,* which produced its own revolution in the field of philosophy of science, was read by scientists, historians, economists, and sociologists, and indeed became one of the most widely discussed intellectual works of the second half of the twentieth century. In this book Kuhn presents a new and striking picture of how science develops.

A traditional picture, which Kuhn rejected, is based on the idea that science is cumulative, that later theories, experiments, and scientific techniques incorporate and extend the results of earlier ones. If we look at the actual history of science, Kuhn declared, we will see that this picture is grossly mistaken. What one actually finds is a "series of peaceful interludes punctuated by intellectually violent revolutions." The peaceful interludes occur during the time when a "paradigm" is accepted by a scientific community. An example is the Newtonian paradigm accepted by the scientific community from the late seventeenth century to the beginning of the twentieth century.

A paradigm contains fundamental theoretical assumptions, rules, and concepts. For example, Newton's paradigm includes his laws of motion and of gravity, his concepts of space, time, motion, and force, and his basic assumption that nature is mechanical (that everything in nature proceeds by bodies exerting forces on other bodies in accordance with laws of motion). The paradigm also determines what problems need to be solved (e.g., the motions of the planets, the tides, bodies on the earth) and what counts and does not count as a legitimate solution. It determines what sorts of observations, experiments, and instruments provide legitimate ways to test problem-solutions in the paradigm. The paradigm also sets the rules of reasoning (e.g., Newton's four rules of causation and induction, and rules of mathematics) that are allowed, as well as scientific values (such as mathematical precision and the need for objectivity in experiments) that are to be satisfied. Solving problems set by the paradigm using conditions set by the paradigm is engaging in what Kuhn calls "normal science." It is what scientists do almost all the time.

However, there are moments in the history of science when a "paradigm shift" occurs, the replacement of one paradigm by another—for example, the paradigm shift from Newtonian to Einsteinian physics in the early twentieth century. The shift involves a radically new way of seeing things, a switch of perspective, like suddenly seeing a meaningful pattern in a set of lines that once appeared disorganized, or seeing a new pattern instead of an old one. This Kuhn calls a "scientific revolution." It does not occur by a process of reasoning or inference. Einstein did not use Newton's four rules of reasoning, or any other set, in arriving at his special or general theories of relativity. Rather, he saw the world, he conceptualized it, in a way that Newton did not.

Critics of Kuhn reacted to different aspects of his views about scientific revolutions, but most strongly to his concept of a "paradigm." The complaint was that Kuhn had not made the idea sufficiently precise. He addresses this criticism in the second edition of the book, published in 1970, in which he adds a "Postscript," a part of which is included below. He replaces the term "paradigm" with the expression "disciplinary matrix," claiming that the latter denotes a system with several components that he proceeds to clarify. One component he calls "symbolic generalizations," which include the formal expressions and equations that members of the group all accept—for example, $f = ma$ (force equals mass times acceleration) in Newtonian physics. Another component is a set of basic ontological commitments shared by the community—for example, that there are bodies in the universe with mass that exert forces on each other. Yet another component is what Kuhn calls "exemplars" or "concrete problem-solutions that students encounter from the start of their scientific education"—the sort found at the ends of textbook chapters. The final component, and the one of special interest to us in connection with rules of method, Kuhn calls the set of scientific "values"—for example, accuracy, quantitative predictions, simplicity, consistency, and plausibility.

Some critics of the first edition complained that Kuhn's position on revolutionary periods of science is much too subjective. If a paradigm shift is not a reasoning process but a matter of seeing things in a new way, a shift of perspective, then it does not conform to objective rules; it is a subjective matter. No, Kuhn now claims, for we must recognize the existence of shared values in science—values shared within a paradigm (or "disciplinary matrix") and values shared by different scientific communities with different paradigms. For example, scientists from very different communities, scientists with different disciplinary matrices, generally agree on the importance and value of predictive accuracy, consistency, and simplicity. Accordingly, Kuhn would reject Feyerabend's idea that in science, "anything goes." There are certain standards that scientists, even within different communities, can and should agree on. However, Kuhn insists, these values do not function as rules that uniquely determine what theory to accept or even work on. The reason is that values can be applied differently by different scientists. Two scientists may agree on the value of simplicity in a theory, yet they may apply the concept of simplicity differently, depending on subjective considerations involving personality and training.

Readers should focus particularly on this part of Kuhn's material and ask whether he has satisfied his critics who accuse him of reducing science to subjectivism, and whether, and if so how, he avoids the anarchism espoused by Feyerabend. An essay by the editor, reprinted at the end of this part of the volume, discusses these issues and suggests ways to interpret Kuhn's views as subjective and ways to avoid that interpretation.

20

Galileo's Refutation of the Tower Argument

From

Dialogue Concerning the Two Chief World Systems / Galileo Galilei

The Second Day

. . . .

Salv. [Salviati] As the strongest reason of all is adduced that of heavy bodies, which, falling down from on high, go by a straight and vertical line to the surface of the earth. This is considered an irrefutable argument for the earth being motionless. For if it made the diurnal rotation, a tower from whose top a rock was let fall, being carried by the whirling of the earth, would travel many hundreds of yards to the east in the time the rock would consume in its fall, and the rock ought to strike the earth that distance away from the base of the tower. This effect they support with another experiment, which is to drop a lead ball from the top of the mast of a boat at rest, noting the place where it hits, which is close to the foot of the mast; but if

---

Reprinted from *Dialogue Concerning the Two Chief World Systems—Ptolemaic and Copernican,* trans. Stillman Drake, 2nd ed. (Berkeley: University of California Press, 1967), 126–27, 139–48.

Translator's notes have been omitted.

the same ball is dropped from the same place when the boat is moving, it will strike at that distance from the foot of the mast which the boat will have run during the time of fall of the lead, and for no other reason than that the natural movement of the ball when set free is in a straight line toward the center of the earth. This argument is fortified with the experiment of a projectile sent a very great distance upward; this might be a ball shot from a cannon aimed perpendicular to the horizon. In its flight and return this consumes so much time that in our latitude the cannon and we would be carried together many miles eastward by the earth, so that the ball, falling, could never come back near the gun, but would fall as far to the west as the earth had run on ahead.

362

They add moreover the third and very effective experiment of shooting a cannon ball point-blank to the east, and then another one with equal charge at the same elevation to the west; the shot toward the west ought to range a great deal farther out than the other one to the east. For when the ball goes toward the west, and the cannon, carried by the earth, goes east, the ball ought to strike the earth at a distance from the cannon equal to the sum of the two motions, one made by itself to the west, and the other by the gun, carried by the earth, toward the east. On the other hand, from the trip made by the ball shot toward the east it would be necessary to subtract that which was made by the cannon following it. Suppose, for example, that the journey made by the ball in itself was five miles and that the earth in that latitude traveled three miles during the flight of the ball; in the shot toward the west, the ball would fall to earth eight miles distant from the gun—that is, its own five toward the west and the gun's three to the east. But the shot toward the east would range no further than two miles, which is all that remains after subtracting from the five of the shot the three of the gun's motion toward the same place. Now experiment shows the shots to fall equally; therefore the cannon is motionless, and consequently the earth is, too. Not only this, but shots to the south or north likewise confirm the stability of the earth; for they would never hit the mark that one had aimed at, but would always slant toward the west because of the travel that would be made toward the east by the target, carried by the earth while the ball was in the air. And not merely shots along the meridians, but even those made to the east or west would not range truly; for the easterly shots would carry high and the westerly low whenever they were aimed point-blank. Since the shots in both directions take the path of a tangent—that is, a line parallel to the horizon—and the horizon is always falling away to the east and rising in the west if the diurnal motion belongs to the earth (which is why the eastern stars appear to rise and the western stars to set), it follows that the target to the east would drop away under the shot, wherefore the shot would range high, and the rising of the western target would make the shot to the west low. Hence in no direction would shooting ever be accurate; and since experience is contrary to this, it must be said that the earth is immovable.

. . . .

SALV. . . . Aristotle says, then, that a most certain proof of the earth's being motionless is that things projected perpendicularly upward are seen to return by the same line to the same place from which they were thrown, even though the movement is extremely high. This, he argues, could not happen if the earth moved, since in the time during which the projectile is moving upward and then downward it is separated from the earth, and the place from which the projectile began its motion would go a long way toward the east, thanks to the revolving of the earth, and the falling projectile would strike the earth that distance away from the place in question. Thus we can accommodate here the argument of the cannon ball as well as the other argument, used by Aristotle and Ptolemy, of seeing heavy bodies falling from great heights along a straight line perpendicular to the surface of the earth. Now, in order to begin to untie these knots, I ask Simplicio by what means he would prove that freely falling bodies go along straight and perpendicular lines directed toward the center, should anyone refuse to grant this to Aristotle and Ptolemy.

363

SIMP. [Simplicio] By means of the senses, which assure us that the tower is straight and perpendicular, and which show us that a falling stone goes along grazing it, without deviating a hairsbreadth to one side or the other, and strikes at the foot of the tower exactly under the place from which it was dropped.

SALV. But if it happened that the earth rotated, and consequently carried along the tower, and if the falling stone were seen to graze the side of the tower just the same, what would its motion then have to be?

SIMP. In that case one would have to say "its motions," for there would be one with which it went from top to bottom, and another one needed for following the path of the tower.

SALV. The motion would then be a compound of two motions; the one with which it measures the tower, and the other with which it follows it. From this compounding it would follow that the rock would no longer describe that simple straight perpendicular line, but a slanting one, and perhaps not straight.

SIMP. I don't know about its not being straight, but I understand well enough that it would have to be slanting, and different from the straight perpendicular line it would describe with the earth motionless.

SALV. Hence just from seeing the falling stone graze the tower, you could not say for sure that it described a straight and perpendicular line, unless you first assumed the earth to stand still.

SIMP. Exactly so; for if the earth were moving, the motion of the stone would be slanting and not perpendicular.

SALV. Then here, clear and evident, is the paralogism of Aristotle and of Ptolemy, discovered by you yourself. They take as known that which is intended to be proved.

SIMP. In what way? It looks to me like a syllogism in proper form, and not a *petitio principii* [a begging of the question].

364

SALV. In this way: Does he not, in his proof, take the conclusion as unknown?

SIMP. Unknown, for otherwise it would be superfluous to prove it.

SALV. And the middle term; does he not require that to be known?

SIMP. Of course; otherwise it would be an attempt to prove *ignotum per aeque ignotum* [something unknown by means of something equally unknown].

SALV. Our conclusion, which is unknown and is to be proved; is this not the motionlessness of the earth?

SIMP. That is what it is.

SALV. Is not the middle term, which must be known, the straight and perpendicular fall of the stone?

SIMP. That is the middle term.

SALV. But wasn't it concluded a little while ago that we could not have any knowledge of this fall being straight and perpendicular unless it was first known that the earth stood still? Therefore in your syllogism, the certainty of the middle term is drawn from the uncertainty of the conclusion. Thus you see how, and how badly, it is a paralogism.

SAGR. [Sagredo] On behalf of Simplicio I should like, if possible, to defend Aristotle, or at least to be better persuaded as to the force of your deduction. You say that seeing the stone graze the tower is not enough to assure us that the motion of the rock is perpendicular (and this is the middle term of the syllogism) unless one assumes the earth to stand still (which is the conclusion to be proved). For if the tower moved along with the earth and the rock grazed it, the motion of the rock would be slanting, and not perpendicular. But I reply that if the tower were moving, it would be impossible for the rock to fall grazing it; therefore, from the scraping fall is inferred the stability of the earth.

SIMP. So it is. For to expect the rock to go grazing the tower if that were carried along by the earth would be requiring the rock to have two natural motions; that is, a straight one toward the center, and a circular one about the center, which is impossible.

SALV. So Aristotle's defense consists in its being impossible, or at least in his having considered it impossible, that the rock might move with a motion mixed of straight and circular. For if he had not held it to be impossible that the stone might move both toward and around the center at the same time, he would have understood how it could happen that the falling rock might go grazing the tower whether that was moving or was standing still, and consequently he would have been able to perceive that this grazing could imply nothing as to the motion or rest of the earth.

Nevertheless this does not excuse Aristotle, not only because if he did have this idea he ought to have said so, it being such an important point in the argument, but also, and more so, because it cannot be said either that such an effect is impossible or that Aristotle considered it impossible. The former cannot be said because, as I shall shortly prove to you, this is not only possible but necessary; and the latter cannot be said either, because Aristotle himself admits that fire moves naturally upward in a straight line and also turns in the diurnal motion which is imparted by the sky to all the element of fire and to the greater part of the air. Therefore if he saw no impossibility in the mixing of straight-upward with circular motion, as communicated to fire and to the air up as far as the moon's orbit, no more should he deem this impossible with regard to the rock's straight-downward motion and the circular motion natural to the entire globe of the earth, of which the rock is a part.

SIMP. It does not look that way to me at all. If the element of fire goes around together with the air, this is a very easy and even a necessary thing for a particle of fire, which, rising high from the earth, receives that very motion in passing through the moving air, being so tenuous and light a body and so easily moved. But it is quite incredible that a very heavy rock or a cannon ball which is dropping without restraint should let itself be budged by the air or by anything else. Besides which, there is the very appropriate experiment of the stone dropped from the top of the mast of a ship, which falls to the foot of the mast when the ship is standing still, but falls as far from the same point when the ship is sailing as the ship is perceived to have advanced during the time of the fall, this being several yards when the ship's course is rapid.

SALV. There is a considerable difference between the matter of the ship and that of the earth under the assumption that the diurnal motion belongs to the terrestrial globe. For it is quite obvious that just as the motion of the ship is not its natural one, so the motion of all the things in it is accidental; hence it is no wonder that this stone which was held at the top of the mast falls down when it is set free, without any compulsion to follow the motion of the ship. But the diurnal rotation is being taken as the terrestrial globe's own and natural motion, and hence that of all its parts, as a thing indelibly impressed upon them by nature. Therefore the rock at the top of the tower has as its primary tendency a revolution about the center of the whole in twenty-four hours, and it eternally exercises this natural propensity no matter where it is placed. To be convinced of this, you have only to alter an outmoded impression made upon your mind, saying: "Having thought until now that it is a property of the earth's globe to remain motionless with respect to its center, I have never had any difficulty in or resistance to understanding that each of its particles also rests naturally in the same quiescence. Just so, it ought to be that if the natural tendency of the earth were to go around its center in twenty-four hours, each of its particles would also have an inherent and natural inclination not to stand still but to follow that same course."

366

And thus without encountering any obstacle you would be able to conclude that since the motion conferred by the force of the oars upon a boat, and through the boat upon all things contained in it, is not natural but foreign to them, then it might well be that this rock, once separated from the boat, is restored to its natural state and resumes its exercise of the simple tendency natural to it.

I might add that at least that part of the air which is lower than the highest mountains must be swept along and carried around by the roughness of the earth's surface, or must naturally follow the diurnal motion because of being a mixture of various terrestrial vapors and exhalations. But the air around a boat propelled by oars is not moved by them. So arguing from the boat to the tower has no inferential force. The rock coming from the top of the mast enters a medium which does not have the motion of the boat; but that which leaves the top of the tower finds itself in a medium which has the same motion as the entire terrestrial globe, so that far from being impeded by the air, it rather follows the general course of the earth with assistance from the air.

SIMP. I am not convinced that the air could impress its own motion upon a huge stone or a large ball of iron or lead weighing, say, two hundred pounds, as it might upon feathers, snow, and other very light bodies. In fact, I can see that a weight of that sort does not move a single inch from its place even when exposed to the wildest wind you please; now judge whether the air alone would carry it along.

SALV. There is an enormous difference between this experience of yours and our example. You make the wind arrive upon this rock placed at rest, and we are exposing to the already moving air a rock which is also moving with the same speed, so that the air need not confer upon it some new motion, but merely maintain—or rather, not impede—what it already has. You want to drive the rock with a motion foreign and unnatural to it; we, to conserve its natural motion in it. If you want to present a more suitable experiment, you ought to say what would be observed (if not with one's actual eyes, at least with those of the mind) if an eagle, carried by the force of the wind, were to drop a rock from its talons. Since this rock was already flying equally with the wind, and thereafter entered into a medium moving with the same velocity, I am pretty sure that it would not be seen to fall perpendicularly, but, following the course of the wind and adding to this that of its own weight, would move in a slanting path.

SIMP. It would be necessary to be able to make such an experiment and then to decide according to the result. Meanwhile, the result on shipboard confirms my opinion up to this point.

367

SALV. You may well say "up to this point," since perhaps in a very short time it will look different. And to keep you no longer on tenterhooks, as the saying goes, tell me, Simplicio: Do you feel convinced that the experiment on the ship squares so well with our purpose that one may reasonably believe that whatever is seen to occur there must also take place on the terrestrial globe?

SIMP. So far, yes; and though you have brought up some trivial disparities, they do not seem to me of such moment as to suffice to shake my conviction.

SALV. Rather, I hope that you will stick to it, and firmly insist that the result on the earth must correspond to that on the ship, so that when the latter is peceived to be prejudicial to your case you will not be tempted to change your mind.

You say, then, that since when the ship stands still the rock falls to the foot of the mast, and when the ship is in motion it falls apart from there; then conversely, from the falling of the rock at the foot it is inferred that the ship stands still, and from its falling away it may be deduced that the ship is moving. And since what happens on the ship must likewise happen on the land, from the falling of the rock at the foot of the tower one necessarily infers the immobility of the terrestrial globe. Is that your argument?

SIMP. That is exactly it, briefly stated, which makes it easy to understand.

SALV. Now tell me: If the stone dropped from the top of the mast when the ship was sailing rapidly fell in exactly the same place on the ship to which it fell when the ship was standing still, what use could you make of this falling with regard to determining whether the vessel stood still or moved?

SIMP. Absolutely none; just as by the beating of the pulse, for instance, you cannot know whether a person is asleep or awake, since the pulse beats in the same manner in sleeping as in waking.

SALV. Very good. Now, have you ever made this experiment of the ship?

SIMP. I have never made it, but I certainly believe that the authorities who adduced it had carefully observed it. Besides, the cause of the difference is so exactly known that there is no room for doubt.

SALV. You yourself are sufficient evidence that those authorities may have offered it without having performed it, for you take it as certain without having done it, and commit yourself to the good faith of their dictum. Similarly it not only may be, but must be that they did the same thing too—I mean, put faith in their predecessors, right on back without ever arriving at anyone who had performed it. For anyone who does will find that the experiment shows exactly the opposite of what is written; that 368 is, it will show that the stone always falls in the same place on the ship, whether the ship is standing still or moving with any speed you please. Therefore, the same cause holding good on the earth as on the ship, nothing can be inferred about the earth's motion or rest from the stone falling always perpendicularly to the foot of the tower.

SIMP. If you had referred me to any other agency than experiment, I think that our dispute would not soon come to an end; for this appears to me to be a thing so remote from human reason that there is no place in it for credulity or probability.

SALV. For me there is, just the same.

SIMP. So you have not made a hundred tests, or even one? And yet you so freely declare it to be certain? I shall retain my incredulity, and my own confidence that the experiment has been made by the most important authors who make use of it, and that it shows what they say it does.

SALV. Without experiment, I am sure that the effect will happen as I tell you, because it must happen that way; and I might add that you yourself also know that it cannot happen otherwise, no matter how you may pretend not to know it—or give that impression. But I am so handy at picking people's brains that I shall make you confess this in spite of yourself.

Sagredo is very quiet; it seemed to me that I saw him move as though he were about to say something.

SAGR. I was about to say something or other, but the interest aroused in me by hearing you threaten Simplicio with this sort of violence in order to reveal the knowledge he is trying to hide has deprived me of any other desire; I beg you to make good your boast.

SALV. If only Simplicio is willing to reply to my interrogation, I cannot fail.

SIMP. I shall reply as best I can, certain that I shall be put to little trouble; for of the things I hold to be false, I believe I can know nothing, seeing that knowledge is of the true and not of the false.

SALV. I do not want you to declare or reply anything that you do not know for certain. Now tell me: Suppose you have a plane surface as smooth as a mirror and made of some hard material like steel. This is not parallel to the horizon, but somewhat inclined, and upon it you have placed a ball which is perfectly spherical and of some hard and heavy material like bronze. What do you believe this will do when released? Do you not think, as I do, that it will remain still?

SIMP. If that surface is tilted?

SALV. Yes, that is what was assumed.

SIMP. I do not believe that it would stay still at all; rather, I am sure that it would spontaneously roll down.

SALV. Pay careful attention to what you are saying, Simplicio, for I am certain that it would stay wherever you placed it.

369

SIMP. Well, Salviati, so long as you make use of assumptions of this sort I shall cease to be surprised that you deduce such false conclusions.

SALV. Then you are quite sure that it would spontaneously move downward?

SIMP. What doubt is there about this?

SALV. And you take this for granted not because I have taught it to you—indeed, I have tried to persuade you to the contrary—but all by yourself, by means of your own common sense.

SIMP. Oh, now I see your trick; you spoke as you did in order to get me out on a limb, as the common people say, and not because you really believed what you said.

SALV. That was it. Now how long would the ball continue to roll, and how fast? Remember that I said a perfectly round ball and a highly polished surface, in order to remove all external and accidental impediments. Similarly I want you to take away any impediment of the air caused by its resistance to separation, and all other accidental obstacles, if there are any.

SIMP. I completely understood you, and to your question I reply that the ball would continue to move indefinitely, as far as the slope of the surface extended, and with a continually accelerated motion. For such is the nature of heavy bodies, which *vires acquirunt eundo* [acquire force as they move]; and the greater the slope, the greater would be the velocity.

SALV. But if one wanted the ball to move upward on this same surface, do you think it would go?

SIMP. Not spontaneously, no; but drawn or thrown forcibly, it would.

SALV. And if it were thrust along with some impetus impressed forcibly upon it, what would its motion be, and how great?

SIMP. The motion would constantly slow down and be retarded, being contrary to nature, and would be of longer or shorter duration according to the greater or lesser impulse and the lesser or greater slope upward.

SALV. Very well; up to this point you have explained to me the events of motion upon two different planes. On the downward inclined plane, the heavy moving body spontaneously descends and continually accelerates, and to keep it at rest requires the use of force. On the upward slope, force is needed to thrust it along or even to hold it still, and motion which is impressed upon it continually diminishes until it is entirely annihilated. You say also that a difference in the two instances arises from the greater or lesser upward or downward slope of the plane, so that from a greater slope downward there follows a greater speed, while on the contrary upon the upward slope a given movable body thrown with a given force moves farther according as the slope is less.

Now tell me what would happen to the same movable body placed upon 370 a surface with no slope upward or downward.

SIMP. Here I must think a moment about my reply. There being no downward slope, there can be no natural tendency toward motion; and there being no upward slope, there can be no resistance to being moved, so there would be an indifference between the propensity and the resistance to motion. Therefore it seems to me that it ought naturally to remain stable. But I forgot; it was not so very long ago that Sagredo gave me to understand that that is what would happen.

SALV. I believe it would do so if one set the ball down firmly. But what would happen if it were given an impetus in any direction?

SIMP. It must follow that it would move in that direction.

SALV. But with what sort of movement? One continually accelerated, as on the downward plane, or increasingly retarded as on the upward one?

SIMP. I cannot see any cause for acceleration or deceleration, there being no slope upward or downward.

SALV. Exactly so. But if there is no cause for the ball's retardation, there ought to be still less for its coming to rest; so how far would you have the ball continue to move?

SIMP. As far as the extension of the surface continued without rising or falling.

SALV. Then if such a space were unbounded, the motion on it would likewise be boundless? That is, perpetual?

SIMP. It seems so to me, if the movable body were of durable material.

SALV. That is of course assumed, since we said that all external and accidental impediments were to be removed, and any fragility on the part of the moving body would in this case be one of the accidental impediments.

Now tell me, what do you consider to be the cause of the ball moving spontaneously on the downward inclined plane, but only by force on the one tilted upward?

SIMP. That the tendency of heavy bodies is to move toward the center of the earth, and to move upward from its circumference only with force; now the downward surface is that which gets closer to the center, while the upward one gets farther away.

SALV. Then in order for a surface to be neither downward nor upward, all its parts must be equally distant from the center. Are there any such surfaces in the world?

SIMP. Plenty of them; such would be the surface of our terrestrial globe if it were smooth, and not rough and mountainous as it is. But there is that of the water, when it is placid and tranquil.

SALV. Then a ship, when it moves over a calm sea, is one of these movables which courses over a surface that is tilted neither up nor down, and if all external and accidental obstacles were removed, it would thus be disposed to move incessantly and uniformly from an impulse once received?

371

SIMP. It seems that it ought to be.

SALV. Now as to that stone which is on top of the mast; does it not move, carried by the ship, both of them going along the circumference of a circle about its center? And consequently is there not in it an ineradicable motion, all external impediments being removed? And is not this motion as fast as that of the ship?

SIMP. All this is true, but what next?

SALV. Go on and draw the final consequence by yourself, if by yourself you have known all the premises.

SIMP. By the final conclusion you mean that the stone, moving with an indelibly impressed motion, is not going to leave the ship, but will follow it, and finally will fall at the same place where it fell when the ship remained motionless.

. . . .

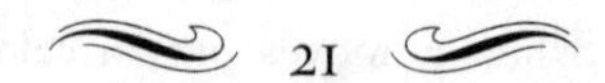

21

Feyerabend's Rejection of Universal Rules

From

Against Method: Outline of an Anarchistic Theory of Knowledge / Paul K. Feyerabend

. . . .

1. Introduction: The Limits of Argument

The idea of a method that contains firm, unchanging, and absolutely binding principles for conducting the business of science gets into considerable difficulty when confronted with the results of historical research. We find, then, that there is not a single rule, however plausible, and however firmly grounded in epistemology, that is not violated at some time or other. It becomes evident that such violations are not accidental events, they are not the results of insufficient knowledge or of inattention which might have been avoided. On the contrary, we see that they are necessary for progress.

Reprinted from *Analyses of Theories and Methods of Physics and Psychology*, ed. Michael Radner and Stephen Winokur, vol. 4 of *Minnesota Studies in the Philosophy of Science* (Minneapolis: University of Minnesota Press, 1970), 21–27, 45–55, 69–70, 91–92.

Feyerabend's lengthy discursive notes (numbers 20–29, 31–35, 38, 40–42, 142, and 184 in the original sequence) have been omitted.

Indeed, one of the most striking features of recent discussions in the history and philosophy of science is the realization that developments such as the Copernican Revolutions, or the rise of atomism in antiquity and recently (kinetic theory; dispersion theory; stereochemistry; quantum theory), or the gradual emergence of the wave theory of light occurred either because some thinkers decided not to be bound by certain "obvious" methodological rules or because they *unwittingly broke* them.

This liberal practice, I repeat, is not just a *fact* of the history of science. It is not merely a manifestation of human inconstancy and ignorance. It is reasonable and *absolutely necessary* for the growth of knowledge. More specifically, the following can be shown: considering any rule, however "fundamental," there are always circumstances when it is advisable not only to ignore the rule, but to adopt its opposite. For example, there are circumstances when it is advisable to introduce, elaborate, and defend ad hoc hypotheses, or hypotheses which contradict well-established and generally accepted experimental results, or hypotheses whose content is smaller than the content of the existing and empirically adequate alternatives, or self-inconsistent hypotheses, and so on.

There are even circumstances—and they occur rather frequently—when argument loses its forward-looking aspect and becomes a hindrance to progress. Nobody wants to assert that the teaching of small children is exclusively a matter of argument (though argument may enter into it and should enter into it to a larger extent than is customary), and almost everyone now agrees that what looks like a result of reason—the mastery of a language, the existence of a richly articulated perceptual world, logical ability—is due partly to indoctrination, partly to a process of growth that proceeds with the force of natural law. And where arguments do seem to have an effect this must often be ascribed to their physical repetition rather than to their semantic content. This much having been admitted, we must also concede the possibility of non-argumentative growth in the adult as well as in (the theoretical parts of) institutions such as science, religion, and prostitution. We certainly cannot take it for granted that what is possible for a small child—to acquire new modes of behavior on the slightest provocation, to slide into them without any noticeable effort—is beyond the reach of his elders. One should expect that catastrophic changes of the physical environment, wars, the breakdown of encompassing systems of morality, political revolutions, will transform adult reaction patterns, too, including important patterns of argumentation. This may again be an entirely natural process and rational argument may but increase the mental tension that precedes and causes the behavioral outburst.

Now, if there are events, not necessarily arguments, which cause us to adopt new standards, including new and more complex forms of argumentation, will it then not be up to the defenders of the status quo to provide, not just arguments, but also contrary causes? (Virtue without terror is ineffective, says Robespierre.) And if the old forms of argumentation turn out to be too weak a cause, must not these defenders either give up or resort to stronger and more "irrational" means? (It is very difficult, and perhaps entirely impossible, to combat the effects of brainwashing by argument.) Even the most puritanical rationalist will then be forced to stop reasoning and to use, say, propaganda and coercion, not because some of his reasons have ceased to be valid, but because the psychological conditions which make them effective, and capable of influencing others, have disappeared. And what is the use of an argument that leaves people unmoved?

Of course, the problem never arises quite in this form. The teaching of standards never consists in merely putting them before the mind of the student and making them as clear as possible. The standards are supposed to have maximal causal efficacy as well. This makes it very difficult to distinguish between the logical force and the material effect of an argument. Just as a well-trained pet will obey his master no matter how great the confusion he finds himself in and no matter how urgent the need to adopt new patterns of behavior, in the very same way a well-trained rationalist will obey the mental image of his master, he will conform to the standards of argumentation he has learned, he will adhere to these standards no matter how great the difficulty he finds himself in, and he will be quite unable to discover that what he regards as the "voice of reason" is but a causal aftereffect of the training he has received. We see here very clearly how the appeal to "reason" works. At first sight this appeal seems to be to some ideas which convince a man instead of pushing him. But conviction cannot remain an ethereal state; it is supposed to lead to action. It is supposed to lead to the appropriate action, and it is supposed to sustain this action as long as necessary. What is the force that upholds such a development? It is the causal efficacy of the standards to which appeal was made and this causal efficacy in turn is but an effect of training, as we have seen. It follows that appeal to argument either has no content at all, and can be made to agree with any procedure, or else will often have a conservative function: it will set limits to what is about to become a natural way of behavior. In the latter case, however, the appeal is nothing but a concealed political maneuver. This becomes very clear when a rationalist wants to restore an earlier point of view. Basing his argument on natural habits of reasoning which either have become extinct or have no point of attack in the new situation, such a champion of "rationality" must first restore the earlier material and psychological conditions. This, however, involves him in "a struggle of interests and forces, not of argument."[1]

That interests, forces, propaganda, brainwashing techniques play a much greater role in the growth of our knowledge and, a fortiori, of science than is commonly believed can also be seen from an analysis of the relation between idea and action. One often takes it for granted that a clear and distinct understanding of new ideas precedes and should precede any formulation and any institutional expression of them. (An investigation starts with a problem, says Popper.) First, we have an idea, or a problem; then we act, i.e., either speak, or build, or destroy. This is certainly not the way in which small children develop. They use words, they combine them, they play with them until they grasp a meaning that so far has been beyond their reach. And the initial playful activity is an essential presupposition of the final act of understanding. There is no reason why this mechanism should cease to function in the adult. On the contrary, we must expect, for example, that the idea of liberty could be made clear only by means of the very same actions which were supposed to create liberty. Creation of a thing, and creation plus full understanding of a correct idea of the thing, very often are parts of one and the same indivisible process and they cannot be separated without bringing the process to a standstill. The process itself is not guided by a well-defined program; it cannot be guided by such a program for it contains the conditions of the realization of programs. It is rather guided by a vague urge, by a "passion" (Kierkegaard). The passion gives rise to specific behavior which in turn creates the circumstances and the ideas necessary for analyzing and explaining the whole development, for making it "rational."

The development of the Copernican point of view from Galileo up to the twentieth century is a perfect example of the situation we want to describe. We start with a strong belief that runs counter to contemporary reason. The belief spreads and finds support from other beliefs which are equally unreasonable, if not more so (law of inertia; telescope). Research now gets deflected in new directions, new kinds of instruments are built, "evidence" is related to theories in new ways until there arises a new ideology that is rich enough to provide independent arguments for any particular part of it and mobile enough to find such arguments whenever they seem to be required. Today we can say that Galileo was on the right track, for his persistent pursuit of what once seemed to be a silly cosmology created the material needed for the defense of this cosmology against those of us who accept a view only if it is told in a certain way and who trust it only if it contains certain magical phrases, called "observational reports." And this is not an exception—it is the normal case: theories become clear and "reasonable" only after incoherent parts of them have been used for a long time. Such unreasonable, nonsensical, unmethodical foreplay thus turns out to be an unavoidable precondition of clarity and of empirical success.

Trying to describe developments of this kind in a general way, we are of course obliged to appeal to the existing forms of speech which do not take them into account and which must be distorted, misused, and beaten into new patterns in order to fit unforeseen situations (without a constant misuse of language there cannot be any discovery and any progress). "Moreover, since the traditional categories are the gospel of everyday thinking (including ordinary scientific thinking) and of everyday practice, [such an attempt at understanding] in effect presents rules and forms of false thinking and action—false, that is, from the standpoint of [scientific] commonsense."[2] This is how dialectical thinking arises as a form of thought that "dissolves into nothing the detailed determinations of the understanding."[3]

It is clear, then, that the idea of a fixed method, or of a fixed (theory of) rationality, arises from too naive a view of man and of his social surroundings. To those who look at the rich material provided by history, and who are not intent on impoverishing it in order to please their lower instincts, their craving for intellectual security as it is provided, for example, by clarity and precision, to such people it will seem that there is only one principle that can be defended under all circumstances, and in all stages of human development. It is the principle: anything goes.

This abstract principle (which is the one and only principle of our anarchistic methodology) must now be elucidated, and explained in concrete detail.

2. Counterinduction I: Theories

It was said that when considering any rule, however fundamental or "necessary for science," one can imagine circumstances when it is advisable not only to ignore the rule, but to adopt its opposite. Let us apply this claim to the rule that "experience," or "the facts," or "experimental results," or whatever words are being used to describe the "hard" elements of our testing procedures, measure the success of a theory, so that agreement between the theory and "the data" is regarded as favoring the theory (or as leaving the situation unchanged), while disagreement endangers or perhaps even eliminates it. This rule is an essential part of all theories of induction, including even some theories of corroboration. Taking the opposite view, I suggest introducing, elaborating, and propagating hypotheses which are inconsistent either with well-established theories or with well-established facts. Or, as I shall express myself: I suggest proceeding counterinductively in addition to proceeding inductively.

There is no need to discuss the first part of the suggestion which favors hypotheses inconsistent with well-established theories. The main argument has already been published elsewhere.[4] It may be summarized by saying that evidence that is relevant for the test of a theory T can often be unearthed only with the help of an incompatible alternative theory T′. Thus, the advice to postpone alternatives until the first refutation has occurred means putting the cart before the horse. In this connection, I also advised increasing empirical contents with the help of a principle of proliferation: invent and elaborate theories which are inconsistent with the accepted point of view, even if the latter should happen to be highly confirmed and generally accepted. Considering the arguments just summarized, such a principle would seem to be an essential part of any critical empiricism.

The principle of proliferation is also an essential part of a humanitarian outlook. Progressive educators have always tried to develop the individuality of their pupils, and to bring to fruition the particular and sometimes quite unique talents and beliefs that each child possesses. But such an education very often seemed to be a futile exercise in daydreaming. For is it not necessary to prepare the young for life? Does this not mean that they must learn one particular set of views to the exclusion of everything else? And, if there should still remain a trace of their youthful gift of imagination, will it not find its proper application in the arts, that is, in a thin domain of dreams that has but little to do with the world we live in? Will this procedure not finally lead to a split between a hated reality and welcome fantasies, science and the arts, careful description and unrestrained self-expression? The argument for proliferation shows that this need not be the case. It is possible to retain what one might call the freedom of artistic creation and to use it to the full, not just as a road of escape, but as a necessary means for discovering and perhaps even changing the properties of the world we live in. For me this coincidence of the part (individual man) with the whole (the world we live in), of the purely subjective and arbitrary with the objective and lawful, is one of the most important arguments in favor of a pluralistic methodology.

. . . .

5. The Tower Argument Stated: First Steps of Analysis

As a concrete illustration and as a basis for further discussion, I shall now briefly describe the manner in which Galileo defused an important counterargument against the idea of the motion of the earth. I say "defused," and not "refuted," because we are dealing with a changing conceptual system as well as with certain attempts at concealment.

According to the argument which convinced Tycho, and which is used against the motion of the earth in Galileo's own Trattato della sfera, observation shows that "heavy bodies . . . falling down from on high, go by a straight and vertical line to the surface of the earth. This is considered an irrefutable argument for the earth being motionless. For if it made the diurnal rotation, a tower from whose top a rock was let fall, being carried by the whirling of the earth, would travel many hundreds of yards to the east in the time the rock would consume in its fall, and the rock ought to strike the earth that distance away from the base of the tower."[5]

In considering the argument, Galileo at once admits the correctness of the sensory content of the observation made, viz. that "heavy bodies . . . falling from a height, go perpendicularly to the surface of the earth."[6]

Considering an author (Chiaramonti) who sets out to convert Copernicans by repeatedly mentioning this fact, he says: "I wish that this author would not put himself to such trouble trying to have us understand from our senses that this motion of falling bodies is simple straight motion and no other kind, nor get angry and complain because such a clear, obvious, and manifest thing should be called into question. For in this way he hints at believing that to those who say such motion is not straight at all, but rather circular, it seems they see the stone move visibly in an arc, since he calls upon their senses rather than their reason to clarify the effect. This is not the case, Simplicio; for just as I . . . have never seen nor ever expect to see the rock fall any way but perpendicularly, just so do I believe that it appears to the eyes of everyone else. It is therefore better to put aside the appearance, on which we all agree, and to use the power of reason either to confirm its reality or to reveal its fallacy."[7] The correctness of the observation is not in question. What *is* in question is its "reality" or "fallacy." What is meant by this expression?

The question is answered by an example that occurs in Galileo's next paragraph, and "from which . . . one may learn how easily anyone may be deceived by simple appearances, or let us say by the impressions of one's senses. This event is the appearance to those who travel along a street by night of being followed by the moon, with steps equal to theirs, when they see it go gliding along the eaves of the roofs. There it looks to them just as would a cat really running along the tiles and putting them behind it; an appearance which, if reason did not intervene, would only too obviously deceive the senses."

In this example we are asked to start with a sensory impression and consider a statement that is forcefully suggested by it. (The suggestion is so strong that it has led to entire systems of belief and rituals as becomes clear from a closer study of the lunar aspects of witchcraft and of other religions.) Now "reason intervenes": the statement suggested by the impression is examined, and one considers other statements in its place. The nature of the impression is not changed a bit by this activity. (This is only approximately true; but we can omit for our present purpose the complications arising from the interaction of impression and proposition.) But it enters new observation statements and plays new, better or worse, parts in our knowledge. What are the reasons and the methods which regulate such exchange?

To start with we must become clear about the nature of the total phenomenon: appearance plus statement. There are not two acts, one, noticing a phenomenon, the other, expressing it with the help of the appropriate statement, but only one, viz. saying, in a certain observational situation, "the moon is following me," or "the stone is falling straight down." We may of course abstractly subdivide this process into parts, and we may also try to create a situation where statement and phenomenon seem to be psychologically apart and waiting to be related. (This is rather difficult to achieve and is perhaps entirely impossible.[8]) But under normal circumstances such a division does not occur; describing a familiar situation is, for the speaker, an event in which statement and phenomenon are firmly glued together.

This unity is the result of a process of learning that starts in one's childhood. From our very early days we learn to react to situations with the appropriate responses, linguistic or otherwise. The teaching procedures both shape the 'appearance' or the 'phenomenon' and establish a firm connection with words, so that finally the phenomena seem to speak for themselves, without outside help or extraneous knowledge. They just are what the associated statements assert them to be. The language they 'speak' is of course influenced by the beliefs of earlier generations which have been held for such a long time that they no longer appear as separate principles, but enter the terms of everyday discourse, and, after the prescribed training, seem to emerge from the things themselves.

Now at this point we may want to compare, in our imagination and quite abstractly, the results of the teaching of different languages incorporating different ideologies. We may even want to consciously change some of these ideologies and adapt them to more 'modern' points of view. It is very difficult to say how this will change our situation, unless we make the further assumption that the quality and structure of sensations (perceptions), or at least the quality and structure of those sensations which enter the body of science, are independent of their linguistic expression. I am very doubtful about even the approximate validity of this assumption which can be refuted by simple examples. And I am sure that we are depriving ourselves of new and surprising discoveries as long as we remain within the limits defined by it. Yet the present essay will remain quite consciously within these limits. (My first task, if I should ever resume writing, would be to explore these limits and to venture beyond them.)

Making the additional simplifying assumption, we can now distinguish between (a) sensations, and (b) those "mental operations which follow so closely upon the senses"[9] and are so firmly connected with their reactions that a separation is difficult to achieve. Considering the origin and the effect of such operations, I shall call them natural interpretations.

6. Natural Interpretations

In the history of thought, natural interpretations have been regarded either as a priori presuppositions of science or else as prejudices which must be removed before any serious examination can proceed. The first view is that of Kant, and, in a very different manner and on the basis of very different talents, that of some contemporary linguistic philosophers. The second view is due to Bacon (who had, however, predecessors, such as the Greek skeptics).

Galileo is one of those rare thinkers who neither wants to forever retain natural interpretations nor wants to altogether eliminate them. Wholesale judgments of this kind are quite alien to his way of thinking. He insists upon critical discussion to decide which natural interpretations can be kept and which must be replaced. This is not always clear from his writings. Quite the contrary, the methods of reminiscence, to which he appeals so freely, are designed to create the impression that nothing has changed and that we continue expressing our observations in old and familiar ways. Yet his attitude is relatively easy to ascertain: natural interpretations are necessary. The senses alone, without the help of reason, cannot give us a true account of nature. What is needed for arriving at such a true account are "the . . . senses, accompanied by reasoning."[10] Moreover, in the arguments dealing with the motion of the earth, it is this reasoning, it is the connotation of the observation terms, and not the message of the senses or the appearance, that causes trouble. "It is therefore better to put aside the appearance, on which we all agree, and to use the power of reason either to confirm [its] reality or to reveal [its] fallacy."[11] "To confirm the reality or reveal the fallacy of appearances" means, however, to examine the validity of those natural interpretations which are so intimately connected with the appearances that we no longer regard them as separate assumptions. I now turn to the first natural interpretation implicit in the argument from falling stones.

According to Copernicus the motion of a falling stone should be "mixed straight-and-circular."[12] By the "motion of the stone" is meant, not just its motion relative to some visible mark in the visual field of the observer, or its observed motion, but rather its motion in the solar system, or in (absolute) space, or its real motion. The familiar facts appealed to in the argument assert a different kind of motion, a simple vertical motion. This result refutes the Copernican hypothesis only if the concept of motion that occurs in the observation statement is the same as the concept of motion that occurs in the Copernican prediction. The observation statement "the stone is falling straight down" must therefore likewise refer to a movement in (absolute) space. It must refer to a real motion.

Now, the force of an "argument from observation" derives from the fact that the observation statements it involves are firmly connected with appearances. There is no use appealing to observation if one does not know how to describe what one sees, or if one can offer one's description with hesitation only, as if one had just learned the language in which it is formulated. An observation statement, then, consists of two very different psychological events: (1) a clear and unambiguous sensation and (2) a clear and unambiguous connection between this sensation and parts of a language. This is the way in which the sensation is made to speak. Do the sensations in the argument above speak the language of real motion?

They speak the language of real motion in the context of seventeenth-century everyday thought. At least this is what Galileo tells us. He tells us that the everyday thinking of the time assumes the "operative" character of all motion.[13] Or, to use well-known philosophical terms, it assumes a naive realism with respect to motion: except for occasional and unavoidable illusions, apparent motion is identical with real (absolute) motion. Of course, this distinction is not explicitly drawn. One does not first distinguish the apparent motion from the real motion and then connect the two by a correspondence rule. Quite the contrary, one describes, perceives, acts toward the apparent motion as if it were already the real thing. Nor does one proceed in this manner under all circumstances. It is admitted that objects may move which are not seen to move; and it is also admitted that certain motions are illusory. . . . Apparent motion and real motion are not always identified. However, there are paradigmatic cases in which it is psychologically very difficult, if not plainly impossible, to admit deception. It is from these paradigmatic cases, and not from exceptions, that naive realism with respect to motions derives its strength. These are also the situations in which we first learn our kinematic vocabulary. From our very childhood we learn to react to them with concepts which have naive realism built right into them, and which inextricably connect movement and the appearance of movement. The motion of the stone in the tower argument, or the alleged motion of the earth, is such a paradigmatic case. How could one possibly be unaware of the swift motion of a large bulk of matter such as the earth is supposed to be! How could one possibly be unaware of the fact that the falling stone traces a vastly extended trajectory through space! From the point of view of seventeenth-century thought and language, the argument is, therefore, impeccable and quite forceful. However, notice how theories ("operative character" of all motion: essential correctness of sense reports), which are not formulated explicitly, enter the debate in the guise of observational terms. We realize again that observational terms are Trojan horses which must be watched very carefully. How is one supposed to proceed in such a sticky situation?

The argument from falling stones seems to refute the Copernican view. This may be due to an inherent disadvantage of Copernicanism; but it may also be due to the presence of natural interpretations which are in need of improvement. The first task, then, is to discover and to isolate these unexamined obstacles to progress.

It was Bacon's belief that natural interpretations could be discovered by a method of analysis that peels them off, one after another, until the sensory core of every observation is laid bare. This method has serious drawbacks. First, natural interpretations of the kind considered by Bacon are not just added to a previously existing field of sensations. They are instrumental in constituting the field, as Bacon says himself. Eliminate all natural interpretations, and you also eliminate the ability to think and to perceive. Second, disregarding this fundamental function of natural interpretations, it should be clear that a person who faces a perceptual field without a single natural interpretation at his disposal would be completely disoriented; he could not even start the business of science. Third, the fact that we do start, even after some Baconian analysis, shows that the analysis has stopped prematurely. It has stopped at precisely those natural interpretations of which we are not aware and without which we cannot proceed. It follows that the intention to start from scratch, after a complete removal of all natural interpretations, is self-defeating.

Furthermore, it is not possible to even partly unravel the cluster of natural interpretations. At first sight the task would seem to be simple enough. One takes observation statements, one after the other, and analyzes their content. However, concepts that are hidden in observation statements are not likely to reveal themselves in the more abstract parts of language. If they do, it will still be difficult to nail them down; concepts, just as percepts, are ambiguous and dependent on background. Moreover, the content of a concept is determined also by the way in which it is related to perception. Yet how can this way be discovered without circularity? Perceptions must be identified, and the identifying mechanism will contain some of the very same elements which govern the use of the concept to be investigated. We never penetrate this concept completely, for we always use part of it in the attempt to find its constituents.[14] There is only one way to get out of this circle, and it consists in using an external measure of comparison, including new ways of relating concepts and percepts. Removed from the domain of natural discourse and from all those principles, habits, and attitudes which constitute its form of life, such an external measure will look strange indeed. This, however, is not an argument against its use. Quite the contrary, such an impression of strangeness reveals that natural interpretations are at work, and it is a first step toward their discovery. Let us explain this situation with the help of the tower example.

The example is intended to show that the Copernican view is not in accordance with 'the facts.' Seen from the point of view of these 'facts,' the idea of the motion of the earth appears to be outlandish, absurd, and obviously false, to mention only some of the expressions which were frequently used at the time, and which are still heard wherever professional squares confront a new and counterfactual theory. This makes us suspect that the Copernican view is an external measuring rod of precisely the kind described above.

We now can turn the argument around and use it as a detecting device that helps us to discover the natural interpretations that exclude the motion of the earth. Turning the argument around, we first assert the motion of the earth and then inquire what changes will remove the contradiction. Such an inquiry may take considerable time, and there is a good sense in which one can say that it is not yet finished, not even today. The contradiction, therefore, may stay with us for decades or even centuries. Still, it must be upheld (Hegel!) until we have finished our examination or else the examination, the attempt to discover the antediluvian components of our knowledge, cannot even start. This, we have seen, is one of the reasons one can give for retaining, and, perhaps, even for inventing, theories which are inconsistent with the facts: Ideological ingredients of our knowledge and, more especially, of our observations, are discovered with the help of theories which are refuted by them. They are discovered counterinductively.

Let me repeat what has been asserted so far. Theories are tested and possibly refuted by facts. Facts contain ideological components, older views which have vanished from sight or were perhaps never formulated in an explicit manner. These components are highly suspicious, first, because of their age, because of their antediluvian origin; second, because their very nature protects them from a critical examination and always has protected them from such an examination. Considering a contradiction between a new and interesting theory and a collection of firmly established facts, the best procedure is, therefore, not to abandon the theory but to use it for the discovery of the hidden principles that are responsible for the contradiction. Counterinduction is an essential part of such a process of discovery. (Excellent historical example: the arguments against motion and atomicity of Parmenides and Zeno. Diogenes of Sinope, the Cynic, took the simple course that would be taken by many contemporary scientists and all contemporary philosophers: he refuted the arguments by rising and walking up and down. The opposite course, recommended here, led to much more interesting results, as is witnessed by the history of the case. One should not be too hard on Diogenes, however, for it is also reported that he beat a pupil who was content with his refutation, exclaiming that he had given reasons which the pupil should not accept without additional reasons of his own.[15])

Having discovered a particular natural interpretation, the next question is how it is to be examined and tested. Obviously, we cannot proceed in the usual way, i.e., derive predictions and compare them with "results of observation." These results are no longer available. The idea that the senses, employed under normal circumstances, produce correct reports of real events, for example reports of the real motion of physical bodies, has now been removed from all observational statements. (Remember that this notion was found to be an essential part of the anti-Copernican argument.) But without it our sensory reactions cease to be relevant for tests. This conclusion has been generalized by some rationalists, who decided to build their science on reason only and ascribed to observation a quite insignificant auxiliary function. Galileo does not adopt this procedure.

If one natural interpretation causes trouble for an attractive view, and if its elimination removes the view from the domain of observation, then the only acceptable procedure is to use other interpretations and to see what happens. The interpretation which Galileo uses restores the senses to their position as instruments of exploration, but only with respect to the reality of relative motion. Motion "among things which share it in common" is "nonoperative," that is, "it remains insensible, imperceptible, and without any effect whatever."[16] Galileo's first step in the joint examination of the Copernican doctrine, and of a familiar but hidden natural interpretation, consists therefore in replacing the latter by a different interpretation, or, considering the function of natural interpretations, he introduces a new observation language.

This is, of course, an entirely legitimate move. In general, the observation language which enters an argument has been in use for a long time and is quite familiar. Considering the structure of common idioms on the one hand, and of the Aristotelian philosophy on the other, neither this use nor the familiarity can be regarded as a test of the underlying principles. These principles, these natural interpretations, occur in every description. Extraordinary cases which might create difficulties are defused with the help of "adjuster words,"[17] such as "like" or "analogous," which divert them so that the basic ontology remains unchallenged. A test is, however, urgently needed. It is needed especially in those cases where the principles seem to threaten a new theory. It is then quite reasonable to introduce alternative observation languages and to compare them both with the original idiom and with the theory under examination. Proceeding in this way, we must make sure that the comparison is fair. That is, we must not criticize an idiom that is supposed to function as an observation language because it is not yet well known and is therefore less strongly connected with our sensory reactions and less plausible than is another and more "common" idiom. Superficial criticisms of this kind, which have been elevated in an entire new "philosophy," abound in discussions of the mind-body problem. Philosophers who want to introduce and to test new views thus find themselves faced not with arguments, which they could most likely answer, but with an impenetrable stone wall of well-entrenched reactions. This is not at all different from the attitude of people ignorant of foreign languages, who feel that a certain color is much better described by "red" than by "rosso." As opposed to such attempts at conversion by appeal to familiarity ("I know what pains are, and I also know, from introspection, that they have nothing whatever to do with material processes!"), we must emphasize that a comparative judgment of observation languages, e.g., materialistic observation languages, phenomenalistic observation languages, objective-idealistic observation languages, theological observation languages, can start only when all of them are spoken equally fluently.

Let me assert at this point that while it is possible to consider and to actively apply various rules of thumb, and while we may in this way arrive at a satisfactory judgment, it is not at all wise to go further and to turn these rules of thumb into necessary conditions of science. For example, one might be inclined to say, following Neurath, that an observation language A is preferable to an observation language B, if it is at least as useful as B in our everyday life, and if more theories and more comprehensive theories are compatible with it than are compatible with B. Such a criterion takes into account that both our perceptions (natural interpretations included) and our theories are fallible, and it also pays attention to our desire for a harmonious and universal point of view. (One always seems to assume that observation languages should be employed not only in laboratories, but also at home, and in the "natural surroundings" of the scientist.) However, we must not forget that we find and improve the assumptions hidden in our observational reports by a method that makes use of inconsistencies. Hence, we might prefer B to A as a starting point of analysis, and we might in this way arrive at a language C which satisfies the criterion even better, but which cannot be reached from A. Conceptual progress like any other kind of progress depends on psychological circumstances, which may prohibit in one case what they encourage in another. Moreover the psychological factors which come into play are never clear in advance. Nor should the demand for practicality and sensory content be regarded as a *conditio sine qua non.* We possess detecting mechanisms whose performance outdistances our senses. Combining such detectors with a computer, we may test a theory directly, without intervention of a human observer. This would eliminate sensations and perceptions from the process of testing. Using hypnosis, one could eliminate them from the transfer of the results into the human brain also, and thus arrive at a science that is completely without experience. Considerations like these, which indicate possible paths of development, should cure us once and for all of the belief that judgments of progress, improvement, etc., are based on rules which can be revealed now and will remain in action for all the years to come. My discussion of Galileo has not, therefore, the aim of arriving at the "correct method." It has rather the aim of showing that such a "correct method" does not and cannot exist.

More especially, it has the limited aim of showing that counterinduction is very often a reasonable move. Let us now proceed a step further in our analysis of Galileo's reasoning!

7. The Tower Argument: Analysis Continued

Galileo replaces one natural interpretation by a very different and as yet (1630!) at least partly unnatural interpretation. How does he proceed? How does he manage to introduce absurd and counterinductive assertions such as the assertion that the earth moves, and how does he manage to get them a just and attentive hearing? One may anticipate that arguments will not suffice—an interesting, and highly important, limitation of rationalism—and Galileo's utterances are indeed arguments in appearance only. For Galileo uses propaganda. He uses psychological tricks in addition to whatever intellectual reasons he has to offer. These tricks are very successful; they lead him to victory. But they obscure the new attitude toward experience that is in the making, and postpone for centuries the possibility of a reasonable philosophy. They obscure the fact that the experience on which Galileo wants to base the Copernican view is nothing but the result of his own fertile imagination, that it has been invented. They obscure this fact by insinuating that the new results which emerge are known and conceded by all, and need only be called to our attention to appear as the most obvious expression of the truth.

. . . .

10. Summary of Analysis of Tower Argument

I repeat and summarize: An argument is proposed that refutes Copernicus by observation. The argument is inverted in order to discover those natural interpretations which are responsible for the contradiction. The offensive interpretations are replaced by others. Propaganda and appeal to distant and highly theoretical parts of common sense are used to defuse old habits and to enthrone new ones. The new natural interpretations which are also formulated explicitly as auxiliary hypotheses are established partly by the support they give to Copernicus and partly by plausibility considerations and ad hoc hypotheses. An entirely new "experience" arises in this way. Independent evidence is as yet entirely lacking, but this is no drawback as it is to be expected that independent support will take a long time appearing. For what is needed is a theory of solid objects, aerodynamics, hydrodynamics, and all these sciences are still hidden in the future. But their task is now well defined, for Galileo's assumptions, his ad hoc hypotheses included, are sufficiently clear and simple to prescribe the direction of future research. Let it be noted, incidentally, that Galileo's procedure drastically reduces the content of dynamics. Aristotelian dynamics was a general theory of change comprising locomotion, qualitative change, generation, and corruption, and it provided a theoretical basis for witchcraft also. Galileo's dynamics and its successors deal with locomotion only, and here again only with the locomotion of matter. The other kinds of motion are pushed aside with the promissory note, due to Democritos, that locomotion will eventually be capable of explaining all motion. Thus, a comprehensive empirical theory of motion is replaced by a much narrower theory plus a metaphysics of motion, just as an "empirical" experience is replaced by an experience that contains strange and speculative elements. Counterinduction, however, is now justified both for theories and for facts. It clearly plays an important role in the advancement of science.

. . . .

Conclusion

The idea that science can and should be run according to some fixed rules, and that its rationality consists in agreement with such rules, is both unrealistic and vicious. It is *unrealistic,* since it takes too simple a view of the talents of men and of the circumstances which encourage, or cause, their development. And it is *vicious,* since the attempt to enforce the rules will undoubtedly erect barriers to what men might have been, and will reduce our humanity by increasing our professional qualifications. We can free ourselves from the idea and from the power it may possess over us (i) by a detailed study of the work of revolutionaries such as Galileo, Luther, Marx, or Lenin; (ii) by some acquaintance with the Hegelian philosophy and with the alternative provided by Kierkegaard; (iii) by remembering that the existing separation between the sciences and the arts is artificial, that it is a side effect of an idea of professionalism one should eliminate, that a poem or a play can be intelligent as well as informative (Aristophanes, Hochhuth, Brecht), and a scientific theory pleasant to behold (Galileo, Dirac), and that we can change science and make it agree with our wishes. We can turn science from a stern and demanding mistress into an attractive and yielding courtesan who tries to anticipate every wish of her lover. Of course, it is up to us to choose either a dragon or a pussycat as our companion. So far mankind seems to have preferred the latter alternative: "The more solid, well defined, and splendid the edifice erected by the understanding, the more restless the urge of life . . . to escape from it into freedom." We must take care that we do not lose our ability to make such a choice.

. . . .

NOTES

1. Leon Trotsky, *The Revolution Betrayed,* trans. M. Eastman (Garden City, N.Y.: Doubleday, 1937), pp. 86–87.

2. H. Marcuse, *Reason and Revolution* (London: Oxford University Press, 1941), p. 130. The quotation is about Hegel's logic.

3. [*Wissenschaft der Logik,* vol. I (Hamburg: Felix Meiner, 1965), p. 6.]

4. "Problems of Empiricism," in *Beyond the Edge of Certainty,* ed. R. G. Colodny (Englewood Cliffs, N.J.: Prentice-Hall, 1965), sections IVff, especially section VI. (The relevant material has been reprinted in P. H. Nidditch, ed., *The Philosophy of Science,* London: Oxford University Press, 1969, pp. 12ff, especially pp. 25–33.) "Realism and Instrumentalism," in *The Critical Approach to Science and Philosophy,* ed. M. Bunge (Glencoe, Ill.: Free Press, 1964). "Reply to Criticism," in *Boston Studies in the Philosophy of Science,* vol. II, ed. R. S. Cohen and M. W. Wartofsky (New York: Humanities, 1965).

5. *Dialogue concerning the Two Chief World Systems,* [trans. Stillman Drake, 2nd ed. (Berkeley: University of California Press, 1967)], p. 126.

6. *Ibid.,* p. 125.

7. *Ibid.,* p. 256.

8. "Problems of Empiricism," pp. 204ff.

9. Bacon, *The New Organon,* Introduction.

10. *Dialogue concerning the Two Chief World Systems,* p. 255. My italics.

11. *Ibid.,* p. 256.

12. *Ibid.,* p. 248.

13. *Ibid.,* p. 171. . . .

14. Cf. "Problems of Empiricism," pp. 204ff.

15. Cf. Hegel, *Vorlesungen uber die Geschichte der Philosophie,* part I, ed. C. L. Michelet (Berlin: Duncker und Humblot, 1840), p. 289.

16. *Dialogue concerning the Two Chief World Systems,* p. 171. . . .

17. J. L. Austin, *Sense and Sensibilia* (New York: Oxford University Press, 1964), p. 74.

22

A Critique of Feyerabend's Anarchism

Proliferation: Is It a Good Thing? / Peter Achinstein

So, I exaggerate
—Paul Feyerabend

I served on several panel discussions with Paul Feyerabend. The most memorable occurred in 1970 at the University of Cincinnati. Paul delivered one of his spellbinding lectures in which he talked about Galileo's tower argument and claimed that Galileo was violating standard rules of logic and methodology. He was practicing "counterinduction" (which, according to Feyerabend, is a good thing to do). He was introducing a hypothesis, namely, that a body falling from a tower has a slanted, not a straight, motion, which is incompatible with highly confirmed theories and indeed is refuted by what we observe when we see the body fall. Galileo was following Feyerabend's favorite principle:

> *Proliferation I:* Invent and elaborate theories that are inconsistent with the accepted point of view, even if the latter should happen to be highly confirmed and generally accepted.[1]

In my response I tried to go through Galileo's argument, with numerous textual references, arguing that Galileo was not doing at all what Feyera-

---

Reprinted from *The Worst Enemy of Science? Essays in Memory of Paul Feyerabend,* ed. John Preston, Gonzalo Munévar, and David Lamb (New York: Oxford University Press, 2000), 37–46.

bend claimed. Feyerabend's reply to me began as follows: "So," he said with a twinkle in his eye, in his charming accent, "I exaggerate".

Feyerabend exaggerated a lot. He espoused, or seemed to, extreme doctrines, the most famous of which he called "anarchism": there are no universal rules of reasoning in science. Any rules of this sort, rules such as the four that Newton proposed at the beginning of Book 3 of the *Principia,* are, and ought to be, frequently violated. Exaggeration, however, suggests that there is a truth there some place, even if it has been distorted or magnified beyond the fact. Is there a truth to proliferation? Perhaps there is, but matters need sorting out.

To begin with, if we understand proliferation in accordance with Feyerabend's explicit formulation above, there is nothing particularly shocking about it. Indeed, scientists and others practice it all the time. Suppose that I want to argue the advantages of a certain theory I favor that explains the facts in some way. Part of my doing so might involve my inventing a different, conflicting theory that also explains these facts but, I demonstrate, is incompatible with other principles that are 'highly confirmed and generally accepted' whereas my theory is not. A favorite theory is frequently defended by "inventing and elaborating" alternative theories in such a way as to satisfy Feyerabend's principle of proliferation. I take this to be straightforward and fairly trivial. Since Feyerabend wants to shock and exaggerate, I doubt that this is what he has in mind. The principle of proliferation, so understood, has no teeth.

Let's give it some teeth. One way to do so is to strengthen Feyerabend's injunction that we are to "invent and elaborate" conflicting theories. Scientists, we might say, don't just aim at "inventing and elaborating" theories for the sheer joy of it. They want to discover true theories, or ones that are probable, or ones that yield reasonably good predictions, or at least theories that are good in some important way (e.g., they are unifying or simple). Accordingly, one might strengthen Proliferation I to say this:

> *Proliferation II:* Believe or accept an "invented and elaborated" theory as true, or probable, or as a good predictor, or as good in some way, if it is inconsistent with the accepted point of view, even if the latter should happen to be highly confirmed.

This version of proliferation has a real bite to it. But it is absurd. First, many such theories can be invented and elaborated. Shall we believe or accept all of them as true, or probable, or good predictors, or good in some other way? That won't do, at least for truth and probability, if such theories are also incompatible with each other. More importantly, why should we regard a theory as true, or probable, or good in some way simply because it is inconsistent with the accepted point of view? I can think of no legitimate reason for doing so.

In any case, to my knowledge Feyerabend never asserts a principle as strong as Proliferation II. So what is he after? To determine this we need to identify his primary interest in formulating the principle of proliferation. What is such a principle supposed to do for us? On several occasions Feyerabend speaks of this principle in the context of the "growth of knowledge" and of obtaining "objective knowledge" (*MS*, pp. 22, 24; *AM*, p. 46). He also speaks of it in the context of "testing a theory" (*MS*, p. 26), and of giving us "evidence that might refute a theory" and thus detect error (*AM*, pp. 29, 41).

Perhaps Feyerabend is committed just to the following thesis, which I take to be more interesting than I and II:

> *Proliferation III:* Suppose we wish to test a theory T by determining whether there is evidence for or against it (so that we may obtain "objective knowledge"). The only way to test T—the only way to obtain evidence for or against it—is to invent and elaborate a conflicting theory T′, even if T′ is inconsistent with an accepted and highly confirmed point of view.

How is the invention of such a conflicting theory T′ supposed to test theory T? One way noted by Feyerabend is that the invention of T′ may enable us to unearth new evidence to test T (*MS*, p. 26). The theory T′ may encompass phenomena that ought to be covered by T but were not thought of or appealed to in defense of T. Discovering such phenomena may then refute T. Unless an alternative such as T′ were conceived, scientists would never have thought of the possibility of such refuting phenomena.

There is no denying this is possible. But this is pretty weak fare. If you have a theory T, then simply imagining a conflicting theory T′ (where there are no constraints on T′ other than perhaps that T′ encompass some phenomena not appealed to in defense of T) does not guarantee or even make it likely that you will discover new evidence to test T. My theory is that O. J. Simpson committed the crimes for which he was charged. How is my inventing the conflicting theory that President Clinton did it going to help me unearth new evidence to test my O. J. theory? (Should I ask special prosecutor Starr to add this to his numerous investigations of Clinton by examining Clinton's logs for the day of the murders?) Simply inventing a contrary theory, without putting more conditions on it than Feyerabend appears to want, isn't likely to do much. I am much more likely to unearth new evidence by continuing to try to identify O. J.'s blood, his motives, his opportunities, his friends and neighbors, and so forth. Without more constraints on the conflicting theory T′—that it have some plausibility or some evidence in its favor—the possibility that it will enable us to unearth new evidence to test T is mere *logical* possibility. Moreover, it is clearly not the only logically possible way to do so. Concentrating on the evidence we have and asking more questions about it is another logical possibility, and it need not involve the invention of incompatible theories.

There is one situation in which the invention of a competing theory T′ (without constraints on its plausibility) is important in testing T. Suppose you defend your theory T hypothetico-deductively by showing that from it a range of facts follows, some of which are known to be true, and others of which are predictions that are later verified. As Isaac Newton and John Stuart Mill pointed out a long time ago, this "consequentialist" test of T is not sufficient by itself to establish truth or even significant probability, since some competing theory T′ may give the same results. Finding such a competitor, then, will serve as a further, and crucial, test of T. If the only argument offered in favor of T is that it generates a set of known and predicted facts, then "inventing and elaborating" a competitor that does equally well is an argument or test that at least blunts the force of the consequentialist argument.

It does not follow, however, that the invention of a competing theory T′ which yields all the facts that T does (and perhaps more) will always test T in this way. This is a point emphasized by Newton in Rule 4 of his *Rules of Reasoning in Philosophy*:

> *Rule 4:* In experimental philosophy we are to look upon propositions inferred by general induction from phenomena as accurately or very nearly true, notwithstanding any contrary hypothesis that may be imagined, till such time as other phenomena occur, by which they may either be made more accurate, or liable to exceptions.[2]

Rule 4 needs to be considered together with the other three. Rule 1 enjoins us to admit no more causes than are true and sufficient to explain the phenomena. Rule 2 insists that we assign the same causes to the same effects. Rule 3 requires us to generalize inductively from properties present in all observed bodies to the presence of those properties in all bodies whatever.[3]

Newton uses his four rules in proving the law of gravity. The planets and their satellites exhibit the same observed effects: they are drawn off from rectilinear motion by forces that vary inversely as the square of the distance of the planets from the sun and the satellites from their planets. By Rule 2, we can infer the same cause or causes in all these cases. By Rule 1, only one cause is needed here, namely, the same force of gravity operating in the case of all the planets and their satellites. By Rule 3, we can generalize this from the observed motions of the planets and satellites to all bodies in the universe. As Newton does in Proposition 7 of Book 3, we can infer "that there is a power of gravity pertaining to all bodies, proportional to the several quantities of matter which they contain."

Now the point of Rule 4 is to say that we can make such an inference and regard it as true (or very nearly so), even if we can imagine some contrary hypothesis that explains the phenomena. If Rules 1–3 are employed to generate a universal proposition such as the law of gravity, then, Newton is claiming, the fact that some contrary hypothesis has been "invented and elaborated" from which known phenomena can be derived does not at all weaken the argument in favor of the universal proposition. So, for instance, suppose you can imagine that there is no universal force of gravity but a different force for each pair of bodies that has imperceptibly different effects in each case; or suppose you can imagine a grue-type universal force that is an inverse-square force until the year 2500 but not thereafter. Such imaginings will in no way test or diminish the effectiveness of the argument in favor of universal gravitation, if that argument is constructed so as to satisfy the first three rules of reasoning. That is Newton's claim.

Let me generalize Newton's claim. Suppose that from what is observed (the "phenomena"), using some acceptable form of reasoning (whether Newton's rules or some others), one argues that some proposition is true or very probable. The fact that you can imagine some contrary proposition that explains the observed facts casts no doubt whatever on the original conclusion. This is a generalization of Newton's claim to which, I believe, he is committed. It is also a claim that Feyerabend seems to deny. How could he? I see at least three possibilities.

1. *Imagining a contrary hypothesis will enable us to challenge the proposition inferred by providing the means to challenge the "observed facts" upon which the inferred proposition is based.* It will enable us to see that some of these "facts" are not really facts. This is what Feyerabend claims about Galileo's tower argument. Galileo imagined a set of hypotheses contrary to the stationary earth hypothesis, namely, that the earth turns on its axis and that a body falling from a tower has two motions, one toward the center of the earth and one in the direction of the earth's motion, which together result in the body's having a slanting motion. These contrary hypotheses challenge not just the original hypothesis (that the earth is stationary), but the observed fact that forms the basis for this hypothesis, namely, that we see the body falling from the top of the tower not in a slanting line but a line that "goes along [the tower] grazing it, without deviating a hairsbreadth to one side or the other" and

landing at its base.[4] Galileo's contrary hypotheses enable us to see things differently and so reject the "phenomena" that serve as a basis for the original hypothesis.

I have two replies. The first is similar to one noted earlier. Simply inventing a contrary theory, without putting constraints on it, isn't guaranteed or even likely to show us that something we took to be an observed fact is not so. This is simply a *logical* possibility, no more. Suppose Galileo had imagined a Cartesian demon that makes it appear that the body is falling straight down; or that a body falls with a slanting motion but when it does so light is refracted so as to produce the appearance of a straight motion; or that there are twenty different motions of the body, resulting in some quite different curve we cannot see. Would, or should, these imaginings show us that what we took to be the observed facts are not really so?

My second point pertains to what Galileo was in fact doing when he proposed his contrary hypotheses. He was not claiming that we don't see the body fall "grazing the tower without deviating". We do see this. Nor was he claiming that the contrary hypotheses he was imagining entail or suggest that we don't see this. What Galileo was proposing was a conflicting explanation for this observed phenomenon, namely, that when we see the body falling what we see is the relative motion of the body with respect to the tower, not its absolute motion (which is slanted). Galileo is not challenging the observed facts but only the standard theory used to explain them.[5]

2. *Imagining a contrary hypothesis will enable us to challenge the mode of reasoning employed to defend the original hypothesis.* As noted earlier, this is the case when a hypothesis is defended hypothetico-deductively solely by reference to consequences. Imagining a contrary hypothesis that yields the same conclusion does or should serve to challenge the mode of reasoning (as Newton and Mill observed). But this is because the mode of reasoning is faulty and deserves to be challenged in just this way. If you claim that your hypothesis is true or probable solely because it entails a range of phenomena that are observed, I can challenge your claim by producing a contrary hypothesis that does the same thing.

Suppose, however, we consider Newton's Rules of Reasoning (at least Rules 1–3), and use them, as Newton did, to infer his universal law of gravity. Does imagining a contrary law not inferred from the phenomena by using these rules cast doubt on the rules themselves? Let's consider two cases. In the first, using Rule 2, from similar inverse-square effects of the planets and their satellites—that they are continually drawn off from rectilinear motions and retained in their orbits—and from the nature of these orbits, Newton infers that the cause that produces these effects in each case is an inverse-square force; and, using Rule 1, he infers that this is the same force (gravity) in each case. To invent a contrary hypothesis, we might suppose that the forces operating between the planets and their satellites, the planets and the sun, objects falling to the earth and the earth itself, etc., are all different in some respects other than being inverse-square forces. Or we might suppose that the force in question is the same in all these cases, but that it is not one force, but several different forces acting together to produce the resultant gravitational force. Doing so would violate Newton's Rule 1, which urges us "to admit no more causes of natural things than such as are both true and sufficient to explain their appearances".

In the second case, using Rule 3, from the fact that the law of gravity is satisfied by all the observed planets and satellites Newton infers that it is satisfied by all bodies whatever. To invent a contrary hypothesis (taking a cue from contemporary quantum mechanics) we might suppose that the law of gravity is satisfied by bodies only when they are observed, and that some different law operates when they are not observed. Or we might suppose that the law is satisfied until the year 2500 but not thereafter.

Does imagining these contrary hypotheses cast doubt on the rules of inference themselves? Newton would emphatically deny this. To be sure, it might turn out that one of these contrary hypotheses, or some other, is true. That is a logical possibility. But even if it does, that would not be enough to impugn the validity of inferring the same cause from the same effects or generalizing from observed instances. Newton is not claiming that such reasoning is *guaranteed* to lead to truth. This point he makes explicit in Rule 4, which says that we may regard propositions inferred by induction from phenomena (and presumably also by causal reasoning, in conformity with Rules 1 and 2) "as accurately or very nearly true, notwithstanding any contrary hypotheses that may be imagined, till such time as other phenomena occur, by which they may either be made more accurate, or liable to exceptions". So Newton explicitly recognizes that the propositions he infers using his rules of inference may be inaccurate, liable to exceptions, and hence false as they stand. But what he is claiming is that this can be shown only by new phenomena, and not by imagining contrary hypotheses. Doing the latter only shows that the original propositions inferred using causal or inductive reasoning or both could turn out false. But that "could" is only a logical possibility and casts no doubt at all on the modes of reasoning (or the propositions inferred).

More generally, with any mode of nondemonstrative reasoning, it is logically possible for the premises in such reasoning to be true and the conclusion false. If this logical possibility is sufficient to cast doubt on the mode of reasoning, then every mode of nondemonstrative reasoning is suspect. This, of course, is Hume's skeptical claim. Is that also what Feyerabend is claiming? If so, one doesn't really need to "proliferate," to "invent and elaborate" contrary hypotheses, to make such a claim. Simply stating that the conclusion of a nondemonstrative argument could be false will do the trick. One need not bother saying how it could be false by imagining specific contraries.

Feyerabend claims that his viewpoint is different from skepticism (*AM*, p. 189). The skeptic, he says, regards every conclusion "as equally good, or equally bad, or desists from making such judgments altogether" (p. 189). Feyerabend, by contrast, calling himself an "epistemological anarchist", wants to be allowed "to defend the most trite, or the most outrageous statement":

> The one thing he [the epistemological anarchist] opposes positively and absolutely are universal standards, universal laws, universal ideas such as 'Truth', 'Reason', 'Justice', 'Love', and the behaviour they bring along, though he does not deny that it is often a good policy to act as if such laws (such standards, such ideas) existed, and as if he believed in them. (*AM*, p. 189)

So in the end Feyerabend's view about rules of reasoning, such as Newton's four rules, seems to be this. Although generally it is "good policy" to act in accordance with them, to act as if we believed in them, they should not be *universally* followed. There are important occasions on which they are, and should be, violated. Rules of reasoning such as Newton's are not universally valid. They are good "rules of thumb," useful "guides" (see *MS*, p. 19). But like other such rules they need to be ignored or flouted on certain occasions.

If this is Feyerabend's view, then an important distinction needs drawing. The distinction is between, on the one hand, flouting, or violating, or ignoring a rule, and on the other, knowing when it is supposed to be applied and when not. As an example of the latter, I might refuse to use inductive generalization when I have very few instances, or when they are all of the same narrow type, or when there are some negative instances or ones that do not fit the generalization very well. Or I might refuse to infer a single cause from similar effects, if I know from other cases that effects of that kind can be produced by quite different causes. In such cases I have not flouted, violated, or ignored Newton's Rules 1–3. Rule 2, for example, does not say "To the same natural effects we must *always* assign the same causes." It explicitly says that we must do so "as far as possible". As is the case with most rules, there are facts about application that are left unstated and that may be unstatable in a way that will cover all possible cases. I have a pretty good idea of how to apply Newton's rules to various cases, when and when not to make the inference from effects to causes or from observed instances to all instances. When I refuse to make such an inference I am not necessarily flouting, or violating, or ignoring the rule, but simply realizing that it is not applicable in the case in question.

What about cases where the rule is flouted, violated, or ignored? There is a fifty-five-mile-per-hour limit on the road I am taking, but a passenger in my car becomes violently ill, so I exceed the speed limit to rush her to the hospital. Here I violate or ignore the fifty-five-mile-per-hour rule in order possibly to save a life. In such a case, I violate a rule to produce some desired state of affairs that perhaps will satisfy some more important rule. There are occasions when the rules can be ignored and violated for a better cause.

Which, if either, of these claims about rules of reasoning is Feyerabend making? If he is making the first, then his position becomes quite plausible. You aren't necessarily violating or ignoring Newton's Rule 2 when you refuse to use it to infer a cause from an effect. You may simply be applying it, as Newton tells you, only "as far as possible", and it may *not* be possible to apply it in a particular case. Similarly, you are not necessarily violating or ignoring Newton's Rule 3 when you refuse to generalize from observed instances. The latter may be too few or too unvaried to permit the inference, or there may be unexplained instances that appear to violate the generalization. Sometimes Feyerabend makes it sound as if this is what he has in mind. For example,

> The limitation of all rules and standards is recognized by *naive anarchism.* A naive anarchist says (a) that both absolute rules and context dependent rules have their limits and infers (b) that all rules and standards are worthless and should be given up. Most reviewers regard me as a naive anarchist in this sense overlooking the many passages where I show how certain procedures *aided* scientists in their research . . . I agree with (a) but I do not agree with (b). I argue that all rules have their limits and that there is no comprehensive "rationality". I do not argue that we should proceed without rules and standards.[6]

Feyerabend also makes the following claim, which might give some support to this interpretation:

> We may start by pointing out that not a single theory ever agrees with all the facts in its domain. And the trouble is not created by rumors, or by the results of sloppy procedure. It is created by experiments and measurements of the highest precision and reliability. (*MS*, p. 36)

Without here attempting to challenge Feyerabend's claim that every theory has counterinstances or at least ones that do not agree with the theory, if he is right, then use of rules such as Newton's in the case of "positive" instances becomes limited, to say the least.

On the other hand, there are many passages in which in addition Feyerabend seems to want to make the second, bolder claim about rules of reasoning, namely, that there are occasions (and they are numerous) when these rules ought to be ignored or violated, presumably for some better cause, where the latter is not simply to deal with known counterinstances or other failures of the theory. Thus,

> [m]ore specifically the following can be shown: considering any rule, however "fundamental", there are always circumstances when it is advisable not only to ignore the rule, but to adopt its opposite. . . . There are even circumstances—and they occur rather frequently—when argument loses its forward-looking aspect and becomes a hindrance to progress. (*MS*, p. 22)

If this view is to be taken seriously, then we must imagine a case of the following sort (unlike the first sort of case). Newton has observed a range of planetary systems in which the gravitational effects are similar. No other phenomena are known indicating a different cause in each case. Newton's Rules 2 and 3 are applicable, in the sense that these rules are supposed to apply to a case of just this sort. Nevertheless, it is permissible, and even a good thing, to ignore or violate these rules by proposing different causes for each observed effect, or perhaps no cause at all, just chance, or by refusing to generalize beyond the observed systems, or by generalizing to a contrary hypothesis. It is permissible to do so in appropriate circumstances for a "better cause". What "circumstances," what "better cause"?

Feyerabend's answer to the "better cause" question seems to be that doing so "is reasonable and absolutely necessary for the growth of knowledge" (*MS*, p. 22). But how will it promote knowledge? The only answer that I can find is that we might discover that what is inferred (a universal gravitational force) does not exist, and that something else is responsible for the effects. We might discover error. Again, there is an appeal to a "logical possibility", which neither Newton nor I would regard as a "better cause."

What is Feyerabend's answer to the question of "circumstances"? In his discussion of Newton's Rule 4 he claims that legitimate criticisms of Newton's "ray" theory of light (according to which white light consists of rays of different refrangibility) can be unearthed only by ignoring Rule 4:

> [Such] criticism can be articulated only if we are allowed to view the success of Newton's theory in the light of "contrary hypotheses" [thus violating Rule 4]. If on the other hand we follow Rule 4 to the letter,

> then contrary hypotheses will not be used and the criticism cannot arise. (IS, p. 401)

A few pages later we find

> Instead of judging theories by a never-examined, mystical, and stable entity, "experience", one should let them compete with each other in the very same manner in which party lines are competing in politics. The invention of "contrary hypotheses" is the first step towards such a competition, and never is their invention more necessary than when it seems that certain ideas have been confirmed beyond doubt and that matters have been settled once and for all. (IS, p. 405)

Finally, in *Against Method,* Feyerabend writes: "Hence it is advisable to let one's inclinations go against reason *in any circumstances,* for science may profit from it" (*AM,* pp. 155–156; italics his). Accordingly, the answer to "under what circumstances" it is legitimate to ignore or violate rules such as Newton's seems to be "always", if error is to be found.

At one point Feyerabend proposes a principle in addition to proliferation, which he calls the "Principle of Tenacity".[7] It urges us "to select from a number of theories the one that promises to lead to the most fruitful results, and to stick to this one theory even if the actual difficulties it encounters are considerable". Perhaps Feyerabend has in mind that in selecting such a theory in accordance with tenacity one is following rules of reasoning, while in proposing contrary theories in accordance with proliferation one is violating those rules (and practicing "counterinduction"). And perhaps he wants to say that while some in the scientific community should attempt to use rules such as Newton's as useful guides (as Newton did in arguing for his law of gravity, thus obeying tenacity), others should ignore those rules by introducing contrary theories that violate those rules (proliferation). Again, the proposed justification would be that such a practice, and only this, can detect error and produce "objective knowledge".

When Feyerabend says that there are circumstances when any rule of reasoning ought to be ignored (and its "opposite" adopted), I think it is fair to interpret him to be making not just the first claim about rules of reasoning, but the second as well. So, recalling his self-confessed penchant for exaggeration, one might say that he exaggerates by turning claim 1 into claim 2. He turns the plausible claim that you need to know how and when to apply rules such as Newton's, which are not applicable in all situations, into the implausible one that there are numerous occasions (perhaps always) on which rules such as these, although applicable, should be violated or ignored for a better cause.

3. *Imagining a contrary hypothesis will cast doubt on the hypothesis inferred even without casting doubt on the facts from which it is inferred, on the mode of inference, or on the claim that the hypothesis follows from the facts in accordance with the mode of reasoning.* So, for example, even if you accept Newton's six "phenomena" from which he infers his law of gravity (phenomena pertaining to the motions of the planets and their satellites), and even if you accept Newton's rules of reasoning (and the earlier propositions of Book 1 of the *Principia*) and agree that using these the law of universal gravitation can be inferred, you cast doubt on that law simply by inventing a contrary hypothesis. You test, and ought to test, Newton's law in this way.

This is just what Newton rejects in his Rule 4, and I think he is right to do so. If you accept the mode of inference, and the claim that the conclusion follows, then how can imagining a contrary conclusion cast doubt? In such a case you cast doubt on the conclusion only if you challenge the facts (the "phenomena"). But, by hypothesis, these are unchallenged. So merely imagining a contrary hypothesis is (again) tantamount to pointing out the logical possibility that the conclusion could be false. And in science as in the law the mere logical possibility that a conclusion is false ought not to cast ("reasonable") doubt on a conclusion that otherwise has support. It is not a test of such a conclusion.

NOTES

1. 'Against Method: Outline of an Anarchistic Theory of Knowledge', in M. Radner and S. Winokur (eds.), *Minnesota Studies in the Philosophy of Science,* Vol. 4 (Minneapolis: University of Minnesota Press, 1970), p. 26. There is a similar formulation in *Against Method* (London: New Left Books, 1975), p. 47. Hereafter I will use the abbreviations *MS* (Minnesota Studies) for the first work, and *AM* for the second.

2. Feyerabend discusses this rule in 'On the Improvement of the Sciences and the Arts, and the Possible Identity of the Two', in R. S. Cohen and M. W. Wartofsky (eds.), *Boston Studies in the Philosophy of Science,* Vol. 3 (Dordrecht: D. Reidel, 1968), pp. 387–415 (hereafter IS). I respond in 'Acute Proliferitis', pp. 416–424 of the same volume.

3. For a discussion of these rules, see my *Particles and Waves* (New York: Oxford University Press, 1991), chap. 2.

4. Galileo, *Dialogue Concerning the Two Chief World Systems* (Berkeley: University of California Press, 1967), p. 139.

5. In the *Dialogue Concerning the Two Chief World Systems,* Simplicio (the Aristotelian defender of the stationary earth theory) claims that "the senses . . . assure us that the tower is straight and perpendicular, and . . . show us that a falling stone goes along grazing it, without deviating a hairsbreadth to one side or the other, and strikes at the foot of the tower exactly under the place from which it is dropped" (p. 139). Salviati (the defender of the moving earth theory) responds by asking this question: "if it happened that the earth rotated, and consequently the tower, and *if the falling stone was seen to graze the side of the tower just the same,* what would its motion have to be?" Salviati's answer is that it would be a compound of two motions resulting in a "slanting". Hence, he concludes, "just from seeing the falling stone graze the tower, you could not say for sure that it described a

straight and perpendicular line, unless you first assumed the earth to stand still" (pp. 139–140). Galileo is claiming that what we see is the same in both theories (we see the falling stone graze the tower and land at its base). The explanation each theory offers for what we see is quite different.

6. *Science in a Free Society* (London: New Left Books, 1978), p. 32. Italics his.

7. 'Consolations for the Specialist', in I. Lakatos and A. Musgrave (eds.), *Criticism and the Growth of Knowledge* (Cambridge: Cambridge University Press, 1970), p. 203.

Kuhn's Rejection of Universal Rules

From

The Structure of Scientific Revolutions / Thomas S. Kuhn

Postscript

. . . .

1. Paradigms and Community Structure

The term 'paradigm' enters the preceding pages early, and its manner of entry is intrinsically circular. A paradigm is what the members of a scientific community share, *and,* conversely, a scientific community consists of men who share a paradigm. Not all circularities are vicious. . . . , but this one is a source of real difficulties. Scientific communities can and should be isolated without prior recourse to paradigms; the latter can then be discovered by scrutinizing the behavior of a given community's members. If this book were being rewritten, it would therefore open with a discussion of the community structure of science, a topic that has recently become a significant subject of sociological research and that historians of science are also beginning to take seriously. Preliminary results, many of them still un-

---

Reprinted from *The Structure of Scientific Revolutions,* 2nd ed., enlarged (Chicago: University of Chicago Press, 1970), 176–87.

published, suggest that the empirical techniques required for its exploration are non-trivial, but some are in hand and others are sure to be developed.[1] Most practicing scientists respond at once to questions about their community affiliations, taking for granted that responsibility for the various current specialties is distributed among groups of at least roughly determinate membership. I shall therefore here assume that more systematic means for their identification will be found. Instead of presenting preliminary research results, let me briefly articulate the intuitive notion of community that underlies much in the earlier chapters of this book. It is a notion now widely shared by scientists, sociologists, and a number of historians of science.

A scientific community consists, on this view, of the practitioners of a scientific specialty. To an extent unparalleled in most other fields, they have undergone similar educations and professional initiations; in the process they have absorbed the same technical literature and drawn many of the same lessons from it. Usually the boundaries of that standard literature mark the limits of a scientific subject matter, and each community ordinarily has a subject matter of its own. There are schools in the sciences, communities, that is, which approach the same subject from incompatible viewpoints. But they are far rarer there than in other fields; they are always in competition; and their competition is usually quickly ended. As a result, the members of a scientific community see themselves and are seen by others as the men uniquely responsible for the pursuit of a set of shared goals, including the training of their successors. Within such groups communication is relatively full and professional judgment relatively unanimous. Because the attention of different scientific communities is, on the other hand, focused on different matters, professional communication across group lines is sometimes arduous, often results in misunderstanding, and may, if pursued, evoke significant and previously unsuspected disagreement.

Communities in this sense exist, of course, at numerous levels. The most global is the community of all natural scientists. At an only slightly lower level the main scientific professional groups are communities: physicists, chemists, astronomers, zoologists, and the like. For these major groupings, community membership is readily established except at the fringes. Subject of highest degree, membership in professional societies, and journals read are ordinarily more than sufficient. Similar techniques will also isolate major subgroups: organic chemists, and perhaps protein chemists among them, solid-state and high-energy physicists, radio astronomers, and so on. It is only at the next lower level that empirical problems emerge. How, to take a contemporary example, would one have isolated the phage group prior to its public acclaim? For this purpose one must have recourse to attendance at special conferences, to the distribution of draft manuscripts or galley proofs prior to publication, and above all to formal and informal communication networks including those discovered in correspondence and in the linkages among citations.[2] I take it that the job can and will be done, at least for the contemporary scene and the more recent parts of the historical. Typically it may yield communities of perhaps one hundred members, occasionally significantly fewer. Usually individual scientists, particularly the ablest, will belong to several such groups either simultaneously or in succession.

Communities of this sort are the units that this book has presented as the producers and validators of scientific knowledge. Paradigms are something shared by the members of such groups. Without reference to the nature of these shared elements, many aspects of science described in the preceding pages can scarcely be understood. But other aspects can, though they are not independently presented in my original text. It is therefore worth noting, before turning to paradigms directly, a series of issues that require reference to community structure alone.

Probably the most striking of these is what I have previously called the transition from the pre- to the post-paradigm period in the development of a scientific field. . . . Before it occurs, a number of schools compete for the domination of a given field. Afterward, in the wake of some notable scientific achievement, the number of schools is greatly reduced, ordinarily to one, and a more efficient mode of scientific practice begins. The latter is generally esoteric and oriented to puzzle-solving, as the work of a group can be only when its members take the foundations of their field for granted.

The nature of that transition to maturity deserves fuller discussion than it has received in this book, particularly from those concerned with the development of the contemporary social sciences. To that end it may help to point out that the transition need not (I now think should not) be associated with the first acquisition of a paradigm. The members of all scientific communities, including the schools of the "pre-paradigm" period, share the sorts of elements which I have collectively labelled 'a paradigm.' What changes with the transition to maturity is not the presence of a paradigm but rather its nature. Only after the change is normal puzzle-solving research possible. Many of the attributes of a developed science which I have above associated with the acquisition of a paradigm I would therefore now discuss as consequences of the acquisition of the sort of paradigm that identifies challenging puzzles, supplies clues to their solution, and guarantees that the truly clever practitioner will succeed. Only those who have taken courage from observing that their own field (or school) has paradigms are likely to feel that something important is sacrificed by the change.

A second issue, more important at least to historians, concerns this book's implicit one-to-one identification of scientific communities with scientific subject matters. I have, that is, repeatedly acted as though, say, 'physical optics,' 'electricity,' and 'heat' must name scientific communities because they do name subject matters for research. The only alternative my text has seemed to allow is that all these subjects have belonged to the physics community. Identifications of that sort will not, however, usually withstand examination, as my colleagues in history have repeatedly pointed out. There was, for example, no physics community before the mid-nineteenth century, and it was then formed by the merger of parts of two previously separate communities, mathematics and natural philosophy (*physique expérimentale*). What is today the subject matter for a single broad community has been variously distributed among diverse communities in the past. Other narrower subjects, for example heat and the theory of matter, have existed for long periods without becoming the special province of any single scientific community. Both normal science and revolutions are, however, community-based activities. To discover and analyze them, one must first unravel the changing community structure of the sciences over time. A paradigm governs, in the first instance, not a subject matter but rather a group of practitioners. Any study of paradigm-directed or of paradigm-shattering research must begin by locating the responsible group or groups.

When the analysis of scientific development is approached in that way, several difficulties which have been foci for critical attention are likely to vanish. A number of commentators have, for example, used the theory of matter to suggest that I drastically overstate the unanimity of scientists in their allegiance to a paradigm. Until comparatively recently, they point out, those theories have been topics for continuing disagreement and debate. I agree with the description but think it no counter-example. Theories of matter were not, at least until about 1920, the special province or the subject matter for any scientific community. Instead, they were tools for a large number of specialists' groups. Members of different communities sometimes chose different tools and criticized the choice made by others. Even more important, a theory of matter is not the sort of topic on which the members of even a single community must necessarily agree. The need for agreement depends on what it is the community does. Chemistry in the first half of the nineteenth century provides a case in point. Though several of the community's fundamental tools—constant proportion, multiple proportion, and combining weights—had become common property as a result of Dalton's atomic theory, it was quite possible for chemists, after the event, to base their work on these tools and to disagree, sometimes vehemently, about the existence of atoms.

Some other difficulties and misunderstandings will, I believe, be dissolved in the same way. Partly because of the examples I have chosen and partly because of my vagueness about the nature and size of the relevant communities, a few readers of this book have concluded that my concern is primarily or exclusively with major revolutions such as those associated with Copernicus, Newton, Darwin, or Einstein. A clearer delineation of community structure should, however, help to enforce the rather different impression I have tried to create. A revolution is for me a special sort of change involving a certain sort of reconstruction of group commitments. But it need not be a large change, nor need it seem revolutionary to those outside a single community, consisting perhaps of fewer than twenty-five people. It is just because this type of change, little recognized or discussed in the literature of the philosophy of science, occurs so regularly on this smaller scale that revolutionary, as against cumulative, change so badly needs to be understood.

One last alteration, closely related to the preceding, may help to facilitate that understanding. A number of critics have doubted whether crisis, the common awareness that something has gone wrong, precedes revolutions so invariably as I have implied in my original text. Nothing important to my argument depends, however, on crises' being an absolute prerequisite to revolutions; they need only be the usual prelude, supplying, that is, a self-correcting mechanism which ensures that the rigidity of normal science will not forever go unchallenged. Revolutions may also be induced in other ways, though I think they seldom are. In addition, I would now point out what the absence of an adequate discussion of community structure has obscured above: crises need not be generated by the work of the community that experiences them and that sometimes undergoes revolution as a result. New instruments like the electron microscope or new laws like Maxwell's may develop in one specialty and their assimilation create crisis in another.

2. Paradigms as the Constellation of Group Commitments

Turn now to paradigms and ask what they can possibly be. My original text leaves no more obscure or important question. One sympathetic reader, who shares my conviction that 'paradigm' names the central philosophical elements of the book, prepared a partial analytic index and concluded that the term is used in at least twenty-two different ways.[3] Most of those differences are, I now think, due to stylistic inconsistencies (e.g., Newton's Laws are sometimes a paradigm, sometimes parts of a paradigm, and sometimes paradigmatic), and they can be eliminated with relative ease. But, with that editorial work done, two very different usages of the term would remain, and they require separation. The more global use is the subject of this subsection; the other will be considered in the next.

Having isolated a particular community of specialists by techniques like those just discussed, one may usefully ask: what do its members share that accounts for the relative fulness of their professional communication and the relative unanimity of their professional judgments? To that question my original text licenses the answer, a paradigm or set of paradigms. But for this use, unlike the one to be discussed below, the term is inappropriate. Scientists themselves would say they share a theory or set of theories, and I shall be glad if the term can ultimately be recaptured for this use. As currently used in philosophy of science, however, 'theory' connotes a structure far more limited in nature and scope than the one required here. Until the term can be freed from its current implications, it will avoid confusion to adopt another. For present purposes I suggest 'disciplinary matrix': 'disciplinary' because it refers to the common possession of the practitioners of a particular discipline; 'matrix' because it is composed of ordered elements of various sorts, each requiring further specification. All or most of the objects of group commitment that my original text makes paradigms, parts of paradigms, or paradigmatic are constituents of the disciplinary matrix, and as such they form a whole and function together. They are, however, no longer to be discussed as though they were all of a piece. I shall not here attempt an exhaustive list, but noting the main sorts of components of a disciplinary matrix will both clarify the nature of my present approach and simultaneously prepare for my next main point.

One important sort of component I shall label 'symbolic generalizations,' having in mind those expressions, deployed without question or dissent by group members, which can readily be cast in a logical form like $(x)(y)(z)\phi(x, y, z)$. They are the formal or the readily formalizable components of the disciplinary matrix. Sometimes they are found already in symbolic form: $f = ma$ or $I = V/R$. Others are ordinarily expressed in words: "elements combine in constant proportion by weight," or "action equals reaction." If it were not for the general acceptance of expressions like these, there would be no points at which group members could attach the powerful techniques of logical and mathematical manipulation in their puzzle-solving enterprise. Though the example of taxonomy suggests that normal science can proceed with few such expressions, the power of a science seems quite generally to increase with the number of symbolic generalizations its practitioners have at their disposal.

These generalizations look like laws of nature, but their function for group members is not often that alone. Sometimes it is: for example the Joule-Lenz Law, $H = RI^2$. When that law was discovered, community members already knew what H, R, and I stood for, and these generalizations simply told them something about the behavior of heat, current, and resistance that they had not known before. But more often, as discussion earlier in the book indicates, symbolic generalizations simultaneously serve a second function, one that is ordinarily sharply separated in analyses by philosophers of science. Like $f = ma$ or $I = V/R$, they function in part as laws but also in part as definitions of some of the symbols they deploy. Furthermore, the balance between their inseparable legislative and definitional force shifts over time. In another context these points would repay detailed analysis, for the nature of the commitment to a law is very different from that of commitment to a definition. Laws are often corrigible piecemeal, but definitions, being tautologies, are not. For example, part of what the acceptance of Ohm's Law demanded was a redefinition of both 'current' and 'resistance'; if those terms had continued to mean what they had meant before, Ohm's Law could not have been right; that is why it was so strenuously opposed as, say, the Joule-Lenz Law was not.[4] Probably that situation is typical. I currently suspect that all revolutions involve, among other things, the abandonment of generalizations the force of which had previously been in some part that of tautologies. Did Einstein show that simultaneity was relative or did he alter the notion of simultaneity itself? Were those who heard paradox in the phrase 'relativity of simultaneity' simply wrong?

Consider next a second type of component of the disciplinary matrix, one about which a good deal has been said in my original text under such rubrics as 'metaphysical paradigms' or 'the metaphysical parts of paradigms.' I have in mind shared commitments to such beliefs as: heat is the kinetic energy of the constituent parts of bodies; all perceptible phenomena are due to the interaction of qualitatively neutral atoms in the void, or, alternatively, to matter and force, or to fields. Rewriting the book now I would describe such commitments as beliefs in particular models, and I would expand the category models to include also the relatively heuristic variety: the electric circuit may be regarded as a steady-state hydrodynamic system; the molecules of a gas behave like tiny elastic billiard balls in random motion. Though the strength of group commitment varies, with non-trivial consequences, along the spectrum from heuristic to ontological models, all models have similar functions. Among other things they supply the group with preferred or permissible analogies and metaphors. By doing so they help to determine what will be accepted as an explanation and as a puzzle-solution; conversely, they assist in the determination of the roster of unsolved puzzles and in the evaluation of the importance of each. Note, however, that the members of scientific communities may not have to share even heurisic models, though they usually do so. I have already pointed out that membership in the community of chemists during the first half of the nineteenth century did not demand a belief in atoms.

A third sort of element in the disciplinary matrix I shall here describe as values. Usually they are more widely shared among different communities than either symbolic generalizations or models, and they do much to provide a sense of community to natural scientists as a whole. Though they function at all times, their particular importance emerges when the members of a particular community must identify crisis or, later, choose between incompatible ways of practicing their discipline. Probably the most deeply held values concern predictions: they should be accurate; quantitative predictions are preferable to qualitative ones; whatever the margin of permissible error, it should be consistently satisfied in a given field; and so on. There are also, however, values to be used in judging whole theories: they must, first and foremost, permit puzzle-formulation and solution; where possible they should be simple, self-consistent, and plausible, compatible, that is, with other theories currently deployed. (I now think it a weakness of my original text that so little attention is given to such values as internal and external consistency in considering sources of crisis and factors in theory choice.) Other sorts of values exist as well—for example, science should (or need not) be socially useful—but the preceding should indicate what I have in mind.

One aspect of shared values does, however, require particular mention. To a greater extent than other sorts of components of the disciplinary matrix, values may be shared by men who differ in their application. Judgments of accuracy are relatively, though not entirely, stable from one time to another and from one member to another in a particular group. But judgments of simplicity, consistency, plausibility, and so on often vary greatly from individual to individual. What was for Einstein an insupportable inconsistency in the old quantum theory, one that rendered the pursuit of normal science impossible, was for Bohr and others a difficulty that could be expected to work itself out by normal means. Even more important, in those situations where values must be applied, different values, taken alone, would often dictate different choices. One theory may be more accurate but less consistent or plausible than another; again the old quantum theory provides an example. In short, though values are widely shared by scientists and though commitment to them is both deep and constitutive of science, the application of values is sometimes considerably affected by the features of individual personality and biography that differentiate the members of the group.

To many readers of the preceding chapters, this characteristic of the operation of shared values has seemed a major weakness of my position. Because I insist that what scientists share is not sufficient to command uniform assent about such matters as the choice between competing theories or the distinction between an ordinary anomaly and a crisis-provoking one, I am occasionally accused of glorifying subjectivity and even irrationality.[5] But that reaction ignores two characteristics displayed by value judgments in any field. First, shared values can be important determinants of group behavior even though the members of the group do not all apply them in the same way. (If that were not the case, there would be no *special* philosophic problems about value theory or aesthetics.) Men did not all paint alike during the periods when representation was a primary value, but the developmental pattern of the plastic arts changed drastically when that value was abandoned. . . . Imagine what would happen in the sciences if consistency ceased to be a primary value. Second, individual variability in the application of shared values may serve functions essential to science. The points at which values must be applied are invariably also those at which risks must be taken. Most anomalies are resolved by normal means; most proposals for new theories do prove to be wrong. If all members of a community responded to each anomaly as a source of crisis or embraced each new theory advanced by a colleague, science would cease. If, on the other hand, no one reacted to anomalies or to brand-new theories in high-risk ways, there would be few or no revolutions. In matters like these the resort to shared values rather than to shared rules governing individual choice may be the community's way of distributing risk and assuring the long-term success of its enterprise.

Turn now to a fourth sort of element in the disciplinary matrix, not the only other kind but the last I shall discuss here. For it the term 'paradigm' would be entirely appropriate, both philologically and autobiographically; this is the component of a group's shared commitments which first led me to the choice of that word. Because the term has assumed a life of its own, however, I shall here substitute 'exemplars.' By it I mean, initially, the concrete problem-solutions that students encounter from the start of their scientific education, whether in laboratories, on examinations, or at the ends of chapters in science texts. To these shared examples should, however, be added at least some of the technical problem-solutions found in the periodical literature that scientists encounter during their post-educational research careers and that also show them by example how their job is to be done. More than other sorts of components of the disciplinary matrix, differences between sets of exemplars provide the community fine-structure of science. All physicists, for example, begin by learning the same exemplars: problems such as the inclined plane, the conical pendulum, and Keplerian orbits; instruments such as the vernier, the calorimeter, and the Wheatstone bridge. As their training develops, however, the symbolic generalizations they share are increasingly illustrated by different exemplars. Though both solid-state and field-theoretic physicists share the Schrödinger equation, only its more elementary applications are common to both groups.

. . .

NOTES

1. W. O. Hagstrom, *The Scientific Community* (New York, 1965), chaps. iv and v; D. J. Price and D. de B. Beaver, "Collaboration in an Invisible College," *American Psychologist,* XXI (1966), 1011–18; Diana Crane, "Social Structure in a Group of Scientists: A Test of the 'Invisible College' Hypothesis," *American Sociological Review,* XXXIV (1969), 335–52; N. C. Mullins, *Social Networks among Biological Scientists* (Ph.D. diss., Harvard University, 1966), and "The Micro-Structure of an Invisible College: The Phage Group" (paper delivered at an annual meeting of the American Sociological Association, Boston, 1968).

2. Eugene Garfield, *The Use of Citation Data in Writing the History of Science* (Philadelphia: Institute of Scientific Information, 1964); M. M. Kessler, "Comparison of the Results of Bibliographic Coupling and Analytic Subject Indexing," *American Documentation,* XVI (1965), 223–33; D. J. Price, "Networks of Scientific Papers," *Science,* CIL (1965), 510–15.

3. [Margaret Masterman, "The Nature of a Paradigm," in *Criticism and the Growth of Knowledge,* ed. Imre Lakatos and Alan Musgrave (Cambridge, 1970).]

4. For significant parts of this episode see: T. M. Brown, "The Electric Current in Early Nineteenth-Century French Physics," *Historical Studies in the Physical Sciences,* I (1969), 61–103, and Morton Schagrin, "Resistance to Ohm's Law," *American Journal of Physics,* XXI (1963), 536–47.

5. See particularly: Dudley Shapere, "Meaning and Scientific Change," in *Mind and Cosmos: Essays in Contemporary Science and Philosophy,* The University of Pittsburgh Series in the Philosophy of Science, III (Pittsburgh, 1966), 41–85; Israel Scheffler, *Science and Subjectivity* (New York, 1967); and the essays of Sir Karl Popper and Imre Lakatos in *Growth of Knowledge.*

24

A Discussion of Kuhn's "Values"

Subjective Views of Kuhn / Peter Achinstein

Many philosophers of science have wrestled with the question of whether there is a universal scientific method, i.e., a set of rules for discovering or testing scientific hypotheses. The rules are universal because they hold for all sciences and all times. Examples might include rules of the sort that Descartes proposed (his 21 rules) in *Rules for the Direction of the Mind,* or the 4 rules of reasoning that begin the third book of Newton's *Principia.*

Two major figures in 20th century philosophy of science, Karl Popper and Paul Feyerabend, whose deaths occurred within a year or two of Kuhn's, offered strikingly different answers to the question of whether universal rules exist. Popper's answer I take to be a resounding, "Yes," at least to the question of whether there are such rules in the case of scientific testing or justification. (Rules of discovery, he believed, are another matter.) Popper's method of corroboration—his version of hypothetico-deductivism—is, I believe, intended to be a universal scientific method for testing hypotheses.

Feyerabend, although a student of Popper, offered a very different answer: There are no universal rules of discovery or testing. Doing science, Feyerabend said in one of his typically provocative moments, is like making love or waging war. There are some useful hints, some rules of thumb, but in general, to quote his most notorious principle, "anything goes." Anarchism, while it may not be a workable political strategy, yields the best science, and is the best philosophy of science, he used to say.

Reprinted from *Perspectives on Science* 9, no. 4 (2001): 423–32.

With Popper on the extreme right and Feyerabend on the extreme left, where does Kuhn fit on this spectrum? One might be tempted to say that Kuhn acknowledges the existence of methodological rules, but they are not universal. They are paradigm-specific. The paradigm supplies rather specific rules for what sort of hypotheses are worth testing, e.g., ones that do not violate conservation of energy, or that do not postulate action at a distance or velocities greater than that of light. And the paradigm supplies specific rules for how to test such hypotheses.

On the other hand, in his *Postscript* written for the second edition of *The Structure of Scientific Revolutions,* Kuhn emphasizes the importance of what he calls scientific *values* as an essential part of the "disciplinary matrix," a term that replaces "paradigm" (Kuhn 1970a, p. 182ff.). Kuhn mentions values such as accuracy, consistency, scope, simplicity, fruitfulness, explanatory power, and plausibility. These are values employed in determining what theory or hypothesis to accept, or to promote, or at least to take seriously and explore. There are passages in Kuhn suggesting that these values are shared by very different scientific communities, that they are shared over time, and indeed that this is the reason why all scientists can, in an important sense, be thought of as forming a single community (Kuhn 1970a, "Postscript," p. 184). So is Kuhn here allied with Popper? Is he a universalist?

I don't think so. Nor is he an anarchist either. Instead, he is delightfully paradoxical. He speaks of values shared within a scientific community and shared even by different communities. But he also says that these shared values are frequently applied differently even by scientists within the same community. Although the criterion of "accuracy" is pretty stable from one application to another, "judgments of simplicity, consistency, plausibility, and so on often vary greatly from individual to individual" (Kuhn 1970a, "Postscript," p. 185). "In short," Kuhn writes, "though values are widely shared by scientists and though commitment to them is both deep and constitutive of science, the application of values is sometimes considerably affected by the features of individual personality and biography that differentiate the members of the group" (Kuhn 1970a, p. 185). So while you and I may both agree that the simplest hypothesis is to be preferred, promoted, tested first, or whatever, we may disagree substantially over what counts as simple. And this disagreement, I take it, may be unresolvable by argumentation, but only explainable by reference to causal features of temperament, training, and so forth. Moreover, to complicate matters further, in determining which of two theories to prefer, promote, or pursue, it may turn out that each has some virtues the other lacks, and that scientists, even from the same community, weight these virtues differently.

Kuhn emphasizes that the criteria for selecting theories "function not as rules, which determine choice, but as values which influence them" (Kuhn

1977, p. 331). The idea seems to be that *rules,* at least if clearly formulated, are, or are supposed to be, applied by different people in the same way under the same circumstances. It wouldn't do if the rule that you must stop your car at a red light could be interpreted by different people in a very different manner. But for Kuhn, *values* can be variably applied, and that is fine and dandy. Moreover, Kuhn thinks this is sufficient to avoid the charge of subjectivity levelled against him by a number of critics. Responding to Scheffler, e.g., Kuhn writes, "it is emphatically *not* my view that 'adoption of a new scientific theory is an intuitive or mystical affair, a matter for psychological description rather than logical or methodological codification'" (Kuhn 1970b, "Reflections on My Critics," p. 261). Kuhn insists that there are good reasons for adopting one theory over another, and these reasons are of the sort philosophers of science typically describe involving accuracy, scope, simplicity, explanatory power, and so on.

So where does Kuhn end up on the spectrum between Popper, the universalist, and Feyerabend, the anarchist? Like Popper, he wants to promote certain values common to scientists generally, from one period to another. He wants to say that scientists generally favor theories satisfying these values. Appeals to these values constitute good reasons for preferring the theories. And it is the acceptance of such values that is at least part of what it is to be a scientist. Like Feyerabend, however, he wants to reject the idea that scientists proceed in accordance with some fixed and universal set of rules which determine what hypothesis it is reasonable to accept.

Let me put this another way which may be more helpful. All three of the philosophers I have mentioned talk about values such as simplicity, consistency, explanatory power, and empirical confirmation (even though Feyerabend at one point calls them "verbal ornaments"). Now there are several views about such values. On one, there are objective rational constraints on how they are to be interpreted and applied in particular circumstances. On another, opposing view, there are no objective rational constraints. What one takes to be simple, consistent, explanatory, or confirmatory, is a personal matter, or perhaps is decided by voting or power politics. I will attribute that view to Feyerabend, at least in his more provocative moments.

As for the rational constraint view, I see two versions: One is a strict constraint view. This is the idea that how a value is applied is strictly constrained by usage, or definition, or practice, so that in accordance with this usage or definition or practice, although there might be borderline cases, what counts, or is to count, as simple, consistent, explanatory, or empirically confirmed, is quite generally determined in individual applications. I think that is Popper's view, or something close to it.

The second rational constraint view is weaker. It is that usage or definition or practice furnish at least some constraints on how scientific values are to be interpreted and applied. Not just any theory a scientist or a community calls simple or confirmed is really so. But these constraints, however universal, do not completely determine how scientific values are to be interpreted and applied. Scientists associated with the same practice can still differ considerably over whether, or to what extent, some theory is simple, confirmed, explanatory, etc. I suspect this weak constraint view is Kuhn's view, or something close to it.

If this is Kuhn's view, there is an interesting similarity between this and the view of someone most people take to be one of Kuhn's major opponents, viz. Rudolf Carnap, the arch-logical positivist. In his *Continuum of Inductive Methods* published in 1952, 10 years before Kuhn's most famous work, Carnap sought to explicate the idea of the degree to which a theory is confirmed or supported by the evidence. He began by setting down all the objective, rational, universal constraints he could think of for this concept, resulting in 11 axioms. However, these axioms do not determine a unique concept of confirmation, but rather an infinity of them. Which concept does one choose, and how is the choice to be made? Carnap proposed that both subjective and objective factors be used. The subjective factors—the ones that can vary from one individual scientist to another—include familiarity and ease of operating with the concept. The main objective factor is success in using that concept, by which Carnap means how frequently highly confirmed hypotheses turn out to be true or yield true predictions.

Carnap's appeal to success in using a particular concept of confirmation as a criterion for selecting it strikes me as circular. It is justifying an inductive method by appeal to induction. If we drop this and keep Carnap's other criteria then we have a view of confirmation that involves some constraints (Carnap's 11 axioms) and some subjectivity. So, on the basis of the same evidence and background information and partially constraining axioms, you can say that a certain hypothesis is highly confirmed by the evidence while I can say that the confirmation is much less. You can say that the degree of confirmation is .96, and I can say it is .69. And we can both be reasonable and even correct. That is, in your confirmation system, which is determined in part by subjective factors, the degree of confirmation is .96, in mine it is .69.

This seems quite subjective.[1] But let me spell this out a bit more. For Carnap there is an important relationship between degree of confirmation and justification of belief. The relationship is this. If the degree of confirmation of h on e is some number r, and if e is all you know, or all you know that is relevant for h, then you are justified in believing h to the degree r. So, for Carnap, it is perfectly okay for there to be two scientists with exactly the same information e, who not only believe h with substantial differences of degree, but who are justified in doing so. These scientists may rightfully have substantially different degrees of belief in the same hypothesis on the same evidence because they have chosen different systems of confirmation for reasons that include subjective considerations, such as familiarity and ease of use.

Is this the kind of subjectivism—call it partial subjectivism—that Kuhn advocates? I don't really know the answer. Instead of saddling Kuhn with one particular view, let me consider three different forms of subjectivism concerning confirming evidence that might possibly be attributed to him.

At one end of the spectrum there is simply the idea that different scientists confronted by the same data, and with roughly the same background knowledge, sometimes, even frequently, come to different conclusions about what to believe or how much to believe it. Putting this in terms of confirming evidence, different scientists sometimes come to different conclusions about whether, or to what extent, the evidence confirms the hypothesis. This is absolutely minimal subjectivism, although I take it to be at least part of what Kuhn is saying. Perhaps you will say that there is not enough here even to be labeled subjectivism.

At the other end of the spectrum there is the idea that not only do scientists with the same background information and the same data sometimes, even frequently, draw different conclusions about whether, or how much, the data support the hypothesis, but that scientists who do so may all be correct or at least reasonable in doing so. The reason is that whether, or to what extent, data support an hypothesis depends in part on subjective "weighting" factors that can vary from one scientist to another. Accordingly, whether, and to what extent, it is reasonable to believe an hypothesis on the basis of some data depends in part on such factors. This I take to be Carnap's position in *The Continuum of Inductive Methods*.

Now let me mention a third, middle position. Let's distinguish confirmation on the one hand from actions to be taken on the basis of confirmation. One might hold the following view. Whether, or the extent to which, some data support an hypothesis is purely objective.[2] It does not depend at all on factors that vary from one scientist to another. Suppose, then, that e confirms h to the degree r. Nothing whatever follows about what actions it is reasonable to engage in concerning hypothesis h. In the sciences, an important question pertaining to action might be whether to "pursue" h, i.e., whether to invest one's time, energy, and resources in developing consequences of h, testing them, extending h to new areas, and so on. If we treat this as a decision theory problem, we have two factors: probabilities and utilities. The probabilities can be thought of objectively, the utilities subjectively. The mere fact that on the basis of the data the probability (or degree of confirmation) of h is low or high tells us nothing about whether to "pursue" h. Suppose h concerns the efficacy of some new drug and e reports a limited study on 15 patients. Even if the degree of confirmation of h is high, whether it is reasonable to pursue h by testing it on more individuals can depend on factors that can vary from one scientific group or even individual to another—for example, on costs, other activities being engaged in, differences in views of how important the research is, etc.

Is this third alternative Kuhn's view? Does this capture his partial subjectivism? It has the advantage of letting him have his cake and eat it too. The view is objective in this sense. It says that whether, or to what extent, data confirm an hypothesis does not depend on subjective facts. It is also subjective in these two senses. Like the minimalist view it allows scientists to disagree about whether, or to what extent, e confirms h. People can disagree about objective matters. More importantly, it allows scientists, on subjective grounds, to reasonably take different actions on the basis of an hypothesis, which is confirmed to some particular degree. Kuhn speaks of "the adoption of a new scientific theory." If "adoption" involves taking certain actions that depend for their rationality on both objective and subjective factors, then this middle position is at least part of Kuhn's view, if not the whole of it. If it is the whole of it, that is fine with me. There is objectivism in the sense I like and subjectivism in a sense that does not bother me at all.

Is that all there is to it? I have my doubts. The first and third subjective views are not particularly exciting. But Kuhn was exciting, if he was anything. So let's say his view was more like Carnap's. And let me delve a bit further into this sort of partial subjectivism. On the Carnapian idea, objective rules of confirmation provide only a partial constraint on what probability or confirmation function to adopt. To take a simple Carnapian example, suppose that we have studied some sample of objects of a certain type, a fraction m/n of which have some simple property P. Suppose we want to determine the probability, i.e., the degree of confirmation, that the next object of that type will have P. According to Carnap, what objective rules of confirmation require is that this probability be a number between the observed relative frequency m/n and what Carnap calls the logical width of the property P, which is a type of a priori probability. To simplify considerably, we toss a coin 100 times, getting 60 heads. Our hypothesis is that the next toss will yield heads. Suppose that "heads" is a simple property whose logical width is 1/2. Then, says Carnap, the objective, rational constraints imply that the probability, or degree of confirmation, of the hypothesis that the next toss will yield heads, given the results of the 100 previous tosses, is between 1/2, the logical width, and 3/5, the observed relative frequency. Suppose Carnap is right that objective rational constraints do not determine a unique probability or degree of confirmation for this hypothesis given the evidence, but only a range. For at least some cases, then, such constraints yield *inexact* probabilities.

Now two options seem possible to me. The first is taken by Carnap. It is to say that inexact probabilities or degrees of confirmation are not as useful to science as exact ones. We must opt for exact ones, but, and this is crucial, the only way we can get them is by introducing an important subjective element. We choose some particular probability function allowed by the constraints on the basis of factors such as familiarity and ease of use—factors that can vary from one person to another.

The second option is simply to stay with inexact probabilities or degrees of confirmation when and if they occur. So if indeed the probability of h is between 1/2 and 3/5 that is what we say, period. Or, to take a non-quantitative assessment, if all that rational constraints allow us to say is that an hypothesis is confirmed to a fairly high degree by the evidence, then that is all we can say. We make our assessment somewhat inexact or vague. But inexactness and vagueness are not the same as subjectivity, nor do they imply or necessarily lead to subjectivity. This second option avoids subjectivism.

So is Kuhn a partial subjectivist in the Carnapian mold, i.e., in accordance with the first option? My feeling is that he may well be so, but that this is not the end of the matter. Nor can I say simply that he is a subjectivist in the two unexciting senses noted earlier, i.e., in the sense that scientists frequently disagree about probabilities and that even when they agree they may rationally take different actions based on subjective differences. My feeling is that Kuhn may want to say that what I have been calling the "rational constraints" can themselves be differently applied by different scientists on subjective grounds.

So let us consider a rational constraint. I pick one that Carnap defends. According to Carnap, the larger the sample the closer the probability that the next individual has the property in question gets to the observed relative frequency of that property in the sample. Perhaps Kuhn wants to say this as well. Suppose this is an objective rational constraint that all scientists accept. Still it can be applied differently. Different scientists can decide differently about how fast the probability in question approaches the relative frequency in the sample. A bold (or very empiricist) scientist may say it does so very quickly. A conservative (more "a priori") scientist may say it does so more slowly. So the rational constraints themselves can be applied differently.

Now one thing an objectivist can say in response is that boldness (as Popper would say), or being conservative (as George W. Bush might say) are *objective* constraints in science no less than in politics. That is, the bold scientist believes in boldness because he thinks it is an objective requirement; likewise for the conservative. And, sticking with boldness, an objectivist can also say that if two bold scientists disagree about probabilities to be inferred from relative frequencies in a sample, at least one of them is wrong. It is not that both are correct and each chooses what probability to infer on subjective grounds such as familiarity or ease of use.

Let me put aside this objectivist response and suppose that boldness and conservativeness do not represent objective constraints, but rather different styles of reasoning that vary with temperament. Some scientists are just bolder in their inferences than others, and that is perfectly okay. At least let me take this to be Kuhn's position.

If this is what Kuhn has in mind, then, although it sounds different from Carnap, I am not sure that it is. Suppose the objective rational constraint is the one expressed earlier, viz. that as the sample size increases, the closer the probability that the next item has the property in question gets to the observed relative frequency. For the sake of argument, suppose we adopt Carnap's view that the probability that the next toss of this coin yields heads is a number between the a priori probability and the observed relative frequency. Finally, suppose we have a relatively small sample of tosses with this coin, say 10, of which 6 result in heads. So, for the sake of argument, rational constraints tell us to infer that the probability of heads on the next toss is between 1/2 and 3/5, but they tell us no more than this. The conservative scientist, let us say, will choose a number closer to 1/2, the bold scientist will choose one closer to 3/5. But if neither conservativism nor boldness is itself an objective rational constraint, but simply a personally subjective style, and if this is what Kuhn's position amounts to, then, I think, Kuhn's position is Carnap's or something like it. Although the constraint that inferred probabilities approach observed relative frequencies is objective, it allows different rates of approach or convergence. How quickly it is taken to converge is, on this view, a personal subjective matter. In our coin tossing example, whether to choose a probability closer to 1/2 or closer to 3/5 varies from one individual to another.

An alternative, of course, is to refuse to play this subjective game. That is, in our example it is to refuse to assign a precise probability to getting heads on the next toss, given the size of the sample. If objective rational constraints yield only that the probability is between 1/2 and 3/5, then that is all we can and should say. I tend to favor this approach. Two versions are possible. On one, precise objective probabilities always exist, but there are times when our epistemic situations allow us only to assign upper and lower bounds and there is no reason to make matters more precise. By analogy, if all the police can say when judging the size of the crowd during the rally is that it was between 1500 and 2000 people, then that is all they should say.

It seems wrong and unnecessary to pick a number between these and say that the crowd was 1759.

On the second version, precise objective probabilities do not always exist; sometimes they are "smeared out." By analogy, if the size of the crowd was constantly changing then there is no precise number denoted by the expression "the size of the crowd," even if there is such a number for the average size of the crowd over the time period of the rally or for the size of the crowd at one particular point in time. However, I won't pursue this here. My claim is only that if Kuhn's point is that the same scientific value can be applied differently by different scientists, and in our confirmation example if this is to be understood in the sense that unique confirmation values, or bold vs. conservative strategies, are to be chosen at least partly on subjective grounds, then Kuhn's subjectivism in this regard is akin to Carnap's.

Finally, there are two more radical alternatives that I would like to resist attributing to Kuhn. One is to say that a constraint such as the one I have cited is not a necessary condition but simply a rule of thumb. Sometimes it is applied, sometimes not, depending on different subjective factors. This sounds like Feyerabend. The second is to say that the constraint is a necessary condition but that it is subject to different interpretations in a more interesting sense than simply one in which the convergence is bolder or more conservative. The claim would be that the very idea of convergence—the very idea that the inferred probability should approach the observed relative frequency as the size of the sample increases—is subject to different meanings. That is, the constraint itself is ambiguous. I won't speculate about what its different possible meanings might be. The idea would be simply that there are different meanings, and which is chosen is a subjective matter. I hope this is not Kuhn's view. If it is, then Kuhn's idea that different scientists share common values is not a good description of what he has in mind. They may share common words, but if these words mean different things, then I would say their values are different. Indeed, on the present alternative, Kuhn's view would seem to amount to Feyerabend's most extreme view that scientific values are mere "verbal ornaments."

So I leave you with three levels of subjectivism one might plausibly attribute to Kuhn. Confining our attention just to the case of confirming evidence, the levels are these.

First, there is the minimal idea that different scientists, confronted with the same data and the same background information, sometimes, perhaps often, come to different conclusions about whether, or to what extent, the data confirm or support an hypothesis.

Second, there is a middle position, which distinguishes confirmation on the one hand from actions (such as pursuing and testing hypotheses) on the other. It says that even though whether, or to what extent, the data confirm the hypothesis is an objective matter, whether it is reasonable to take certain actions with respect to that hypothesis is at least in part subjective. It depends on, and varies with, individual desires and resources.

Third, there is the view that objective rational constraints exist but in many cases do not determine a unique probability or degree of confirmation or strength of evidence. To arrive at such unique judgments, rather than simply leaving the matter somewhat imprecise, decisions need to be made that depend on subjective factors, e.g., conservativeness or boldness in judgments, that can vary from one person to another. This type of subjectivism, which can be found in Carnap as well, is the strongest and most interesting of the three.

NOTES

1. An even more vigorous subjectivism than Carnap's is that of the subjective Bayesians for whom the only objective constraint on rational degree of belief is satisfaction of the mathematical rules of probability. I will assume that this is more subjective than Kuhn wishes to be.

2. For a defense of the idea that there are one or more such concepts employed in the sciences, see Achinstein 2001.

REFERENCES

Achinstein, Peter. 2001. *The Book of Evidence.* New York: Oxford University Press.

Kuhn, Thomas S. 1977. *The Essential Tension.* Chicago: University of Chicago Press.

———. 1970a. *The Structure of Scientific Revolutions.* 2nd ed. Chicago: University of Chicago Press.

———. 1970b. "Reflections on My Critics." In *Criticism and the Growth of Knowledge.* Edited by Imre Lakatos and Alan Musgrave. Cambridge, U.K.: Cambridge University Press.

Suggested Further Reading

PART I. DESCARTES' RATIONALISM AND LAWS OF MOTION

1. Desmond M. Clarke, *Descartes' Philosophy of Science* (University Park: Pennsylvania State University Press, 1982).

2. Dennis Des Chene, *Physiologia: Natural Philosophy in Late Aristotelian and Cartesian Thought* (Ithaca, N.Y.: Cornell University Press, 1996).

3. Daniel Garber, *Descartes' Metaphysical Physics* (Chicago: University of Chicago Press, 1992). (Fourth selection in Part I is from this book.)

4. Stephen Gaukroger, John Schuster, and John Sutton, eds., *Descartes' Natural Philosophy* (London: Routledge, 2000).

5. Gary Hatfield, *Routledge Philosophy Guidebook to Descartes and the Meditations* (London: Routledge, 2003).

PART II. NEWTON'S INDUCTIVISM AND THE LAW OF GRAVITY

1. Peter Achinstein, *Particles and Waves* (New York: Oxford University Press, 1991), essay 2.

2. I. B. Cohen, *Introduction to Newton's "Principia"* (Cambridge, Mass.: Harvard University Press, 1978).

3. I. B. Cohen, *The Newtonian Revolution* (Cambridge: Cambridge University Press, 1980).

4. John Herivel, *The Background to Newton's "Principia"* (Oxford: Oxford University Press, 1965).

5. Ernan McMullin, *Newton on Matter and Activity* (Notre Dame: Notre Dame University Press, 1978).

PART III. HYPOTHETICO-DEDUCTIVISM, THE MILL–WHEWELL DEBATE, AND THE WAVE THEORY OF LIGHT

1. Peter Achinstein, *Particles and Waves* (listed above), essays 3 and 4.

2. G. N. Cantor, *Optics after Newton* (Manchester: University of Manchester Press, 1983).

3. Carl G. Hempel, *Philosophy of Natural Science* (Englewood Cliffs, N.J.: Prentice-Hall, 1966).

4. Larry Laudan, "The Medium and Its Message," in G. N. Cantor and M. J. Hodge, eds., *Conceptions of the Ether* (Cambridge: Cambridge University Press, 1981).

5. Laura J. Snyder, "The Mill–Whewell Debate: Much Ado about Induction," *Perspectives on Science* 5 (1997): 159–98.

PART IV. REALISM VS. ANTIREALISM, AND MOLECULAR REALITY

1. Arthur Fine, "The Natural Ontological Attitude," in Leplin (listed below).

2. Philip Kitcher, "Real Realism: The Galilean Strategy," *Philosophical Review* 110 (April 2001):151–97.

3. Jarrett Leplin, ed., *Scientific Realism* (Berkeley: University of California Press, 1984).

4. Mary Jo Nye, *Molecular Reality* (London: Macdonald, 1972).

5. Stathis Psillos, *Scientific Realism* (New York: Routledge, 1999).

PART V. GALILEO'S TOWER ARGUMENT, AND REJECTIONS OF UNIVERSAL RULES OF METHOD

1. Paul Horwich, ed., *World Changes: Thomas Kuhn and the Nature of Science* (Cambridge, Mass.: M.I.T. Press, 1993).

2. Paul Hoyningen-Huene, *Reconstructing Scientific Revolutions: Thomas Kuhn's Philosophy of Science* (Chicago: University of Chicago Press, 1993).

3. Imre Lakatos and Alan Musgrave, eds., *Criticism and the Growth of Knowledge* (Cambridge: Cambridge University Press, 1970).

4. Peter Machamer, ed., *The Cambridge Companion to Galileo* (Cambridge: Cambridge University Press, 1998).

5. John Preston, *Feyerabend: Philosophy, Science, and Society* (Cambridge, U.K.: Polity Press, 1997).

Index

Achinstein, Peter
on hypothetico-deductivism vs. inductivism, 235–37
on Kuhn's "values," 412–21
on principle of proliferation, 389–400
scientific realism, 327–52
on wave theory of light, 237–47
Anti-realism
Duhem's, 258–80
van Fraassen's, 281–97
Avogadro's number, 256, 308, 313–14, 327, 329–30

Brownian motion, 255–56, 300–305, 314–15, 328

Carnap, Rudolf, 341–42, 415–19
Cohen, I. Bernard, 99–111
Coherence
See Whewell
Common-cause argument, 319–26
See Salmon
Consilience of inductions, 160–62
See Whewell
Constructive empiricism
See van Fraassen

Descartes, René
deductions, 10, 19–20, 32
intuitions, 10, 19
laws of motion, 11–14, 40–47, 54–66
methodological rules, 9–11, 17–34, 48–54
ontological proof of God, 11, 35–39
simple natures, 29–33
visible world, 45–47
Disciplinary matrix
See Kuhn
Duhem, Pierre, 258–80
natural classification, 273–78
physical theory
as economy of thought, 269–72
rejections of explanations, 258–68, 336
physics of a believer, 278–79

Eliminative-causal argument, 331

Falsificationism
See Popper
Feyerabend, Paul, 355–58, 372–88
natural interpretations, 380–86
principle of proliferation, 376–77
on tower argument, 377–87
Fine, Arthur, 342, 347

Galileo
tower argument, 356–58, 361–71
Garber, Daniel, 14, 15, 48–66

Hypotheses
Mill, 215–228
Newton, 72–74, 97–99, 104–7
Whewell, 156–66
Hypothetico-deductivism
Mill's critique, 133–36

Hypothetico-deductivism (*continued*)
Popper's version, 132–33
simple hypothetico-deductive view, 130
Whewell's version, 130–32

Induction
Mill, 173–82
Newton, 70–75, 77–80, 97–98
Popper's rejection, 168–70
Whewell, 150–55
Inference to the best explanation
van Fraassen, 293–96
Whewell's version, 159–66
Intuition
See Descartes

Kuhn, Thomas
disciplinary matrix, 407
exemplars, 410
paradigms, 402–7
subjectivism, 414–21
symbolic generalizations, 407–8
values in science, 408–10, 412–21

Light
wave theory, 128–30, 137–49, 237–47

Method of hypothesis
See hypothetico-deductivism
Mill, John Stuart
debate with Whewell, 133–36, 182–90, 225, 228–33, 276–77
deductive method, 207–15
on hypotheses, 215–28
induction, 173–82
methods of experimental inquiry, 190–207

Natural interpretations
See Feyerabend
Newton, Isaac
Cohen's discussion, 99–111
on God, 95–97
"hypotheses," 72–74, 97–99, 104–07
law of gravity, 75–76, 85–94
methodological rules, 70–75, 78–80, 100–101, 113–23, 392–95
"phenomena," 72–73, 81–85, 101–2
Whewell's critique, 112–23

Ontological proof of God
See Descartes

Paradigms
See Kuhn
Perrin, Jean
argument for molecules, 255–57, 300–311, 328–33
inductive method, 300
intuitive method, 300
scientific realism, 346
See also Achinstein; Salmon
Popper, Karl, 132–33
deductive testing of theories, 171–72
rejection of induction, 168–70
Principle of proliferation, 376, 389–400
See also Feyerabend

Rationalism
See Descartes
Realism
See scientific realism

Salmon, Wesley
empirical defense of realism, 255, 312–26
Scientific method, rules of
Descartes' rationalism, 9–11, 17–34, 48–59
Mill's empiricism
inductivism, 173–90
methods of experimental inquiry, 190–207
deductive method, 207–28, 236
Newton's inductivism, 70–75, 78–80, 100–101, 113–23
Popper's falsificationism, 168–72
Rejection of universal rules
Feyerabend, 355–58, 372–77
Kuhn, 358–60
Whewell's hypothetico-deductivism, 130–32, 150–66, 235–37
Scientific realism, 251–52
Achinstein's defense, 327–52
vs. antirealism, 252–54
Perrin's argument for reality of molecules, 255–57, 300–311
Salmon's defense, 312–26

types of scientific realism, 340–43
van Fraassen's formulations of realism and antirealism, 282–88
Scientific values
See Kuhn
Subjectivism
in Kuhn, 414–21

Tower argument
Feyerabend, 377–87
Galileo, 356–58, 361–71

van Fraassen, Bas, 281–97, 336–37, 341
constructive empiricism, 286–88
inference to the best explanation, 293–96
scientific realism, 282–84
theory-observation distinction, 288–92

Wave theory of light, 128–30, 137–49, 237–47
Whewell, William
coherence, 162–66
consilience of inductions, 160–62
critique of Newton's methodology, 112–23
debate with Mill, 133–36, 182–90, 225
hypothetico-deductivism, 130–32, 150–66, 235–37
induction, 150–55

Young, Thomas
wave theory of light, 127–30, 137–49